The Historical Development of Quantum Theory

Springer
New York
Berlin
Heidelberg
Barcelona
Hong Kong
London
Milan
Paris
Singapore
Tokyo

Jagdish Mehra
Helmut Rechenberg

The Historical Development of Quantum Theory

VOLUME 2

The Discovery of Quantum Mechanics 1925

Springer

Library of Congress Cataloging-in-Publication Data
Mehra, Jagdish.
The discovery of quantum mechanics, 1925
(The historical development of quantum theory/
Jagdish Mehra and Helmut Rechenberg; v. 2)
Bibliography: p.
Includes index.
1. Quantum theory—History.
I. Rechenberg, Helmut, joint author. II. Title.
III. Series: Mehra, Jagdish.
The historical development of quantum theory; v. 2.
QC173.98.M44 vol. 2 530.1'2'09s 82-5445
[530.1'2'09] AACR2

Printed on acid-free paper.
First softcover printing, 2001.

Production managed by Christina Torster; manufacturing supervised by Jacqui Ashri.
Printed and bound by Berryville Graphics, Inc., Berryville, VA.
Printed in the United States of America.

9 8 7 6 5 4 3 2 1

ISBN 0-387-95176-8 SPIN 10789460

Springer-Verlag New York Berlin Heidelberg
A member of BertelsmannSpringer Science+Business Media GmbH

Contents

Foreword

This volume deals with Werner Heisenberg's intellectual development and work on the problems of quantum theory (anomalous Zeeman effects, the helium problem and perturbation theory, resonance fluorescence, dispersion theory, etc.) —beginning in fall 1920, and continuing with full vigour and enthusiasm after Niels Bohr's Wolfskehl lectures in Göttingen in June 1922—culminating in his discovery of the quantum-theoretical reformulation of the mechanical equations of atomic theory in summer 1925.

From the early 1950s onwards I obtained many opportunities for discussing the historical development of quantum theory with Heisenberg and the other architects of quantum mechanics. Helmut Rechenberg and I have used the relevant parts of these conversations to illuminate the human aspects of the historical background and the individual contributions to various problems that constituted the fabric of quantum theory. We have also drawn upon the rich source materials of the *Archives for the History of Quantum Physics*, especially the systematic interviews with the quantum physicists (cited here as 'AHQP Interviews').* In addition, we have made extensive use of the scientific correspondence of Niels Bohr, Werner Heisenberg, Wolfgang Pauli and Arnold Sommerfeld.

I wish to express my gratitude to the memory of Werner Heisenberg, who had allowed me in February 1975 to quote from his scientific correspondence and my conversations and other interviews with him. I also thank Mrs. Elisabeth Heisenberg for confirming the permission to make use of these sources.

I am grateful to Aage Bohr for giving me complete access to the *Bohr Archives* and the *Archives for the History of Quantum Physics* (the latter deposited, among other places, at the Niels Bohr Institute in Copenhagen).

I am indebted to Mrs. Franca Pauli and Victor F. Weisskopf for permission to quote from Pauli's scientific correspondence.

I also wish to thank the heirs of the literary and scientific estates of numerous other quantum physicists, whose work is cited here, for permission to do so.

JAGDISH MEHRA

*An inventory of these sources is contained in: T. S. Kuhn, J. L. Heilbron, P. Forman and L. Allen, *Sources for History of Quantum Physics*, The American Philosophical Society, Philadelphia, 1967.

The Discovery of Quantum Mechanics 1925

Introduction

During the second half of the year 1924 the speed of the development towards achieving a consistent quantum theory increased considerably. The main drive behind the efforts was what one called the sharpening ('*Verschärfung*') of the correspondence principle; that is, one attempted seriously to put the requirements of the correspondence principle into mathematical equations or to build them into particular mechanical models that described selected processes in atomic physics. These models lost more and more the property of visualizability ('*Anschaulichkeit*'), but they worked well beyond expectation. Hence the majority of the quantum theorists believed that they were moving in the right direction. One of the most active contributors to the theoretical development at that time was Werner Heisenberg. He was a young man, extremely ambitious, who worked very hard and fast, and never gave up before he had obtained the solution of a problem he had undertaken to investigate. He also had the advantage of having gone through three of the foremost schools of atomic theory: Munich, Göttingen and Copenhagen. These three places determined Heisenberg's scientific development, and he characterized many years later their respective influences in the following words: 'From Sommerfeld I learned optimism, in Göttingen mathematics, from Bohr physics' (Heisenberg, Conversations; see also Hermann, 1976, p. 28: '*Bei Sommerfeld hab' ich den Optimismus gelernt, bei den Göttingern die Mathematik, bei Bohr die Physik.*'). All these influences served to sharpen Heisenberg's scientific abilities and to enable him to take the first great step towards the discovery of quantum mechanics.

Chapter I
Werner Heisenberg's Scientific Development

I.1 Early Influences

Werner Karl Heisenberg was born on 5 December 1901 in Würzburg, an old university town in Lower Franconia, Bavaria, Germany, as the son of August Heisenberg and his wife Anna (*née* Wecklein). August's father, August Wilhelm Heisenberg, who had been a blacksmith in Osnabrück, Westfalia, respected religion and education and adored music. He purchased a piano for his son out of his meagre earnings and arranged music lessons. The protestant pastor of Osnabrück advised August Wilhelm to send his son, born 13 November 1869 at Osnabrück, to school and, if possible, to the university. August did well at school and also began to love music, especially the opera. He became very fond of the music of Richard Wagner, and, after a period of study at Marburg and Leipzig, he decided to attend the University in Munich because Wagner's music and a variety of operas were regularly performed there. At the University of Munich, August Heisenberg studied theology, philosophy, and the classics, and completed a doctoral dissertation (1894) on medieval Greek under Karl Krumbacher. In 1892 he passed the state examination (*Staatsprüfung*) to qualify for teaching classical languages and history in Bavarian secondary schools. He began his teaching career in schools at the renowned *Maximilian Gymnasium* in Munich and then continued in Landau (*Rheinpfalz*) and Lindau. In 1898 he travelled to Italy and in 1899 to Greece on an archaeological fellowship. On his return to Munich in spring 1899 he married Anna, daughter of Nikolaus Wecklein—then director of the *Maximilian Gymnasium*.[1] In 1901 August Heisenberg went to Würzburg to become a teacher at the *Alte Gymnasium* and, simultaneously, *Privatdozent* for medieval and modern Greek philology at the University of Würzburg. The Heisenbergs had two sons, Erwin and Werner, the latter younger

[1] Nikolaus Wecklein (1843–1926), the son of a farmer from Gänheim, Unterfranken, Germany, studied classical philology at the University of Würzburg and obtained his doctorate with a thesis on Plato and the Sophists (1865). After working as a high school teacher for several years, he continued his studies in Berlin (1868–1869), and in 1869 obtained his *Habilitation* at the University of Munich. At the same time he became a teacher in the *Maximilian Gymnasium* in Munich; in 1883 he moved to the *Gymnasium* in Bamberg, and in 1882 he was called to Passau as head of the *Gymnasium* there. Four years later he returned to the *Maximilian Gymnasium* as head and remained in this position until his retirement. In addition to his work in school administration, Wecklein edited Greek classics, especially the works of Aeschylus, Euripides and Homer. He wrote essays on various topics of classical philology, including those on Plato. (For his obituary, see Rehm, 1926.)

by about one and a half years.[2] Upon Krumbacher's death, August Heisenberg was invited in January 1910, to take the Chair of Classical Philology and Greek and Byzantine Studies (*Byzantinistik*) at the University of Munich, and the family moved to the Bavarian capital later that year.

August Heisenberg, who held the only Chair of Byzantine Studies in Germany, continued to sponsor the development of this field. He headed the Munich Seminar for medieval and new Greek philology; he edited the *Byzantinische Zeitschrift* and was also responsible for other related publications and activities. Basing his approach on philology, he emphasized the unity of Byzantine studies, considering history, history of art, and language as inseparable parts of the same field. In contrast to the tendency of the times, he opposed increasing specialization and tried to solve, by combining philological and historical methods, the problems of his research by examining the details of all aspects at the same time. He remained a very active man throughout his life; he could do enormous work and always showed great enthusiasm and indestructible optimism. He passed on many of these qualities to his son Werner. August Heisenberg died on 22 November 1930 in Munich.[3] His widow survived him by fifteen years.

The two boys, Erwin and Werner, grew up in an atmosphere of respect for learning in their parents' home. They not only became interested in arts and languages, but in science and technology as well. They shared the enthusiasm of many boys for making their own technical toys. 'Emulating the example of a fellow pupil, or perhaps because of a present received at Christmas—or even through school lessons—they begin to have a desire to handle small engines, and perhaps even to build one. This is precisely what I did with great enthusiasm during the first five years of my life at high-school' (Heisenberg, 1958c, p. 55). The Heisenberg boys once constructed, for example, the model of a battleship which was steered by radio waves, an unusual achievement at that time (Hermann, 1977, p. 34).

After attending the elementary schools in Würzburg and later in Munich, Werner Heisenberg became a pupil of *Maximilian Gymnasium*, his grandfather's school, on 18 September 1911. He immediately proved to be a very good student, and after one year his teachers pronounced a judgment about him: 'He has an eye for what is essential, and never gets lost in details. His thought processes in grammar and mathematics operate rapidly and usually without mistakes. Spontaneous diligence, great interest and thoroughness, and ambition' (Hermann, 1976, pp. 7–8). Thus, at school he developed an early interest in mathematics and languages that was encouraged by his father. His father possessed numerous Sanskrit books and a dictionary, which Werner began to study with a school friend, and he learned to write in Sanskrit characters. August Heisenberg would

[2] Erwin Heisenberg (1900–1965), the elder of August and Anna Heisenberg's two sons, studied chemistry and worked in industry, later on becoming the leader of a research department (*Abteilung für Synthese*) of *Vereinigte Glanzstoff Fabriken*, a German company.

[3] An obituary of August Heisenberg was written by Franz Dölger, his successor in the chair at the University of Munich (Dölger, 1933).

play all kinds of games with his two sons. Being a good teacher he employed games as an educational tool. When the youngsters received mathematical problems as homework for school, August used them as a game to find out who could do them most quickly. Werner discovered that he was quick in solving these problems and he began to develop a special interest in mathematics. At the age of twelve or thirteen he encountered differential calculus at school and was fascinated. He asked his father to bring home some mathematics books from the university library. Not having much idea of mathematics himself, August Heisenberg would bring home whatever he could find. It occurred to him that the study of mathematics and Latin could be encouraged simultaneously, and he started to bring home books on mathematics written in Latin for young Werner to read. When he was fifteen August brought him Leopold Kronecker's dissertation on the theory of numbers, written in Latin.[4]

As a result of the study of Kronecker's thesis, which he did not fully comprehend at the time, Werner Heisenberg acquired a strong interest in the theory of numbers. He was particularly excited by an essay, written by one of his teachers at the *Gymnasium*, on Pell's equation.[5] He wrote a paper dealing with this problem and sent it to a journal for publication. It was, however, an ambitious but poor attempt and the paper was rejected.

The *Maximilian Gymnasium* laid a great emphasis on the classics, but Chris-

[4]Leopold Kronecker (1823–1891) had been uniformly brilliant and many-sided at school. In addition to the Greek and Latin classics which he mastered with ease and for which he retained a lifelong liking, he shone in Hebrew, philosophy and mathematics. Kronecker's mathematical talent appeared early under the expert guidance of Ernst Eduard Kummer, his teacher at the *Gymnasium*, who in 1855 succeeded P. G. Lejeune Dirichlet when the latter moved from Berlin to Göttingen as the successor of Carl Friedrich Gauss. Kronecker's teachers at the University of Berlin (1841–1845) were Dirichlet, Jakob Steiner and Carl Gustav Jacob Jacobi. Kronecker submitted his famous dissertation, '*De Unitatibus Complexis*' ('On Complex Units'), to the Faculty of Philosophy at Berlin on 30 July 1845, defended it in his oral examination on 14 August 1845, and was awarded the doctorate on 10 September at the age of twenty-two. The particular complex units he discussed were those in algebraic number fields arising from the Gaussian problem of the division of the circumference of a circle into n equal parts. Dirichlet, his professor and examiner, was to remain one of Kronecker's closest friends, as was Kummer, his first mathematics teacher.

[5]In a letter to Frénicle in February 1657, Fermat stated the theorem that $x^2 - Ay^2 = 1$ has an unlimited number of solutions when A is positive and not a perfect square (Pierre de Fermat, *Oeuvres, 2*, pp. 333–335). Leonhard Euler, in a paper of 1732/33, erroneously called the equation Pell's equation and the name stuck. Euler became interested in this equation because he needed its solutions to solve $ax^2 + bx + c = y^2$ in integers; he wrote several papers on this theme. In 1759 Euler gave a method of solving Pell's equation by expressing $\sqrt{A}$ as a continued fraction (Euler, *Opera*, (1), 3, pp. 73–111). He failed to prove that his method always gave solutions and that all its solutions were given by the continued fraction development of $\sqrt{A}$. The existence of solutions of Pell's equation was shown in 1766 by Lagrange (*Oeuvres, 1*, pp. 671–731) and then more simply in later papers in 1770 (*Oeuvres, 2*, pp. 377–535 and pp. 655–726; also in Lagrange's additions to his translation of Euler's *Algebra*).

In the letter to Frénicle mentioned earlier, Fermat had also challenged mathematicians to find an infinity of integral solutions. Lord Brouncker gave solutions, but he did not prove that there was an infinity of them. Wallis did solve the full problem and gave his solutions in letters of 1657 and 1658 (Fermat, *Oeuvres, 3*, pp. 457–80, 490–503, and in Chapter 98 of his *Algebra*). Fermat also asserted that he could show when $x^2 - Ay^2 = B$ is solvable for given A and B and could solve it. We do not know how Fermat solved either equation, though he said in a letter in 1658 that he used the method of descent for the former. The more general equation, $x^2 - Ay^2 = B$, was solved by Lagrange (1770).

tian Wolff, the mathematics teacher, made a genuine effort to make the learning of mathematics attractive to his students. He encouraged Heisenberg in the study of mathematics by assigning him special problems. On one occasion he gave a problem on the diffraction of light in a vessel of water; Heisenberg wrote a long paper on it, employing elliptic functions.

Heisenberg's interest in mathematics continued to develop throughout his studies in the *Gymnasium*, and was stronger in mathematics than in physics. The theory of numbers attracted him the most. Having been introduced to Kronecker's thesis, he began to study other books on the theory of numbers and was fascinated by Paul Bachmann's books on this subject.[6] Heisenberg became interested in Fermat's *Last Theorem*[7] and tried to prove it. He also made an effort to prove Goldbach's conjecture.[8] He enjoyed this kind of mathematics more than, say, the differential calculus, because it seemed to him clearer, one which he could seek to delve into and understand. He even contemplated the pursuit of pure mathematics on leaving school.

Heisenberg also developed an interest in geometry, the elementary steps of which he learned at school. 'At first,' he recalled, 'I felt this to be very dry stuff; triangles and rectangles do not kindle one's imagination as much as do flowers and poems. But then our outstanding mathematics teacher, Wolff by name, introduced us to the idea that one could formulate generally valid propositions from these figures, and that some results, quite apart from their demonstrable geometric properties, could also be proved mathematically' (Heisenberg, 1958c, p. 55–56). The thought that mathematics might somehow be useful in describing one's experience struck him as remarkably strange and exciting.

> At first, stimulated by Herr Wolff's lessons, I tried out this application of mathematics for myself, and I found that this game between mathematics and immediate perception was at least as amusing as most other games. Later on, I discovered that geometry alone was no longer adequate for this mathematical game which had given me so much pleasure. From some books I gleaned that the behaviour of quite a few of my home-made instruments could also be described mathematically and I now began to read voraciously in somewhat primitive mathematical textbooks, in order to acquire the mathematics needed for the description of physical laws, i.e., the differential and integral calculus. In all this I saw the achievements of modern times, of Newton and his successors, as the immediate consequence of the efforts of the Greek mathematicians and philosophers, and never once did it occur to me to consider the science and technology of our times as belonging to a world basically

[6]Paul Gustav Heinrich Bachmann (1837–1920) had given a complete survey of the state of number theory in *Zahlentheorie: Versuch einer Gesamtdarstellung dieser Wissenschaft in ihren Hauptteilen*, 5 vols. (Leipzig, 1892–1923), *Niedere Zahlentheorie*, 2 vols. (Leipzig, 1902, 1910), and *Das Fermat-Problem in seiner bisherigen Entwicklung* (Leipzig, 1919).

[7]For an account of Fermat's *Last Theorem,* see Volume I, Section III.1, footnote 439.

[8]Christian Goldbach (1690–1764) conjectured that all even numbers may be expressed as the sum of two primes (taking 1 as a prime where necessary). Euler agreed with the assertion but could offer no proof, nor has any proof of 'Goldbach's conjecture' yet been found. Goldbach also stated that every odd number may be expressed as the sum of three primes; in the form given it by Edward Waring (which excludes 1 as a prime) this assertion also remains an unproved conjecture. (See Leonard E. Dickson, *History of the Theory of Numbers*, 2nd ed., New York, 1925.)

different from that of the philosophy of Pythagoras or Euclid. (Heisenberg, 1958c, pp. 56–57)

Thus, the study of geometry prepared Heisenberg to consider the other aspect of mathematics: its application to physics.

During 1918, the last year of World War I, times were particularly hard. Food was scarce and Heisenberg's parents worried about feeding the two boys. The family agreed that Werner should temporarily work as a farmhand; this would make it easier to obtain food. He worked on a farm in Miesbach, about thirty miles south of Munich, from spring to fall. There were several youths of his age who ran the farm together, since the men had all gone to war.[9] Their chores would begin before dawn and continue until late in the evening. At the farm Heisenberg acquired the habit of hard work and the enjoyment of physical exercise, and he returned home much healthier for it.

Heisenberg was not able to pursue any scientific interests during the several months which he spent on the farm at Miesbach. He did have with him a copy of Kant's *Kritik der reinen Vernunft* (*Critique of Pure Reason*) which he had intended to study. The hard physical labor did not do much to promote the study of Kant, though a growing interest in philosophy remained.

On his return home Heisenberg came across Hermann Weyl's *Raum-Zeit-Materie* (*Space, Time, Matter*), which had been published recently.[10] The prob-

[9]The service on the farm, which was called '*Kriegshilfsdienst*', was more or less obligatory for boys of Heisenberg's age.

[10]Hermann Weyl's *Raum, Zeit, Materie* (Weyl, 1918c) was based on the lectures which he gave during the summer semester of 1917 at the *Eidgenössische Technische Hochschule* (E.T.H. or the Swiss Federal Institute of Technology) in Zurich. In 1919 Weyl revised the book thoroughly; he included T. Levi-Civita's conception of the infinitesimal displacements and his own attempt to extend general relativity so as to derive from world-geometry not only gravitational but also electromagnetic phenomena (see, e.g., Weyl, 1920). The fourth extended edition (Weyl, 1921) was translated as *Space–Time–Matter* by H. L. Brose (Methuen & Company, London, 1922) and reprinted by Dover Publications, New York, 1952.

Heisenberg read the first edition of Weyl's book; it contained a discussion of the theories of special and general relativity and their mathematical tools. It is of interest to note that this book dealt with the basic mathematical apparatus of matrix mechanics, such as matrix algebra and calculus; however, the noncommutativity of the multiplication of two quadratic schemes was not displayed explicitly. This mathematical treatment does not seem to have influenced Heisenberg.

In some other respects Weyl's book did influence Heisenberg. He learned the important steps involved in Einstein's treatment of special relativity. Weyl's discussion of this was very lively. For example, he wrote about the ether: 'Aether mechanics has thus to account not only for Maxwell's laws but also for this remarkable interaction between matter and aether [as revealed by the Michelson–Morley experiment]. It seems that the aether has betaken itself to the land of the shades in a final effort to elude the inquisitive search of the physicist' (Weyl, 1952, p. 172) [The German original was even more picturesque: '*Es wäre also die Aufgabe der Äthermechanik, nicht nur die Maxwellschen Gesetze zu erklären, sondern auch diese merkwürdige Wirkung auf die Materie, die so erfolgt, als hätte der Äther sich ein für allemal vorgenommen: Ihr verflixten Physiker, mich sollt ihr nicht kriegen!*' (Weyl, 1918c, p. 134)] . . . 'The only reasonable answer that was given as to why a translation in the aether cannot be distinguished from rest was of Einstein, namely, that there *is no aether!*' Similarly, he remarked about the concept of simultaneity: 'The physical purport of this [Einstein's geometry and the null result of the Michelson–Morley experiment] is that *we are to discard our belief in the objective meaning of simultaneity; it was the great achievement of Einstein in the field of the theory of knowledge that he banished this dogma from our minds*' (Weyl, 1952, p. 174). From Weyl's book, Heisenberg, at an impressionable age, obtained a deep insight into Einstein's thinking about relativity theory.

lems of the theory of relativity interested him at once and he made an effort to understand the significance of the Lorentz transformations and Einstein's theory from Weyl's book. It was a challenging thought that the familiar concept of time had to be revised, which both worried and excited Heisenberg; he felt deeply that the theory of relativity had introduced a really fundamental and radically new way of looking at the physical universe. The problem of space, time, and matter appeared to Heisenberg as primarily a philosophical problem, and Albert Einstein appealed to him as a profound natural philosopher. Weyl's book did not turn his thoughts towards physics in particular.

There was much disorder in Munich in the spring of 1919. The *Räterepublik*, a kind of 'Soviet Republic,' had been established in the city, and lawlessness was rampant.[11] When at last a new Bavarian government, formed outside Munich, sent its troops into the city to restore order, the young people joined the advancing troops as guides. Heisenberg, who had turned seventeen the previous December, served from April to June 1919 in the Cavalry Rifle Command No. 11 (*Kavallerie-Schützenkommando* 11), with headquarters in the Theological Seminary building opposite the University of Munich in Leopoldstrasse (Hei-

[11]The Bavarian *Räterepublik* was mainly localized in the City of Munich and grew out of the activities of one man, Kurt Eisner. He belonged to the so-called 'Independent Social Democrats' ('*Unabhängige Sozialdemokraten*'), a small minority to the left of the Bavarian Social Democrats.

Eisner, alias Kosmanowski, was born in Berlin on 14 May 1867, the son of a manufacturer. He became a journalist and worked successfully on the staffs of *Vorwärts* (the main programmatic organ of the German Social Democrats) and of other socialistic newspapers in Nuremberg and Munich. He was frequently imprisoned because of the socialistic tendencies of his writings, which included the books entitled *Psychopathia Spiritualis* (1892), *Eine Junkersrevolte* (1899), *Wilhelm Liebknecht* (1900), *Feste der Festlosen* (1903), and *Die Neue Zeit* (1919). Eisner was a brilliant journalist, renowned for his critical articles on art in the newspaper *Münchner Post*. After the outbreak of World War I, he turned against the majority of the German Social Democrats, the so-called 'Majority Social Democrats' ('*Mehrheitssozialdemokraten*'), attacking them for their support of the war. He held discussion meetings on politics in the backroom of a Munich inn without receiving, however, large audiences or support. In January 1918 he was convicted of treason, but released later because of his candidacy for the *Reichstag*, the German Parliament in Berlin.

In the last days of the war—the armistice was signed on 11 November 1918—the political events in Germany became dramatic. The Bavarian revolution started on the evening of 7 November 1918 with a mass meeting on the *Theresienwiese* (where the *Oktoberfest* is held normally) in Munich. At this meeting, intended as a demonstration for peace, Eisner took the occasion to excite public opinion against the Government. A crowd of people went through the city, bearing red flags and marching to the *Residenz*, seizing military camps and supply offices on the way. At *Bierhaus Mathäser*, where soldiers usually congregated, a 'Workers and Soldiers' Council' ('*Arbeiter- und Soldatenrat*') was established under Eisner's chairmanship, and Bavaria was declared a 'democratic and socialist republic.' During the night of 7–8 November, the military and police headquarters as well as the seat of the *Landtag* (the Bavarian Parliament) were taken over by the revolutionaries. A government was formed by the Independent Social Democrats, which the Majority Social Democrats joined later. King Louis III of Bavaria fled from Munich to Anif, close to Salzburg, Austria. After the new republic had been proclaimed throughout Germany, Eisner became a protagonist of Bavarian rights; he was particularly opposed to the re-establishment of the federal system in the German *Reich*, the election of the National Constituent Assembly, and the admission of German war-guilt. The elections in Bavaria on 12 January 1919 ended in a total defeat for Eisner and his Independent Social Democrats, who won 3 out of 188 seats, while the bourgeois parties obtained a two-thirds majority. On his way to open the Bavarian Assembly on 21 February 1919, Eisner was shot dead by Count Toni Arco-Vally, a young lieutenant of the former *Infanterie-Leibregiment* of the King. A new

senberg to Sommerfeld, 18 January 1936; Heisenberg 1971, p. 7). Although this was his military service, it was often adventurous. Heisenberg's duties consisted of writing reports for the officer in charge and of running errands like delivering guns somewhere. During the early weeks of the fighting it seemed at times like playing a game of 'cops and robbers' (Heisenberg, Conversations, p. 4). Soon, however, the military service became monotonous. Heisenberg would often spend the whole night on duty at the telephone exchange and get the following day free. To catch up on his schoolwork, which had been neglected under the circumstances, he would retire to the solitude of the roof of the Seminary. There, 'lying in the wide gutter,' he would pursue his studies. He brushed up on his Greek by reading a school edition of Plato's *Dialogues*. One night early in June 1919, he was unable to fall asleep and went up to the roof at about 4 a.m. There it was warm and comfortable, and he began to read Plato:

> I came to the *Timaeus*, or rather to those passages in which Plato discusses the smallest particles of matter. Perhaps that section captured my imagination only because it was so hard to translate into German, or because it dealt with mathemati-

---

outbreak of the revolution followed this event, and the new socialist government of Bavaria, led by Johannes Hoffmann, left Munich after a few weeks, moving to Bamberg. However, in the Bavarian capital the leftist groups declared the formation of the *Räterepublik* ('Soviet Republic') on 7 April 1919.

In the beginning, the leaders of the *Räterepublik* were *Schwabing* writers like Gustav Landauer, Erich Mühsam and Ernst Toller. Toller, a student and modern (expressionist) poet, who, after Eisner's death, played a leading role in the *Räte* ('Soviets'), was soon replaced by Bolshevist revolutionaries, including Eugen Leviné, Max Levien, and the sailor Eglhofer, who was City Commander and responsible for the murder of hostages at the *Luitpold Gymnasium*. Munich became the scene of operations of the so-called 'Red Terror.' The lack of law and order, in turn, led to the formation of citizens' defense corps ('*Einwohnerwehren*') and the free corps ('*Freikorps Epp*'). They helped the units of the *Reichswehr* (the German Army), which were called in by the Bamberg Government; they recaptured Munich on 2 May after brutal fighting in the streets of the city on 1 and 2 May 1919. The ruthlessness of the invaders (the army) made people speak of the 'White Terror.' Only very slowly did law and order return to Munich. It was during spring and summer 1919 that Heisenberg served with the troops.

In August 1919 the Bamberg Parliament passed the new constitution which made Bavaria a parliamentary republic. The following February the Socialist Government resigned. Gustav von Kahr became the new Prime Minister (*Ministerpräsident*); he was supported by the strongest party, the *Bayerische Volkspartei*—a separate Bavarian branch of the conservative *Zentrum* party in the *Reichstag*. The quarrels with the *Reichs-Regierung* during later years and the conflicting views of various population groups made Munich a place that was easily affected by revolutionary turmoil. The economic and political situation in the German Republic, including inflation and the French occupation of the Ruhr region, influenced the turn to the right. Thus, Adolf Hitler and General Erich Ludendorff easily won support for their march to the *Feldherrnhalle* during the night of 8–9 November 1923. Only after the defeat of this enterprise, and the subsequent Hitler–Ludendorff trial, did the situation in Munich, and Bavaria generally, become stable again, the conservative *Volkspartei* staying in power until March 1933. (For an account of the Revolution of 1918 see: The article on 'Bavaria' in *Encyclopaedia Britannica*, 1967, Volume 3, p. 305; B. Hubensteiner, *Bayerische Geschichte,* Fourth Edition, R. Pflaum Verlag, Munich, 1964; M. Pittock, *Ernst Toller*, Twayne Publishers, Boston, 1979; A. Mitchell, *Revolution in Bavaria, 1918–1919. The Eisner Regime and the Soviet Republic*, Princeton University Press, Princeton, 1965.)

> cal matters, which had always interested me. In any case, I worked my way laboriously through the text, even though what I read seemed completely nonsensical. The smallest particles of matter were said to be right-angled triangles which, after combining in pairs into isosceles triangles or squares, joined together into the regular bodies of solid geometry: cubes, tetrahedrons, octahedrons and icosahedrons. These four bodies were said to be the building blocks of the four elements, earth, fire, air and water. I could not make out whether these regular bodies were associated with the elements merely as symbols—for instance, the cube with the element earth so as to represent the solidity and balance of that element—or whether the smallest parts of the earth were actually supposed to be cube-shaped. In either case, the whole thing seemed to be wild speculation, pardonable perhaps on the ground that the Greeks lacked the necessary empirical knowledge. Nevertheless, it saddened me to find a philosopher of Plato's critical acumen succumbing to such fancies. I looked for a principle that might help me to find some justification for Plato's speculation, but, try though I might, I could discover none. Even so, I was enthralled by the idea that the smallest particles of matter must reduce to some mathematical form. After all, any attempt to unravel the dense skein of natural phenomena is dependent upon the discovery of mathematical forms, but why Plato should have chosen the regular bodies of solid geometry, of all things, remained a complete mystery to me. They seemed to have no explanatory value at all. If I nevertheless continued reading the *Dialogues*, it was simply to brush up on my Greek. Yet I remained perturbed. The most important result of it all, perhaps, was the conviction that, in order to interpret the material world we need to know something about its smallest parts. Moreover, I knew from textbooks and popular writings that modern science was also inquiring into atoms. Perhaps, later in my studies, I myself might enter this strange world. (Heisenberg, 1971, pp. 8–9)[12]

Plato's thoughts about atoms were in marked contrast to what Heisenberg had learned in physics at school. He recalled later on:

> At that time we were using a rather good textbook of physics in which, quite understandably, modern physics was treated in a somewhat off-hand manner. However, the last few pages dealt briefly with atoms and I distinctly remember an illustration depicting a large number of them. The picture was obviously meant to represent the state of a gas on a large scale. Some of the atoms were clustered in groups and were connected by means of hooks and eyes supposedly representing their chemical bonds. On the other hand, the text itself stated that according to the concepts of the Greek philosophers atoms were the smallest indivisible building-stones of matter. I was greatly put off by this illustration, and I was enraged by the fact that such idiotic things should be presented in a textbook of physics, for I thought that if atoms were indeed such crude structures as this book made out, if their structure was complicated enough for them to have hooks and eyes, then they could not possibly be the smallest indivisible building-stones of matter. (Heisenberg, 1958c, pp. 58–59)

[12] The passages from Plato, to which Heisenberg refers, may be found in *Timaeus and Critias*, Penguin Classics, 1965, pp. 69–82.

His reading of *Timaeus* introduced him to another view of atoms, which he found to be much more adequate for describing the elementary constituents of matter, a picture which he also found more poetic and beautiful. Heisenberg would return to Plato's thoughts about the atom again and again during his later investigations in atomic and elementary particle physics, discovering in Plato an expression of the philosophical background of his own mature scientific work.[13] In the meantime, however, he continued with his military duty and his homework in Greek. In summer 1919 order returned to Munich, the government and parliament returned to the Bavarian capital, and the students were sent back to school.

During his last year at the *Gymnasium*, which began on 15 September 1919, Heisenberg performed really well. He obtained the mark of 'very good' in all subjects (religion, Latin, Greek, French, mathematics, physics, history and gymnastics) except German, in which he received a 'good.' Because of the excellent result of his *Abitur* (or higher-secondary examination), he was allowed to compete in the scholarship examinations of the Maximilianeum Foundation (*König-Max II-Stiftung*), which he passed successfully. Since he wished to stay with his parents during his student days at the University, he did not live at the house of the Foundation (with free room and board) with the other students who had won the same Foundation Award. In fall 1920 Heisenberg entered the University of Munich with the intention of studying mathematics. The assorted knowledge of mathematics which he had acquired at school contained various gaps and there were numerous elementary things of which he had no idea. Yet his familiarity with the calculus[14] and the theory of numbers had given him enough confidence to wish to study pure mathematics.

I.2 Escape from Mathematics and the Encounter with Sommerfeld

It occurred to Heisenberg that rather than just attending the available lectures on mathematics he should also seek to get admitted in the Mathematical Seminar of

[13] In contrast to Plato, Heisenberg never read Ernst Mach very seriously, certainly not in the early days, though he became familiar with his writings later in his career. He found Mach 'a bit formal.' He also found the scope of what Mach sought in his scientific method and philosophy rather modest and limited, if not too negative. He also found Mach's approach 'too little poetical'; Mach was anything but a poet. Heisenberg thought that Mach's dryness was perhaps the bane of all positivists, with the exception of Wittgenstein, 'who, in some ways, was also a poet' (Heisenberg, Conversations, p. 7; also AHQP Interview).

[14] Heisenberg had obtained a good knowledge of differential calculus by giving lessons when he was a lad of sixteen or seventeen. A chemist friend of his parents, a young lady of about twenty-four, wanted to take her examination in chemistry, but she also had to be examined in mathematics. She had to learn differential calculus in order to pass her examinations, and Heisenberg coached her. He also learned integral calculus at the same time.

a professor.[15] He thought that the opportunity of listening to discussions in the Seminar would introduce him to the current problems and methods of mathematical research.

At his urging his father, August Heisenberg, requested the mathematician Ferdinand von Lindemann to let his ambitious son visit him to discuss his plans of study at the University. Lindemann, a distinguished mathematician and professor, a university administrator and a crusty old man, did not relish the idea of receiving a mere freshman student but accepted it as a favour to his colleague.[16] Heisenberg went to see Lindemann at his office in the University's administration building, where his pet, a little black dog, would sit on the desk, examining the visitors who came into the office. When Heisenberg came in, the dog started to bark and would not stop, thus making it difficult for Heisenberg to talk and for Lindemann, who was hard of hearing, to listen. He was incensed to learn that the object of young Heisenberg's visit was to obtain permission to participate in Lindemann's Seminar in his first semester. He wanted to know about the books on mathematical subjects that Heisenberg had been reading. When the latter proudly told him that he had read Hermann Weyl's *Raum-Zeit-Materie*, Lindemann replied, to the accompaniment of the yapping of the little dog, 'In that case you are completely lost to mathematics' (Heisenberg, 1971, p. 16). The failure to obtain encouragement of his plan dejected Heisenberg. It was just as well that, instead of dwelling too long on his unhappiness at the demise of his short-lived ambition to become a pure mathematician, he looked for another avenue.

August Heisenberg, Professor of Greek and Byzantine Studies, and Arnold Sommerfeld, Professor of Theoretical Physics at the University of Munich, were good friends. August advised Werner to try to get himself admitted in Sommerfeld's Seminar, and made the appointment with Sommerfeld to see him.

The visit to Sommerfeld, in his bright study with windows overlooking a courtyard, where groups of students sitting on benches beneath an acacia tree could be seen, had an effect opposite to that of Lindemann's gloomy quarters. 'The small squat man with his martial dark mustache looked rather austere to

[15] It was indeed rather ambitious of Werner Heisenberg to seek admittance to the 'Mathematical Seminar' right away in his first semester. The '*Seminar*' at a German university denotes a special class attended by a number of advanced students, who meet once a week for about two hours to hear talks on specialized topics in a given field. Every participant of a 'Seminar' is supposed, as a rule, to give one talk per semester. In general, each professor directs a 'Seminar,' the members of which are his doctoral candidates and other students seeking the candidacy.

[16] Carl Louis Ferdinand von Lindemann was probably the most famous mathematician at the University of Munich. In 1882 he had proved the transcendence of π; hence the quadrature of the circle cannot be proved in an elementary way. Lindemann was born on 12 April 1852 in Hanover, and obtained his education at the Universities of Göttingen, Erlangen, Munich, Paris and London. In 1879 he became Professor of Mathematics in Freiburg-im-Breisgau; in 1883 he was called to Königsberg, where he became the doctoral supervisor of David Hilbert. Finally, from 1893 until his retirement in 1923, he was Professor of Mathematics at the University of Munich; he died on 7 March 1939 in Munich.

Besides the nature of π, Lindemann worked mainly on geometrical problems—on which he wrote two volumes (*Untersuchungen über Geometrie*, 1875, 1891)—and on Abelian functions. In 1892 he proposed a method of solving equations of any degree by the use of transcendental functions.

me. But his very first sentences revealed his benevolence, his genuine concern for young people, and in particular for the boy who had come to ask for his guidance and advice' (Heisenberg, 1971, p. 16). The question again arose about the mathematical studies which young Heisenberg had pursued as a hobby at school, and he mentioned Weyl's *Space, Time, Matter*. Sommerfeld's reaction was quite different from Lindemann's:

> You are much too demanding [he said]. You can't possibly start with the most difficult part and hope that the rest will automatically fall into your lap. I gather that you are fascinated by relativity theory and atomic problems. But remember that this is not the only field in which modern physics challenges basic philosophical attitudes, in which extremely exciting ideas are being forged. To reach them is much more difficult than you seem to imagine. You must start with a modest but painstaking study of traditional physics. And if you want to study science at all, you must first make up your mind whether you want to concentrate on experimental or theoretical research. From what you have told me, I take it that you are much keener on theory. But didn't you do experiments and dabble with instruments at school? (Heisenberg, 1971, pp. 16–17)

Heisenberg told him about the small engines, motors, and induction coils which he used to build at school, but he had never liked to work with instruments and apparatus and had found the care necessary in making even simple measurements 'as being sheer drudgery.'

Arnold Sommerfeld, being the great teacher that he was, came right to the point. 'Still, even if you study theory, you will have to pay particular attention to what may appear trivial tasks. Even those who deal with the larger issues, issues with profound philosophical implications—for instance, with Einstein's relativity theory or with Planck's quantum theory—have to tackle a great many petty problems. Only by solving these can they hope to get an over-all picture of the new realms they have opened up' (Heisenberg, 1971, p. 17).

Heisenberg protested modestly that he was much more interested in the underlying philosophical ideas of physics than in detailed problems and the techniques of their solution, but Sommerfeld would have none of this. He reminded Heisenberg of what Schiller said about Kant and his interpreters: '*Wenn die Könige bauen, haben die Kärrner zu tun.*'[17] ('When kings go a-building, wagoners have more work.') 'At first,' said Sommerfeld, 'none of us are anything but wagoners. But you will see that you, too, will get pleasure from performing minor tasks carefully and conscientiously and, let's hope, from achieving decent results' (Heisenberg, 1971, p. 17).

It was a different attitude than Ferdinand von Lindemann's. Sommerfeld had won over the young man who would become his most distinguished disciple. He gave him advice about the courses he should take at the University and told him that he would be welcome in his Seminar.

[17] The full quotation is: '*Wie doch ein einziger Reicher so viele Bettler in Nahrung setzt. Wenn die Könige bauen, haben die Kärrner zu tun.*' (The *Xenien* of Schiller and Goethe, No. 53, entitled '*Kant und seine Ausleger*.')

I.3 Sommerfeld's Seminar

In Arnold Sommerfeld, Heisenberg had met an authentic scientist who had himself made important contributions to the solutions of problems in relativity and quantum theory. By being allowed to join Sommerfeld's Research Seminar in his first term at the University, Heisenberg had come close to achieving his initial goal—that of being introduced to the current fundamental scientific problems, and the opportunity of discussing them in a creative milieu.

The Institute for Theoretical Physics, of which Sommerfeld was Director, was situated in a court in one of the university buildings off Ludwigstrasse.[18] The Institute had several rooms at its disposal. In the large Seminar Room there were several desks for young physicists who were engaged on research problems under Sommerfeld. At a given time there would be five or six such people working in the room, some on their research for the doctorate. The Institute had its own small library containing reference books and new publications, certain important journals, and reprints of scientific papers which Sommerfeld regularly received from all over.

When Heisenberg first joined Sommerfeld's Seminar, Gregor Wentzel and Wolfgang Pauli were the seniors among the young people. Wentzel acted as Sommerfeld's assistant and Pauli as a kind of second assistant,[19] and if someone had a question he would go to either of them for help or opinion. The Seminar Room was the place where one passed on the news, exchanged views and discussed the latest developments in physics, especially the quantum theory. Such discussions took place every day, and the occasion for it could be almost any problem of interest, a new publication in one of the journals, information brought by a visitor, or a letter from one of the great physicists, such as Einstein and Bohr, to Sommerfeld that he would pass on to Pauli for opinion. Pauli would

[18]The official postal address of Sommerfeld's Institute was: *Institut für Theoretische Physik, Universität, Ludwigstraße 17, München*. Within the old University building, the Institute's rooms lay close to the *Amalienstraße*; they faced a court next to the street, and were separated by a corridor from the outer façade of the University at *Amalienstraße*. Below the Institute's rooms there was the mechanical workshop and laboratory, the same where in 1912 the first X-ray diffraction diagrams had been obtained. The Institute for Theoretical Physics remained at the same place under Sommerfeld's successor until the early 1960s; then it was transferred first to the new *Dreierinstitut* in the *Schellingstraße* and, ultimately, in the early 1970s to the new buildings in the *Theresienstraße*.

[19]Sommerfeld's assistants around 1920 were: Wilhelm Lenz (since 1911), who left in 1920 to become associate professor in Rostock; Paul Ewald (since 1914), who was called to a professorship at the Technical University of Stuttgart in 1921. Of these, Lenz was about to leave when Heisenberg joined the Institute. Gregor Wentzel became Sommerfeld's official assistant after obtaining his doctorate in spring 1921. Wentzel was two years older than Wolfgang Pauli (who passed his doctoral examination at about the same time); therefore, he acted even before his doctorate as the assistant-in-charge. As a rule, the position of an assistant was given to a post-doc; a graduate student before receiving his doctorate could, however, perform some of the duties of an assistant (he was then called '*Verwalter einer Assistentenstelle*,' i.e., 'in-charge of an assistant's position'). When Heisenberg arrived, Wentzel was Sommerfeld's most advanced doctoral candidate in atomic theory; hence he acted as the assistant-in-charge. Pauli, who was younger, and worked in the same field, also did not yet qualify for an assistant's position; he acted as a kind of deputy assistant (*Hilfsassistent*). Upon graduation in July 1921 Pauli became Sommerfeld's private assistant (being paid from grants rather than university funds) before going to Max Born as his assistant in fall 1921.

most often analyze the problems on the blackboard and the others would get into the discussion and offer comments.

Heisenberg soon fell into the pattern of his new life at the Seminar and the University. At the Seminar he established rather early a rapport with Wolfgang Pauli. His acquaintance with Pauli began as soon as he started to attend Sommerfeld's lectures. As he recalled:

> A few days later, when I walked into the hall where Sommerfeld usually gave his lectures, I spotted a dark-haired student with a somewhat secretive face in the third row. Sommerfeld had introduced us during my first visit and had then told me that he considered this boy to be one of his most talented students, one from whom I could learn a great deal. His name was Wolfgang Pauli, and for the rest of his life he was to be a good friend, though often a very severe critic. I sat down beside him and asked him if, after the lecture, I might consult him about my preparatory studies. Sommerfeld now entered the hall, and as soon as he started to address us Wolfgang whispered in my ear: "Doesn't he look the typical old Hussar officer?" After the lecture, we went back to the Institute of Theoretical Physics, where I asked Wolfgang two questions. I wanted to know how much experimental work had to be done by someone interested chiefly in theory, and what he thought of the respective importance of relativity and atomic theory. (Heisenberg, 1971, pp. 24–25)[20]

Pauli answered Heisenberg's questions frankly by saying that although Sommerfeld laid great emphasis on the importance of experimental work, he personally did not care much for it. Pauli thought that once the empirical results have been obtained, the theoretical description in modern physics becomes too difficult to be grasped by most experimentalists because of the abstract mathematical language involved. He himself had found this language easy to learn and to apply. As for the relative importance of relativity and atomic theory, Pauli thought that special relativity was already an accomplished part of modern physics but one could still find new things to do in general relativity. However, he considered atomic theory as the more promising field to pursue in the near future.

Wolfgang Pauli was about to complete his article on the theory of relativity for the *Encyklopädie der mathematischen Wissenschaften* (Pauli, 1921b) when Heisenberg first met him. Pauli was only a year and a half older than Heisenberg but he possessed great brilliance and virtuosity in mathematical physics. He had a caustic wit and was very critical. With undiminished regularity, and with little or no provocation, Pauli would declare to Heisenberg, or to anyone else for that

[20]On another occasion Heisenberg recalled his first encounter with Pauli as follows:

I had seen him sitting there at a desk in the Seminar but never spoke a word with him. At that time I was interested in general relativity. Sommerfeld had talked about a letter he had received from Einstein about general relativity, concerning whether there was a red shift or no red shift, etc. I had studied Weyl's *Raum, Zeit, Materie* again, and there was a formula I couldn't understand. So I thought, "Well, there is Pauli." I knew that Pauli had written on general relativity, and he was writing a book on it [Pauli, 1921b]. So I asked him and he explained the thing to me. I think that was definitely our first conversation. He always knew everything; I was impressed that a man of my own age already knew all about general relativity, and I was at once interested in this fellow student who knew so much. From that time on we started to discuss things again and again. (Heisenberg, Conversations, p. 19; also AHQP Interview).

matter, 'You are a complete fool.' Strangely, however, 'that helped quite a lot' (Heisenberg, Conversations, p. 18). Heisenberg and Pauli became good friends and were not offended when they criticized each other. However, throughout their friendship, which lasted thirty-eight years until Pauli's death, the acidic remarks, the bile, usually came from Pauli.[21] Pauli, by his strong personality and critical genius, at times exerted a decisive influence on Heisenberg, and continued to do so, in some strange way, even after his death.[22]

Otto Laporte was another of Heisenberg's friends at the Institute. Laporte had attended the University of Frankfurt for one year and joined Sommerfeld's Seminar in summer 1921.[23] He was almost the same age as Heisenberg. His approach to the problems of physics was 'sober and pragmatic,' and he often served as a mediator in arguments between Pauli and Heisenberg (Heisenberg, 1971, p. 28).

Heisenberg also became acquainted with Gregor Wentzel and Adolf Kratzer.

[21] Pauli's advice and criticism accompanied Heisenberg in nearly all the decisive parts of his scientific work, with the exception of the years during World War II, which Pauli spent in Princeton, New Jersey, and Heisenberg in Germany. Heisenberg always used to present his ideas first to Pauli for criticism. He would say: 'If Pauli agrees then the thing is all right. If not, it may still be all right, but one has to be careful' (Heisenberg, Conversations).

Only twice did Heisenberg and Pauli collaborate closely on scientific problems, and only the papers on quantum field theory came out of their joint work (Heisenberg and Pauli, 1929, 1930). In 1957 and early 1958 they again worked together on a unified theory of elementary particles, but Pauli withdrew from this project in late spring 1958.

[22] The work on the unified quantum field theory, which Pauli joined in during the last year of his life, sought to bring together Heisenberg and Pauli's knowledge and experience of fundamental particle theory going back thirty years. The original programme was developed by Heisenberg. Pauli was particularly attracted by the doubling process which seemed to occur there. He remarked in a letter to Heisenberg: '*Er [der Pudel] hat seinen Kern enthüllt: "Zweiteilung und Symmetrieverminderung."*' ('He [the poodle] has revealed its core (interior): "Doubling and reduction of symmetry"' (Pauli to Heisenberg, 29 December 1957). The doubling came about in the nonlinear spinor theory when one had to introduce the isospin, which was done with the help of the Pauli–Gürsey group. Particles like the nucleons, having spin and isospin at the same time, could thus be described, for the fundamental equation of motion was invariant under the transformations of the rotation group and the Pauli–Gürsey group separately. Now the hyperons, Λ and Σ, had the 'wrong' connection between spin and isospin properties from this point of view since they exhibited half-integral spin but integral isospin. In order to account for this situation one had to assume the degeneracy of the vacuum, so that isospin could be taken from the nonunique vacuum state.

Thus the sentence quoted above described Pauli's last contribution to particle physics. For about two months Pauli experienced a certain euphoria with this project, and at one point even declared: 'Ich habe das Gefühl wir sind nun *nahe* einer Lösung.' ['I have the feeling that we are now *close* to a solution.'] (Pauli to Heisenberg, 12 December 1957). But then he withdrew his collaboration and participation entirely, and during the International High Energy Physics Conference, Geneva, July 1958, he severely criticized every detail of the analysis (as well as Heisenberg personally). Yet a few weeks later, at Varenna, Italy, he encouraged Heisenberg to proceed further with his nonlinear spinor theory. With Pauli's death on 15 December 1958, Heisenberg lost his best scientific friend and critic. Yet he remained for him in the background of his thinking as a silent witness, and Heisenberg, during his later work, would frequently ask what Pauli would have thought about the further development of the theory.

[23] Otto Laporte, born on 23 July 1902 at Mainz, Germany, went to Munich after spending a year in Frankfurt (1920–1921) under Max Born's guidance. Upon his arrival in Munich he was asked to give a seminar on the paper of Einstein and de Haas (1915). Pauli helped him to understand this work (Conversation with Laporte). After obtaining his doctorate in 1924, Laporte went to the United States; he first worked at the U.S. National Bureau of Standards and later joined the University of Michigan. He died on 28 March 1971 at Ann Arbor, Michigan.

During the summer semester of 1921 four students, including Pauli and Wentzel, completed their doctorate under Sommerfeld, and Kratzer obtained his *Habilitation* to become a *Privatdozent* (Sommerfeld to Einstein, 10 August 1921). Kratzer had served as Sommerfeld's assistant from 1918 to 1920, when he obtained the doctorate with a dissertation on band spectra (Kratzer, 1920a).[24]

Walther Kossel, who had been around Sommerfeld since 1913, was still in Munich when Heisenberg joined the Seminar. Kossel used to maintain close contact with Sommerfeld and he would attend the colloquium regularly. His interest primarily was in the periodic system of elements. He was amiable and kind, and Heisenberg would see him often.

Regular seminars and colloquia were organized at the Institute for Theoretical Physics under Sommerfeld's leadership. In general, a particular theme would be continued in the seminars during a semester. Impromptu seminars would take place, too, in which the topics varied considerably. Sommerfeld used to keep himself informed about the new developments not only by reading the literature and keeping up a vast scientific correspondence, but also by asking the young scholars in his Seminar to give talks on published papers of interest or to analyze the results contained in a letter which he had received.[25] A constant tempo of scientific stimulation was thus maintained at the Institute. Just before Heisenberg joined the Institute, Lenz, Kratzer, Wentzel and Pauli mainly used to take part in

[24]Adolf Kratzer was also not the official university assistant of Sommerfeld at that time. After completing his doctorate he left Munich to become for one year (1920–1921) Hilbert's private assistant for physics. On his return to Munich in 1921 he obtained his *Habilitation* and served as a *Privatdozent* until he was called to a professorship at the University of Münster the following year. It was during this period (1921–1922) that Heisenberg became acquainted with Kratzer.

[25]At about this time Einstein and Sommerfeld wrote to each other quite frequently. In the published correspondence (*Einstein und Sommerfeld, Briefwechsel*, 1968) there are Einstein's letters dated 6 September 1920, 20 (?) December 1920, 4 January 1921, 9 March 1921, 13 July 1921, 27 September 1921, 9 October 1921, 14 (?) January 1922, 28 January 1922, and 16 September 1922. Sommerfeld's letters during this period (11 September 1920, 7 October 1920, 18 December 1920, 29 December 1920, 14 March 1921, 4 July 1921, 2 August 1921, 10 August 1921, 17 October 1921, and 11 January 1922) dealt mainly with the attacks by certain people, including the physicists Philipp Lenard and Ernst Gehrcke, on Einstein's relativity theory. Sommerfeld also invited Einstein to give a talk on relativity at the University of Munich, but Einstein eventually declined because of an article that appeared in the weekly journal *Die Weltbühne* (see *Einstein und Sommerfeld, Briefwechsel*, 1968, Einstein's letter dated 27 September 1921, pp. 89–90). There occurred only a few remarks about physical problems, and the first were connected with an explanation of the result of the Einstein–de Haas experiment that the ratio of the angular momentum to the magnetic moment of the metal electrons was only half of the value predicted by Bohr's theory. This explanation was proposed by K. F. Herzfeld, *Privatdozent* at the University of Munich, who had assumed that the missing half of the angular momentum was taken away by radiation (Sommerfeld to Einstein, 18 and 29 December 1920). Einstein did not accept this interpretation and raised objections, while Herzfeld tried to prove his hypothesis experimentally at the laboratory in Sommerfeld's Institute without success (Sommerfeld to Einstein, 14 March 1921 and 4 July 1921). Later, Sommerfeld mentioned his progress in dealing with the spectroscopic data, including the second-order Stark effect (Sommerfeld to Einstein, 14 March 1921), the anomalous Zeeman effect (Sommerfeld to Einstein, 10 August 1921), and the inner quantum numbers (Sommerfeld to Einstein, 17 October 1921). In his letter of 11 January 1922 Sommerfeld mentioned to Einstein the progress which his student Heisenberg had achieved in dealing with the anomalous Zeeman effect. Einstein, on the other hand, informed him about a canal ray experiment that he had proposed to decide the nature of light, and which Geiger expected to carry out (Einstein to Sommerfeld, 27 September 1921 and 9 October 1921). On 28 January 1922 Einstein informed Sommerfeld that the experiment had not been conclusive. (See also Einstein, 1922b.)

the seminars. Soon after his arrival Lenz and Kratzer left but Laporte and, somewhat later, Karl Bechert came in.[26]

Sommerfeld devoted much time and attention to every research student who was associated with him in the Seminar, thereby fulfilling 'the first condition of being a good teacher' (Heisenberg, Conversations, p. 16). He would call every one of them by turn to his office to discuss things, and Heisenberg was summoned at least every alternate morning for an hour or two. During these sessions Sommerfeld would inquire about the student's progress in his research work or his opinion of somebody else's results. He would encourage them to propose problems on which they might work. Sometimes he would induce them to take interest in new problems by declaring, 'I cannot solve this problem, why don't you try it?' His approach appealed to the young people, and they would then put all their resources to work on the new ideas and difficulties. From the very beginning Heisenberg developed a great enthusiasm for his work in such an encouraging atmosphere.

I.4 Sommerfeld's Mathematical Approach

Since Heisenberg worked in Sommerfeld's Seminar from the very beginning of his studies at Munich, he came immediately under the influence of his eminent teacher. This was of great importance for his education, for normally the students went much later to a professor to work for a doctorate under him. As a rule, the new students of physics would take a more or less standard curriculum consisting of lectures and exercises in physics and mathematics and some laboratory work.[27] Only after about two years of training in the more elementary parts of physics and basic mathematical tools, the advanced student would either join the Theoretical Seminar—in case he wanted to specialize in theory—or an institute of experimental physics. Thus, even for the early 1920s, Heisenberg's curriculum was an exception rather than a rule. However, he was not the only one who turned to Sommerfeld immediately as a freshman. Wolfgang Pauli had also joined the Theoretical Seminar immediately after graduating from high school in 1918. More than any other scientist in Germany, Sommerfeld allowed young students to participate in his Seminar, which meant that they could work

[26]Karl Richard Bechert was born on 23 August 1901 in Nuremberg, Germany, and studied at the University of Munich from 1920 to 1925, when he received his doctorate under Sommerfeld. He spent the year 1925–1926 at the Physics Institute, University of Madrid, as a Rockefeller Foundation Fellow. From 1926 to 1933 he was one of Sommerfeld's assistants; he also became a *Privatdozent* in 1930. In 1933 he was appointed to the chair of theoretical physics at the University of Gießen, from where he was called to the University of Mainz in 1946. He died on 1 April 1981.

[27]This curriculum corresponded to the courses of undergraduate studies at good American universities. In general, undergraduate studies at German universities were much less regulated than at other places in Europe or America; the students enjoyed considerable freedom in selecting and pursuing their courses of studies.

on independent research problems before they had gone through the regular curriculum. He was able to do so because of his intense yet relaxed manner of teaching and advising, and his method was particularly successful with brilliant students like Heisenberg and Pauli.

Sommerfeld was precise in his instructions to his students as to what had to be done. Because of his early training as a mathematician, Sommerfeld always placed the greatest emphasis on the mathematical aspects of a physical problem. He liked 'clean' mathematics, but also did not want to have any mathematical formulae written down that could not be understood in terms of physics. This was the essence which characterized Sommerfeld's own work and the training he imparted: emphasis on mathematical rigour according to the standard methods and a complete explanation in terms of physical significance. He was not interested in strict mathematical rigour, such as proofs of convergence or existence, etc. Sommerfeld's approach to the mathematical formulation of physical problems was pragmatic; he enjoyed connecting mathematics to physics, but not in obtaining mathematical proofs alone. He took it for granted that the mathematics would somehow work and did not at times even worry about the inconsistency of the mathematical scheme he employed. He felt that if a given mathematical scheme did not work, then another could be found without altering the physical context.

Sommerfeld sought to ensure that his students would learn the mathematical techniques and acquire the knowledge of physics at the same time. He employed this method both in his courses on theoretical physics and in his personal guidance of students. It seemed that in discussing physical problems he always had a bagful of mathematical tricks, and he taught these tricks as appropriate tools for solving problems. For instance, in his course on electrodynamics Sommerfeld would teach how one could solve the linear and partial differential equations by means of exponentials, but he would not discuss the general principles, methods and rules of solving differential equations.[28] In classical mechanics he would show how to solve the equation of motion of a system and how to derive, say, energy conservation: namely, by taking the equation of motion, multiplying it by the velocity and integrating it, he would arrive at the equation denoting energy conservation. He would, however, not discuss the

[28]The method of presenting the solution of problems, less in a systematic manner than as a summary of nice, workable and, at times, beautiful tricks, came through especially in Sommerfeld's lectures on the partial differential equations of physics, which he gave as the last part of his course-cycle on theoretical physics. The book based on these lectures, though it covers the principal mathematical tools, collects the main examples of partial differential equations in physics and shows how to obtain their solutions (Sommerfeld, 1945b).

This method of treating the subject might well be appropriate for a physicist, but other theoreticians followed different styles. For example, Max Born, who grew up in David Hilbert's school, was interested in the mathematical methods themselves. Thus, when he examined a problem and decided upon the method to use, he would carefully investigate the existence of a solution and the question whether the solution was unique. 'For Born, once a question had been formulated mathematically, the examination of the mathematics of the question itself would assume basic importance' (Heisenberg, Conversations, p. 36).

underlying group-theoretical problem. As Heisenberg recalled: 'Sommerfeld would say, "Now we have a problem which has rotational symmetry." But he would not say, "Now we consider the group of rotations." He would say, "Since we have rotational symmetry, then, of course, it's a nice trick to introduce . . . for the coordinates; then you will see that things work out"' (Heisenberg, Conversations, p. 38). In other words, Sommerfeld considered group theory less as a systematic mathematical tool than as another mathematical trick. It did not matter much to him as to what lay behind the trick, or why certain tricks worked and others did not. Therefore, his students did not receive from him strict and infallible prescriptions for attacking a given problem or set of problems. They had to go through a large number of examples before they acquired a feeling for various mathematical methods and devices. After they had learned these, they knew the importance of many mathematical tools, some of which had not yet even entered any mathematical textbook.[29]

Sommerfeld, by his insistence and example, also passed on to his students his own attitude regarding the mathematical methods of physics. Wolfgang Pauli, in particular, took over his attitude. 'Pauli was not interested in mathematical proofs,' Heisenberg recalled. 'In this sense he was a true disciple of Sommerfeld. On the other hand, John von Neumann, an excellent mathematician, was very interested in rigorous mathematical proofs. Von Neumann once told Pauli: "I can prove this and this." To which Pauli replied, "Well, if a proof was important in physics, you would be a great physicist." This exchange, although rather strongly expressed by Pauli's comment, symbolizes the two approaches dealing with the connections between physics and mathematics. For Sommerfeld and his school, the goal was to describe nature in mathematical terms; the mathematical foundations of proofs were not important. The most important thing was the mathematical representation of nature' (Heisenberg, Conversations, p. 37; also AHQP Interview).[30]

I.5 Lectures and Courses

The most important of Heisenberg's courses were the lectures on theoretical physics. At the University of Munich, Arnold Sommerfeld used to give a

[29] As Heisenberg recalled later:

> There was always an interest in mathematical techniques at Sommerfeld's Institute, questions such as solving integral equations, how to go to the complex plane, etc. For instance, Sommerfeld would have been delighted about the development of Regge-pole theory. In fact, the Regge-pole theory goes back to a paper of Sommerfeld and Hopf, which is still quoted [Hopf and Sommerfeld, 1911]. They actually did make such a transformation [the so-called Sommerfeld–Watson transformation] from the angular momentum of the Regge poles. (Heisenberg, Conversations, pp. 41–42; also AHQP Interview).

[30] On another occasion Pauli expressed the same attitude even more definitely by saying to George Uhlenbeck: '*All diese Weierstrass'sche Konvergenz, das ist doch keine Physik.*' [All these Weierstrassian convergence [proofs], that's not physics!'] (Mehra, Conversations with Uhlenbeck and Kac, 1973, p. 157).

course-cycle on theoretical physics extending over six semesters. He would start with classical mechanics and go through, successively, the mechanics of deformable media, thermodynamics and statistical mechanics, electrodynamics, optics, and mathematical methods of theoretical physics (the partial differential equations of physics). The cycle would then commence again. There was a separate course on atomic theory.[31] Before joining the first course on mechanics in their third or fourth term at the University, the students were expected to have devoted themselves sufficiently to mathematical preparation in the calculus and differential equations. A student could enter the cycle of Sommerfeld's courses on theoretical physics at an intermediate point, but he would have had to complete his preparation with the help of textbooks. Later, Sommerfeld started giving two course-cycles of six terms, but with the difference in phase, so that every third term the cycle would start again, and the students wishing to begin the cycle would not have to wait long.

There used to be eighty to a hundred students in Sommerfeld's lectures. Sommerfeld would assign them some problems once a week to be worked out at home, and the solutions would then be turned in to his assistant who would go through them. From his homework Sommerfeld would obtain some impression of how the student was doing, but he would not give any grades for this work. The problems would then be explained on the blackboard, sometimes by his assistant and at other times by himself. Often one of the students would be called upon to explain the problem on the blackboard, and if things were not quite right, then either Sommerfeld or his assistant would interrupt and point out the better way of doing it. In this way a continuing discussion would develop between Sommerfeld, his assistant, and the students. They would spend about fifteen or twenty minutes on a problem, and then go on to the next one, taking up three or four problems every week. With these problems solved, the students knew all about them, and they would receive new problems for the following week, and the process would go on from one semester to the next.

Heisenberg's first semester at the University of Munich was in winter 1920–1921. Since he had already learned the calculus at school, he attended Sommerfeld's course on mechanics right away, and learned a lot from his efficient method of teaching. Heisenberg often gave very long and detailed solutions to the problems, at times running to ten or twenty pages, and he had to learn to solve problems more concisely.

In his first semester, in addition to Sommerfeld's lectures, Heisenberg attended Wilhelm Wien's course on experimental physics, dealing with mechanics and optics, but he did not learn much from it. 'It was a kind of a show. Wien's

[31] Sommerfeld later published, between 1943 and 1952, his course on theoretical physics: Vol. I, *Mechanics*, 1943; Vol. II, *Mechanics of Deformable Bodies*, 1945; Vol. III, *Electrodynamics*, 1948; Vol. IV, *Optics*, 1950; Vol. V, *Thermodynamics and Statistical Mechanics*, 1952; Vol. VI, *Partial Differential Equations in Physics*, 1945. (Sommerfeld: 1943; 1945a, b; 1948; 1950; 1952, and other editions.)

The contents of the lectures on atomic theory appeared in *Atombau und Spektrallinien* (*Atomic Structure and Spectral Lines*) (Sommerfeld, 1919, 1921e, 1922d, 1924d, and other editions.)

demonstrations were designed for effect, but not much information was conveyed. Some spectacular experiments were performed, but it was not clear what they were about' (Heisenberg, Conversations, p. 44). For a while he also attended Leo Graetz's course on a general survey of physics.[32] Occasionally he attended the Seminar on X-rays and Bohr's Atomic Model, which Sommerfeld had organized jointly with Wilhelm Lenz, Paul P. Ewald and Karl F. Herzfeld, and where the questions of spectral lines and anomalous Zeeman effects were also discussed.

During the summer semester 1921, Heisenberg attended Sommerfeld's course on continuum mechanics (including hydrodynamics and elasticity), which turned out to be the starting point of his later work on turbulence. The lectures on hydrodynamics were attended by a large number of students, principally those who intended to pursue careers in industry or engineering. Heisenberg was probably Sommerfeld's only student who worked seriously on hydrodynamics. During this semester he worked on a paper on turbulence (Heisenberg, 1922b), and from this early interest grew up his doctoral dissertation on the stability of laminar flow (Heisenberg, 1924c).

Karl Herzfeld, a *Privatdozent* at the University of Munich, used to give a course on the 'Mathematical Introduction to Physical Chemistry' which Heisenberg attended in summer 1921. This course primarily dealt with equilibrium thermodynamics, Van't Hoff's law, chemical reactions, etc., rather than kinetic theory and statistical mechanics. Heisenberg found that, although Herzfeld was rather poor as a lecturer, he was a good teacher and one could learn a lot from him.[33]

[32] Wilhelm Wien, the famous experimentalist and discoverer of Wien's radiation law, went to the University of Munich in 1920 as Wilhelm Conrad Röntgen's successor in the Chair of Experimental Physics. He was director of the Physics Institute (*Physikalisches Institut*). Leo Graetz was the second man at Wien's Institute. Graetz, born on 26 September 1856 in Breslau, studied physics and mathematics at the Universities of Berlin and Strasbourg. After receiving his doctorate at Strasbourg he became an assistant there in 1881. In 1883 he joined the University of Munich as *Privatdozent*; he was promoted to an associate professorship (*Extraordinarius*) in 1893, and to professorship in 1908. Graetz worked in many fields of experimental physics, including heat conduction and radiation, electromagnetic radiation and X-rays, and he wrote several textbooks on modern physics. He died on 12 November 1941 in Munich.

For the winter semester of 1920–1921 Wien had announced lectures on *Experimental Physics II*, covering mechanics and optics, while Graetz gave a course entitled *Experimental Physics I*, which included heat and electricity (see *Phys. Zs.* **21**, 1920, p. 559). In summer 1921 Wien's course (again announced as *Experimental Physics II*, see *Phys. Zs.* **22**, 1921, p. 287) dealt with electromagnetism. In his lectures, Wien—who as director of the Institute for Experimental Physics gave the main courses that were attended not only by the students of physics, but also by the students of chemistry, biology, mathematics and medicine—covered the entire field in a rather general way. He showed many dramatic effects in experimental demonstrations. However, Heisenberg found that he did not learn much from Wien's courses; he considered them as a kind of show. He felt that not much information was conveyed and, though spectacular experiments were performed, it was not clear what they were about.

[33] Karl Ferdinand Herzfeld, born on 24 February 1892 in Vienna, Austria, attended the *Schottengymnasium* in his hometown (1902–1910), and then studied at the Universities of Vienna (1910–1912), Zurich (1912–1913) and Göttingen (1913–1914). After obtaining his doctorate from Vienna in 1914, he served during World War I as a first lieutenant in the Austrian Army. In 1920 he became a

During the following winter semester (1921–1922), besides Sommerfeld's course on electrodynamics and electron theory, Heisenberg attended Hugo von Seeliger's lectures on theoretical astronomy (the three-body problem) but he was not able to follow them too well.[34] He also attended Herzfeld's lectures on '*Quantenmechanik der Atommodelle*' (Quantum Mechanics of Atomic Models) and the introduction to kinetic gas theory.

During his first four semesters at the University of Munich, Heisenberg variously attended the lectures of the mathematicians Voss, Rosenthal, Pringsheim and Lindemann. Aurel Voss[36] lectured on the theory of functions, and Artur Rosenthal taught analytical and differential geometry. Rosenthal was small, lively, and 'a very nice gentleman'; he was an excellent teacher who sought

---

Privatdozent at the University of Munich, and in 1923 was promoted to extraordinary (i.e., associate) professor, representing there the field of theoretical physical chemistry. In 1926 Johns Hopkins University invited Herzfeld as Speyer Guest Professor, and he remained there as a professor of physics until 1936, when he joined the Catholic University of America, Washington, D.C., as Head of the Physics Department. Herzfeld became known for his research on thermodynamics, solid state physics, kinetic theory of gases, ultrasonics, electronic structure of molecules, reaction velocities, interior ballistics and semiconductors, and as an author of books on kinetic theory and ultrasonic waves. He received numerous honours, including honorary degrees, from various universities in the United States and Germany. He died in June 1978.

[34]Hugo von Seeliger lectured on celestial mechanics during the winter semester of 1921–1922, dealing with the theory of planetary perturbations. In the following summer term he announced a course on 'Selected Topics in the Three-Body Problem.' Heisenberg attended his courses both for his interest in astronomy and because he had selected this field as one of his minors (*Nebenfächer*) for his doctoral examination. He was examined by von Seeliger in July 1923 and received the mark of 'good.'

Hugo von Seeliger, for forty-two years Professor of Astronomy at the University of Munich and Director of the Munich Observatory, was born on 23 September 1849 in Bielitz-Biala, Austria. He studied at the Universities of Heidelberg and Leipzig, became Observer at the Bonn Observatory from 1873 to 1877 (participating in the Transit of Venus Expedition to Auckland Isles) and *Privatdozent* at the Universities of Bonn and Leipzig, before he was appointed Director of the Gotha Observatory in 1881. The following year he accepted the call to the chair at Munich, which he occupied until his death on 2 December 1924.

Von Seeliger performed pioneering research on stellar distribution between 1884 and 1911, working both on observational data and theoretical explanation; in 1898 he proposed the first systematic mathematical theory of star statistics. In 1893 he suggested a test of Maxwell's theory of Saturn's rings, which led to its empirical confirmation. He was very interested in the study of cosmic dust masses, and in connection with it he suggested a prerelativistic explanation of the perihelion motion of Mercury. In his obituary notice of von Seeliger, Arthur S. Eddington said:

> He himself was an uncompromising opponent of Einstein's theory. Indeed in his latter years he seems to have had little sympathy with the new ideas growing up in physics and astronomy. His writings do not give the impression of a stalwart conservative, and he could speculate on the nature of Novae as rashly as the rest of us. But he was evidently a man not easily moved in his opinions, and new proposals were regarded chiefly as a challenge to his keen critical faculties. (Eddington, 1925, p. 318)

Von Seeliger was a great teacher; Karl Schwarzschild had been one of his students. He made Munich an important centre for training in astronomy. He served as President of the *Astronomische Gesellschaft* from 1897 to 1921.

[35]The title of Herzfeld's course in the winter semester of 1921–1922 was one of the earliest occasions when the name 'quantum mechanics' was used. (See *Phys. Zs.* **22**, 1921, p. 591.)

[36]Aurel Voss, born on 7 December 1845 in Altona, Germany, was Professor of Mathematics at the Universities of Würzburg and Munich (1902–1923). He became Emeritus in 1923 and died in Munich on 19 April 1931.

to make the learning of mathematics attractive to his students.[37] Like Sommerfeld, Rosenthal used to hold problem-solving sessions in which Heisenberg took part regularly. He competed in solving the assigned problems with a fellow student, Robert Sauer, the two of them succeeding with most of the problems.[38]

Alfred Pringsheim lectured on number theory and the theory of functions.[39] Both Pringsheim and Lindemann gave rigorous proofs of convergence theorems *ad infinitum*, and Heisenberg found it all boring and unprofitable. Although he still thought now and then about his earlier desire of pursuing pure mathematics, his lack of enthusiasm and success in the lectures of Pringsheim and Lindemann —apart from the disappointment with his encounter with the latter—had con-

[37] Artur Rosenthal was born on 24 February 1887 in Fürth, Bavaria. He began his mathematical career at the University of Munich as *Privatdozent* in 1912 and was promoted in 1920 to extraordinary (i.e., associate) professor. In 1922 he went to the University of Heidelberg as an Extraordinary Professor, becoming a full professor in 1932. He was made Professor Emeritus (forced) in 1935; he emigrated to the United States, where he started a second career. He became a research fellow and lecturer at the University of Michigan, Ann Arbor, in 1940; then lecturer (1942–1943), assistant professor (1943–1946) and associate professor (1946–1947) at the University of New Mexico, Albuquerque. From 1947 to 1957 Rosenthal served as Professor of Mathematics at Purdue University, Lafayette, Indiana; he died in Lafayette on 15 September 1959.

Rosenthal worked especially on measure and integration theory, which he also applied to physical problems like ergodic theory. Thus he proved that the so-called 'ergodic' hypothesis of Ludwig Boltzmann could not be satisfied by a physical system, but that the 'quasi-ergodic' hypothesis of P. and T. Ehrenfest might apply to the molecules of a free gas (Rosenthal, 1913, 1914).

[38] Robert Max Friedrich Sauer, born on 16 September 1898 in Pommersfelden, became *Privatdozent* at the Technical University of Munich in 1926; he was given the title of 'professor' (*Außerplanmäßiger Professor*) in 1932. The same year he was appointed an Extraordinary Professor of Mathematics at the Technical University in Aachen, and was made a full professor in 1937. He moved in 1944 to the Technical University, Karlsruhe, and, finally, in 1948, to the Technical University, Munich, where he remained as Professor of Mathematics and Mechanics until his retirement. Sauer worked on problems of pure and applied mathematics, especially on differential geometry, differential equations, gas dynamics and ballistics. He died in Munich on 22 August 1970.

Heisenberg remembered that either he or Sauer would solve practically all the problems in Arthur Rosenthal's class, and they had great fun competing with each other. Heisenberg loved those courses in which he was required to calculate things. He argued:

> It was extremely important that one should try to solve problems. Just listening [to lectures] is of little use. Particularly if the lecture is good, then everything is too smooth. That's the same in music; if the performance is too good, you really don't enjoy it, because it just goes by and you can never penetrate into the heart of it. Sometimes a poor performance is better for enjoyment, because you can look at those things that were wrong and analyze them. (Heisenberg, Conversations, pp. 43–44)

There used to be problem sessions in the courses of Rosenthal and Pringsheim, but not in those of Lindemann and Voss.

[39] Alfred Pringsheim was born on 2 September 1850 in Ohlau, Silesia. As a young man, Pringsheim, who was acquainted with Richard Wagner, hesitated a while before deciding whether he should turn to music or to mathematics; he chose the latter. He obtained his doctorate in 1872 from the University of Heidelberg and became a *Privatdozent* at the University of Munich in 1877. In 1886 he was appointed an extraordinary professor at the University of Munich, and was made a professor of mathematics in 1901; he retired from this position in 1922.

Pringsheim was a productive mathematician who devoted himself to the theory of functions in the spirit of Weierstrass; he proved many fundamental theorems in this field. Among his mathematical hobbies were the investigation of convergence of infinite processes involved in the summation, products, etc., of series, and the study of Fourier series. Frequently he picked up previous proofs of

vinced him that he was really not suited for a career in mathematics.[40] The decision to study physics under Sommerfeld, which had been made automatically, had been a wise one after all.

During the early semesters in Munich, Heisenberg would study in the mornings in the Seminar Room, concentrating on the textbooks and problems related to the lectures he was attending at a given time.[41] Sometimes he would try to read a newly published paper, often finding it too difficult to understand. In order to understand the papers he would go back to the textbooks. He also took part in the physics colloquia. The colloquia used to be on diverse subjects, some of which—such as spectroscopy or experimental physics—he had not learned yet and were difficult to follow. His education thus far had been one-sided, as he had always specialized in modern theoretical problems. His training was especially incomplete in experimental physics. For example, like every physics student, he had to participate in a laboratory course ('*Physikalische Übungen*'), where groups of two students performed more or less elementary experimental exercises. Heisenberg did so in the summer semester, 1921, and his partner was Wolfgang Pauli. But instead of working earnestly on the given problems, the two preferred to discuss some topic of atomic theory that was more interesting to them. Once they had to determine the frequency of a tuning fork; when the time was nearly up, Heisenberg, who was well-trained in music, just guessed the

---

mathematical theorems and removed gaps and errors in them; at the same time he tried to simplify the argumentation. Pringsheim was also interested in the history of mathematics; thus he became an excellent writer of review articles, especially for the *Encyklopädie der mathematischen Wissenschaften*.

Pringsheim was a brilliant and witty lecturer who loved clarity and accuracy combined with a fine critical irony. He often gave popular mathematical talks, and on general public occasions he would sprinkle his speeches with poems that he had composed himself. He was a man of great culture; his fine Munich home was visited by artists like the painter Franz Lenbach, the poet Paul Heyse, and, later on, the novelist Thomas Mann, who married his daughter Katja. Following the political events in Germany after 1933, Alfred Pringsheim was obliged to sell his house and finally to leave Munich with his wife. He died in Zurich on 25 June 1941. His son, the physicist Peter Pringsheim (1881–1963), escaped from France to the United States with the help of Thomas Mann.

[40] Heisenberg never had much fun in mathematics where one had to prove things. 'This proving of such and such I found to be almost like cheating. You start somewhere, and then you go into a dark tunnel, and then you come out at another place. You find that you have proved what you wanted to prove, but in the tunnel you don't see anything' (Heisenberg, Conversations, p. 20).

In Munich the rigorous mathematical methods were taught in the courses of Pringsheim and Lindemann. Heisenberg's lack of success with these two mathematical authorities, on the one hand, and the encouragement he received from Sommerfeld, on the other, made him go smoothly into theoretical physics. As he said, he was too 'sloppy' ('*schlampig*') for mathematics (Heisenberg, Conversations, p. 18). However, he continued to have good experience with mathematical examinations; for his performance in the oral examination in mathematics connected with his doctorate, he earned the mark of 'very good,' having been examined by Oskar Perron.

[41] Since Sommerfeld's lectures on theoretical physics were not yet available as books, Heisenberg had to use the literature available. For mechanics, he first used the relevant parts of Müller–Pouillet's *Lehrbuch der Physik und Meteorologie*, and later the new textbook on mechanics by Müller and Prange (1923); for electrodynamics he studied the textbook of Abraham (1914, 1918). For certain specific parts of mechanics, Heisenberg could also consult Sommerfeld's *Atombau und Spektrallinien* (1919, 1921e) because Sommerfeld had treated this subject in some detail in the appendix of that book.

correct result. This lax attitude towards experimental physics nearly led to a catastrophe in his doctoral examination. For the moment, however, he did not care. He immensely enjoyed working on the theoretical problems. And his first four semesters at Munich became most exciting because of the discussions with Sommerfeld, Pauli, and Wentzel, from whom he learned a great deal. In the fall of 1922 Sommerfeld went to the United States, where he spent the first part of the academic year 1922–1923 as Karl Schurz Memorial Professor at the University of Wisconsin, Madison, and lectured on atomic structure and the theory of relativity. In Sommerfeld's absence from Munich, Heisenberg spent the winter semester, 1922–1923, with Max Born in Göttingen. He returned to Munich again to complete the work on his doctorate during summer 1923.

I.6 Introduction to Research in Quantum Theory

Just four weeks after Heisenberg had been in the Seminar, Sommerfeld gave him a problem which immediately excited him: to explain the experimental data on the anomalous splitting of spectral lines in a magnetic field. Sommerfeld had, for some time, been interested in deriving the so-called anomalous Zeeman effects on the basis of Bohr's theory of atomic structure. He had available certain pictures showing the observed patterns in the case of specific multiplet lines, and he wanted Heisenberg to determine explicitly what the initial and final states of the emitting atom were; that is, by using the selection rules, to find the quantum numbers by which the states had to be labelled. Heisenberg was delighted. This was a problem particularly suited for him because it contained elements of number theory which he loved from school. Moreover, Sommerfeld had earlier characterized the situation in the anomalous Zeeman effects as a 'mystery of numbers' ('*Zahlenmysterium*,' see Sommerfeld, 1920a, p. 64), and this mystery attracted Heisenberg.

The problem which Heisenberg received had already puzzled Sommerfeld for a while. In 1916 he had been able to derive the normal Zeeman effect, i.e., the splitting of hydrogen lines into three components in a magnetic field—such that the middle component was not shifted at all and the outer ones were displaced by an amount proportional to H, the strength of the magnetic field, the factor of proportionality being $a = e/4\pi m_e c$—from his extension of Bohr's theory (Sommerfeld, 1916d). Three years later he had turned also to consider the anomalous Zeeman effects on the same basis, starting out from Runge's empirical rule. By analyzing the complex magnetic splitting of the spectral lines of neon, Carl Runge had discovered long ago that $\Delta\nu$, the frequency differences, satisfied the relation

$$\Delta\nu = \frac{q}{r} aH, \tag{1}$$

where a $(= e/4\pi m_e c)$ is the constant factor mentioned above, H denotes the magnetic field strength, and q and r are integers, usually lying between 1 and 10 but not much larger (Runge, 1907). This observation had then been confirmed in many examples of spectra that showed the anomalous Zeeman effect, and Eq. (1) was called *Runge's rule*, the numbers q and r being named, respectively, 'Runge's numerator' and 'Runge's denominator.' What Sommerfeld had done in late 1919 was to separate the factors q/r, connected with the frequencies, into the difference of two such factors, q_1/r_1 and q_2/r_2, each of which should be associated with one of the two terms whose combination yields the transition frequencies in atomic theory. However, he did not get very far with this task. All he had succeeded in obtaining was a table suggesting the values of the Runge denominators for the terms: these values were 1 for all terms of atoms emitting singlet lines (which exhibit the normal Zeeman effect); for triplet atoms they increased from 1 for the s-term, to 2 for the p-term, to 3 for the d-term, etc., while for doublet atoms they were 1 for the s-term, 3 for the p-term, 5 for the d-term, etc. It was this table which Sommerfeld called a 'number mystery' (Sommerfeld, 1920a, p. 64). However, about the main question, namely, how to interpret the denominator scheme in the case of doublet and triplet series (in the absence of external magnetic fields), he had not made any progress.

This situation remained practically unchanged in 1920, as represented in the extended paper on the magneto-optical decomposition rule ('*Zerlegungssatz*')—which Sommerfeld submitted in March to *Annalen der Physik*, and which appeared in October (Sommerfeld, 1920f). There he stated the problem in the following words:

> Our consideration, which is based on the combination principle, would be completed only when we succeed in associating with the terms s, p_i, d_i [where the suffix i runs from 1 to 2 in the case of doublets, from 1 to 3 in the case of triplets] separately [i.e., with each of the terms individually] certain magnetic splittings [i.e., additional energy terms involving quotients of numerators and denominators] and in deriving from their combination the observed splitting patterns and their polarizations . . . According to the example of the derivation of the normal Zeeman effect in the case of the hydrogen atom model, the following procedure appears to me as being ideal: One associates with the different terms the following energy levels of the magnetic splitting (suppressing the common factor $(e/4\pi m_e c)H$):
>
> $$\left.\begin{array}{lllllll} s: & 0, & \pm 1, & \cdots & & & \\ p: & 0, & \pm\frac{1}{2}, & \pm 1, & \cdots & & \\ d: & 0, & \pm\frac{1}{3}, & \pm\frac{2}{3}, & \pm 1, & \cdots & \end{array}\right\} \text{(triplet systems)}$$
>
> $$\left.\begin{array}{lllllll} s: & 0, & \pm 1 & \cdots & & & \\ p: & 0, & \pm\frac{1}{3}, & \pm\frac{2}{3}, & \pm 1, & \cdots & \\ d: & 0, & \pm\frac{1}{5}, & \pm\frac{2}{5}, & \pm\frac{3}{5}, & \pm\frac{4}{5}, & \pm 1, \cdots \end{array}\right\} \text{(doublet systems)}$$
>
> and combines these splittings by taking into account suitably chosen selection rules . . . However, I have not succeeded in describing theoretically the empirical splitting patterns in this manner. (Sommerfeld, 1920f, pp. 253–254)

It was Heisenberg's first task at the Seminar to find in the case of doublet spectra the quantum numbers that gave rise to the observed magnetic splittings.[42]

Heisenberg did not have to work very long on the problem. He quickly acquainted himself with the scheme which Sommerfeld had invented for describing the states of atoms emitting doublet and triplet lines, including the recently postulated 'inner quantum number' (Sommerfeld, 1920f, p. 232), whose values characterized the different components of a doublet or triplet. He applied his new knowledge to disentangle the doublet data, returning to Sommerfeld a couple of weeks later with a complete scheme. He made a statement, however, which he almost did not dare to speak out aloud: 'The whole thing works only if one uses half-integral quantum numbers.' This rather shocked Sommerfeld, because at that time—i.e., in late 1920—nobody spoke about half-integral quantum numbers, and his immediate reaction was: 'That is absolutely impossible. The only fact we know about quantum theory is that we have integral numbers, and not half numbers' (Heisenberg, Conversations, p. 14). On the other hand, Sommerfeld was certain that one could disentangle the data on the basis of his '*Zerlegungssatz*' ('Decomposition Rule'), though he had already employed the integral quantum numbers unsuccessfully. Since Heisenberg was a complete novice in atomic physics at that time, he did not have Sommerfeld's prejudices, and he said to himself: 'Why not try half-integral quantum numbers?' The old master and the young pupil had a long discussion about the problem, whether half-integral quantum numbers would be allowed or not, and they finally agreed that the half quantum numbers were perhaps all right.

The idea of half-integral quantum numbers became the subject of many discussions in Sommerfeld's Seminar, and the judgment was mostly negative. People thought that it was interesting for a beginner to entertain such an idea, but it was surely wrong. Wolfgang Pauli remarked to Heisenberg that if he used half quantum numbers, '[You] would soon have to introduce quarters and eighths as well, until finally the whole quantum theory would crumble to dust in [your] capable hands' (Heisenberg, 1971, p. 35).[43] The only real support came more than a year later. Early in 1922 Adolf Kratzer became interested in half-integral quantum numbers and had many discussions with Heisenberg on

[42] Heisenberg mentioned to me that Sommerfeld gave him the experimental values of the anomalous Zeeman effects, but he did not remember to what exactly they referred. From the published paper it appears that he might have worked first on the doublets of sodium or lithium (Heisenberg, 1922a). (J. Mehra)

[43] At that time Heisenberg was not aware of the fact that half-integral quantum numbers had been introduced earlier in connection with the zero-point energy. He recalled, however, that once Pauli told him, 'You may be right anyhow with your half quantum numbers. There is an old paper on the specific heat problem—something about the equilibrium between the solid state and a liquid—where you can actually see that there is a zero-point energy' (Heisenberg, Conversations, p. 93). The old paper, which Pauli had in mind, clearly referred to the work of Otto Stern on the chemical constant (Stern, 1919). Heisenberg remembered that Pauli mentioned the same argument to Sommerfeld, who then said: 'Well, there may be something in it' (Heisenberg, Conversations, p. 93).

In any case, when Heisenberg introduced the half-quantum numbers he was not aware of the earlier discussions about this question, which had started as early as 1911 with Planck's 'second quantum hypothesis' (Planck, 1911a).

this question. He told Heisenberg that the concept of half quantum numbers was useful in band spectra and that they probably fitted better with the experiments than integral quantum numbers; he also used them in a published paper (Kratzer, 1922). Heisenberg felt that his idea had been confirmed and he would use it again.[44] For the moment—i.e., in December 1920—however, nothing about half-integral quantum numbers was published from Sommerfeld's Institute, especially not Heisenberg's result on the anomalous Zeeman effects.

The situation changed a few months later when Alfred Landé's work on the same subject became known in Munich. Landé had begun to treat the anomalous Zeeman effect about the same time as Heisenberg; he obtained the first results in February 1921, about which he soon informed Sommerfeld.[45] Sommerfeld's initial response was very positive, but he soon tried to persuade Landé to wait for the publication of his results until Ernst Back, who had supplied important data, had come out with his paper. 'Don't be angry with me,' he wrote to Landé, 'that I seem to influence you in this matter; but the trustworthy collaboration with experiment, especially the one with Paschen's institute [of which Back was a member] should not be upset for the sake of theory. And I fear, because of Paschen's sensitive character and Back's tender nature, that a permanent disharmony may result if we do not tread here most delicately' (Sommerfeld to Landé, 3 March 1921).[46] Landé, however, obtained the consent

[44] The next occasion for employing half-integral quantum numbers occurred very soon. In his work on the energy states of helium, which he carried out mainly in Göttingen during the winter semester of 1922–1923, Heisenberg again suggested using half-integral quantum numbers for the orbits of the electron in order to fit the data. (See Section II.3.)

[45] Landé's first letter to Sommerfeld is not contained in the collection of Sommerfeld's scientific correspondence at the *Deutsches Museum* in Munich. However, its content was probably identical to his letters to Niels Bohr in which he discussed the rules for doublet and triplet lines (Landé to Bohr, 16 February 1921, 23 February 1921). Sommerfeld's first letter to Landé in the above collection is dated 25 February 1921.

Landé recalled later that he had already been working on the Zeeman effects before he saw Sommerfeld's paper in *Annalen der Physik* (Sommerfeld, 1920f), but the reading of the latter provided a major hint (Landé, AHQP Interview, Third Session). From this one may conclude that Landé began his investigation of the anomalous Zeeman effect after his visit to Copenhagen in October 1920.

[46] The background of Sommerfeld's request to Landé was a rather complex one. On the one hand, the collaboration between Paschen's Experimental Institute in Tübingen and Sommerfeld's Theoretical Institute in Munich was extremely close and successful; Sommerfeld was thus able to obtain experimental data long before publication, and this led to rapid progress in experimental spectroscopy *and* atomic theory between 1915 and 1920. On the other hand, in publishing the results of investigations each party, especially the theoreticians who based their work on not-yet-published data, had to respect the wishes of the other. Now Ernst Back, whose publications had been delayed because of his service in World War I, was about to write up his research work for his *Habilitation*, and Sommerfeld felt that one should not take away the priority of his results before he had had the opportunity of publishing them. (Sommerfeld also wrote a letter to Max Born, who was Landé's superior at Frankfurt, dated 8 March 1921, in which he stated these arguments very clearly.) That Sommerfeld's treatment of the situation was correct follows from a letter of Back, who wrote:

> This simultaneousness of both our efforts [i.e., those of Landé and Back] in explaining the Zeeman effect has anyhow already hit me gravely, of course entirely without the fault of Mr. Landé or of anybody else. I had to extend my *Habilitation* thesis by including totally different things, since these number games are no longer in my sole possession and the distinction of discovering them—after sharing it with Mr. Landé—does not suffice anymore for a *Habilitation*. (Back to Sommerfeld, 7 June 1921)

of Back and Friedrich Paschen to make free use of their data (Back to Landé, 7 March 1921; Paschen to Landé, 8 March 1921) and pushed forward. On 17 March 1921 he wrote letters both to Niels Bohr and Arnold Sommerfeld, in which he reported a simplified version of his rules for the anomalous Zeeman effects of doublet lines. He said:

> A state with the inner quantum number j and the term denominator r_j should assume in a magnetic field the $2j$ (additional magnetic) energy levels, $\pm 1\, j/r_j$, $\pm 3\, j/r_j, \ldots, \pm(2j-1)\, j/r_j$, with which the ("equitorial" quantum) numbers $\pm \frac{1}{2}$, $\pm \frac{3}{2}, \ldots, \pm(2j-1)/2$, might be associated. Then all doublet separations are obtained correctly when one applies the single *rule*: $\left.\begin{smallmatrix} p- \\ s- \end{smallmatrix}\right\}$ components arise from a change of $\left\{\begin{smallmatrix} 0 \\ +1 \end{smallmatrix}\right.$ in the equitorial quantum number (Landé to Sommerfeld, 17 March 1921).

To which Sommerfeld replied: 'The beginning of your paper is striking proof of my assertion that your Zeeman effect considerations were not yet ripe for publication. Your new representation agrees well with what had been found by one of my students (of the first semester [Heisenberg]), but which had *not* been published' (Sommerfeld to Landé, 31 March 1921).[47] But Landé could not be stopped anymore: on 16 April 1921 his paper entitled '*Über den anomalen Zeemaneffekt*' ('On the Anomalous Zeeman Effect') was received by *Zeitschrift für Physik* (Landé, 1921c). It came out in print the following June and contained rules to describe the anomalous Zeeman effect, which Sommerfeld had declared 'not yet ripe for publication.'[48] Later the same year Landé published two more articles on anomalous Zeeman effects, thus becoming the leading expert on the subject (Landé, 1921d, 1921f).

Heisenberg naturally had some reason to be unhappy, for Landé had taken away his priority. But Sommerfeld asked him not to feel bad, for the theoretical problem was by no means solved, and declared: 'The electrodynamic mechanism of the anomalous Zeeman effects is still hidden from us' (Sommerfeld and Back, 1921, p. 913). Indeed, Sommerfeld recognized the principal difficulty with the theoretical derivation of the observed data: the conventional theory of radiation, which had been employed, e.g., by his disciple Adalbert Rubinowicz to derive the selection rules for atomic radiation, only gave the normal Zeeman effect (Rubinowicz, 1918b). This failure, together with Bohr's apparent progress obtained at the same time in explaining the periodic structure of atoms from the correspondence principle (Bohr, 1921a), caused Sommerfeld to write to Bohr: 'I am completely convinced that your path is the right one . . . However, I am heretical enough to believe that this [i.e., the explanation of the length of periods in the

[47] Sommerfeld's letter to Landé is not to be found in the latter's scientific correspondence; however, a stenographic draft of this letter was written on Landé's letter to Sommerfeld of 17 March 1921.

[48] Heisenberg remembered later that Sommerfeld reacted with displeasure to Landé's letters concerning the anomalous Zeeman effect. He would say, 'This Landé always comes up with these complicated things and he is always so unclear. I don't know what the man means' (Heisenberg, Conversations, p. 98).

periodic system of elements] will be found in another, unified and less formal way. To be sure, this will happen only after the continuous electrodynamics has been replaced by a needle-type electrodynamics' (Sommerfeld to Bohr, 25 April 1921). As an immediate consequence, Sommerfeld changed his previously rather reserved attitude towards the correspondence principle; from now on he would consider the latter, in contrast to his earlier statements, as more general than Rubinowicz's selection rules.[49] With respect to the anomalous Zeeman effect, on the other hand, he continued to work on another suggestion of Niels Bohr: namely, to reformulate Woldemar Voigt's classical theory of the anomalous Zeeman effect (Voigt, 1913a, d) in quantum-theoretical language, in which he more or less applied the correspondence idea. And, in this work, he also engaged Heisenberg.

The approach to the quantum theory of the anomalous Zeeman effects, starting from Voigt's classical theory, was different from the previous theoretical work of Sommerfeld, Heisenberg and, in particular, of Landé, who had only used the regularities of the empirical data. Voigt's idea had been to consider, e.g., the doublet of sodium D-lines, as arising from quasi-elastically bound electrons, two for the D_2-line and one for the D_1-line (because the former was twice as intense as the latter); the electrons were independent when no external magnetic field was present (yielding the respective frequencies and intensities of the lines of free sodium atoms), and coupled together if the magnetic field were switched on, the coupling being linear with different strengths for parallel and perpendicular oscillations. From these assumptions Voigt had derived coupled differential equations for the motion of the electrons, whose solution led to a third-order equation in the magnetically induced frequency shift $\Delta\nu$. The solutions for $\Delta\nu$ and the corresponding intensities did fit the observed magnetic splitting of the sodium doublet very well; they also represented perfectly the transition of the anomalous Zeeman components into the normal triplet in very strong magnetic fields.

Since, in quantum theory, the lines must be interpreted as differences of terms, one had to find formulae for the magnetic splitting of terms, which were analogous to the ones obtained by Voigt for the frequencies themselves. Sommerfeld and Heisenberg thus set out in 1921 to perform the quantum translation. They worked together for some time on this problem; then Sommerfeld worked alone, and he finally submitted a paper in December of that year (Sommerfeld, 1922a). By systematic guessing he arrived at a formula describing the magnetic shift of a sodium term, ΔW:

$$\Delta W = \frac{h}{2}\left(m \pm \sqrt{1 + \frac{2m}{2n-1}v + v^2}\right)\Delta\nu_{\text{normal}}, \tag{2}$$

[49]This change in Sommerfeld's attitude occurred in spring 1921. In November 1920 he had still written to Bohr: 'Nevertheless, I must confess that the origin of your [correspondence] principle, alien as it is to quantum theory, still distresses me, however much I recognize that it reveals a most important connection between the quantum theory and classical electrodynamics' (Sommerfeld to Bohr, 11 November 1920).

where n and m are the azimuthal and spatial (magnetic) quantum numbers, respectively, v the ratio of the (field-free) doublet splitting over the normal Zeeman splitting, $\Delta\nu_{\text{normal}}$, and h is Planck's constant. This formula, together with the selection rule, $\Delta m = 0, \pm 2$, fitted the data very well. The same was true of the reformulated intensity of Voigt. Sommerfeld was very pleased, and in his acknowledgement at the end he noted: 'I do not wish to omit to thank sincerely my student Mr. W. Heisenberg for his successful collaboration on the entire problem of the anomalous Zeeman effects' (Sommerfeld, 1922a, p. 272).

Meanwhile, Heisenberg's work on the atomic theory of anomalous Zeeman effects progressed very rapidly. This can be seen from the intensive correspondence between Heisenberg and Landé from October to December 1921. These letters provide the earliest evidence of Heisenberg's approach to a physical problem, of his quick and goal-oriented activity, and of his constant push towards a desired result. The young man reported to Landé about every step relating to what he called the '*Atomrumpf*' ('*Rumpf*' = 'trunk' or 'core') model, dealing at the same time with Landé's objections to his frequently bold assumptions on the path to the first systematic explanation of the anomalous Zeeman effects from the point of view of the Bohr–Sommerfeld theory of atomic structure.[50]

Heisenberg started out by discussing a formula of Landé for the maximum magnetic energy of the triplet states, which was proportional to $n + 1$ (Landé, 1921c, p. 236, Eq. (12), with n replacing Landé's k). He interpreted this result as being due to the sum of the angular momenta of the valence electron, n (in units of $h/2\pi$), and of the core, 1, respectively (Heisenberg to Landé, 3 October 1921). From there he proceeded to specific models of atoms (emitting doublet and triplet series spectra), based on the assumption that the term structures of these atoms arose from an internal Zeeman effect, which was produced by the action of the external valence electrons, viz., their magnetic field, on the rest of the atom, i.e., the core. Heisenberg arrived at these models in November (doublet model: Heisenberg to Landé, 5 November 1921; triplet model: Heisenberg to Landé, 9 November 1921). He showed that these models allowed one to obtain Landé's formulae for the anomalous Zeeman effects (Landé, 1921c, f); but they did much more. Sommerfeld had suggested that Heisenberg also investigate the dependence of the energy levels of doublet and triplet atoms on the external magnetic field and seek to establish formulae similar to Eq. (2), his own quantum translation of Voigt's equation for the doublet D-lines. Heisenberg succeeded with this programme, and on 19 November he wrote a letter to his friend Pauli, reporting all the results he had obtained with the core model.

[50] About fifteen letters and postcards of Heisenberg to Landé have been preserved (Heisenberg to Landé, 3, 11, 16, 23, 29 October 1921; 5, 9, 16, 19, 28 November 1921; 1, 8, 15 December 1921). Landé's scientific correspondence is kept at: *Staatsbibliothek Preußischer Kulturbesitz, Handschriften-Abteilung*, Berlin (West), Federal Republic of Germany. Landé's letters to Heisenberg, as most of the letters which Heisenberg had received in the 1920s and 1930s, were lost during World War II. Heisenberg's correspondence with Landé was the most intensive exchange of letters in a short period of time that he had with a physicist on scientific matters until his correspondence with Pauli from the beginning of 1957 to April 1958.

According to Heisenberg's core model of November 1921 a doublet atom (i.e., an atom emitting doublet spectral lines like sodium) possessed in the ground state the angular momentum 1 (in units of $h/2\pi$), which was equally distributed between the core ($\frac{1}{2}$) and the valence electron ($\frac{1}{2}$). In the excited state the core still had a momentum $\frac{1}{2}$, but the electron assumed momentum values $n - \frac{1}{2}$ ($n = 1, 2, \ldots$). The magnetic energy in the absence of an external field became

$$\Delta W_{H=0} = \tfrac{1}{2} H_i \frac{e}{4\pi m_e c} \cos\theta \, \frac{h}{2\pi}, \tag{3}$$

where H_i is the strength of the internal magnetic field acting on the core (H_i being calculated from the classical nonrelativistic electrodynamics), $-e$ and m_e are the charge and the mass of the electron, respectively, and c is the velocity of light. The angle θ between the directions of the magnetic moment of the core and the internal field obeyed what Heisenberg called the 'quantum condition,'

$$\cos\theta = \pm 1. \tag{4}$$

Hence the states of the atom formed doublets, which were denoted by different total angular momenta, n and $n - 1$, provided the electron carried the momentum $n - \frac{1}{2}$. If an external magnetic field was switched on, the moments of the core and the electron changed their directions in order to make the energy a minimum. Heisenberg calculated ΔW, the total magnetic energy, according to the laws of classical mechanics, obtaining the result

$$\Delta W = h\left(m^* \pm \frac{1}{2}\sqrt{1 + 2\frac{m^*}{n^*}v + v^2}\right)\nu_{\text{normal}}, \tag{5}$$

where the quantum numbers m^* and n^* ($= n - \frac{1}{2}$) assumed half-integral values, $\frac{1}{2}, \frac{3}{2}, \frac{5}{2}$, etc. The value of m^* determined the projection of the momentum of the valence electron (in units of $h/2\pi$) along the direction of the external field, hence $|m^*| \leqslant n^*$. Equation (5) could be identified with Sommerfeld's Eq. (2) if $2m^*$ were put equal to m and n^* put equal to $n - \frac{1}{2}$. The quantity v obtained a simple physical interpretation in Heisenberg's model as being equal to H_i/H, the ratio of the internal to the external magnetic field strength.

Heisenberg immediately drew attention to a numerical success: if he used Eqs. (3) and (4) to calculate $\Delta\nu_d$, the doublet width of the lithium $2p$-state, he obtained 0.32 cm^{-1} as compared to the experimentally observed 0.34 cm^{-1}. He also noted that the dependence of $\Delta\nu_d$ on the quantum number n^* of the valence electron was in agreement with available data and that he was able, of course, to reproduce the anomalous Zeeman patterns for all values of the external field by applying the selection rules $\Delta m^* = 0, \pm 1$, respectively.

If one takes a doublet atom, i.e., one consisting of a core, a valence electron with the properties described above and a second (inner) valence electron with angular momentum 1 ($h/2\pi$), and distributes the momentum of the latter among the core and the electron, then one arrives at Heisenberg's model for atoms emitting both singlet and triplet lines. In the ground state the core might either assume the angular momentum zero, and the valence electrons the momenta $+\frac{1}{2}$ and $-\frac{1}{2}$, such that the level will be the singlet state 1s; or the core might assume the angular momentum $+1$, and both electrons the momenta $+\frac{1}{2}$, the resulting level being a 1*s*-triplet state. In the excited atom the first (outer) valence electron carries the angular momentum $n-\frac{1}{2}$ (with $n = 2, 3, \ldots$); hence the total angular momentum of the atom is $n-1$ ($n = 1, 2, 3, \ldots$) for the singlet states. In the triplet states the core and the inner valence electron form a unit of momentum $\frac{3}{2}$. Heisenberg argued that the lowest triplet term was practically a singlet because both the valence electrons of momenta $\frac{1}{2}$ penetrated into the atomic core and, as a result, their momenta would be aligned.[51] However, in the excited states of the outer electron, denoted by the angular momentum $n-\frac{1}{2}$ (with $n = 2, 3, \ldots$), the terms of the free atoms become real triplets due to the following different situations: the angular momentum of the outer valence electron being parallel, perpendicular, or antiparallel to the angular momentum of the combined core–inner electron system. Hence the 'quantum condition' for the triplet atom becomes

$$\cos\theta = +1, 0, -1, \tag{6}$$

where θ is the angle between the angular momenta under consideration. In the presence of an external magnetic field the singlet lines (emitted in transitions between two different singlet states) are split into a normal Zeeman triplet, while the triplet lines (emitted in transition between triplet states) exhibit complex anomalous separations. For calculating the magnetic shifts of the energy levels Heisenberg retained the assumption that the core and the inner electron always remain strongly coupled; hence he could describe both of them by the same set of coordinates. Then he followed the same method as in the case of doublet atoms, deriving a third-order equation for a quantity X, which determines the magnetic energy according to the equation

$$\Delta W = h(m + X)\Delta\nu_{\text{normal}}, \tag{7}$$

where m is the magnetic or spatial quantum number of the triplet state. Each of the three roots of X for small external fields, that is,

$$X = \begin{cases} \frac{3}{2}v + \dfrac{m}{n} \\ -\dfrac{\frac{3}{2}v}{2(n-\frac{1}{2})} + \dfrac{m}{n(n-1)} \\ -\frac{3}{2}v - \dfrac{m}{n-1}, \end{cases} \tag{8}$$

[51] Heisenberg employed the same argument to explain why the *s*-term of doublet atoms was not split.

is associated with one of the triplet states in which the outer electron has the angular momentum $n - \frac{1}{2}$. The magnetic quantum number m is found to assume absolute values smaller than or equal to n, $n - 1$, and $n - 2$, respectively. This means that the 'inner quantum number' of Sommerfeld, which denotes the splitting of the triplet states in a magnetic field, is not identical with their total angular momenta, which are $n + 1$ (i.e., $n - \frac{1}{2} + \frac{3}{2}$, for parallel momenta of the outer electron and the core–inner electron system), n, and $n - 1$, respectively. For large external magnetic fields, X assumes the values $+1$, 0, and -1, respectively. By using Eqs. (6)–(8), together with the selection rules $\Delta m = 0, \pm 1$, Heisenberg was able to describe the experimental data completely. For example, for the energy differences of the triplet components belonging to the quantum number n of a free atom, he obtained the result

$$\frac{W_1 - W_2}{W_2 - W_3} = \frac{n}{n-1}, \tag{9}$$

this ratio assuming the value 2:1 for the p-states in agreement with observation. Moreover, he pointed out that, although for high external magnetic fields the normal Zeeman triplet is approached, each of the triplet components should exhibit—according to his theory—the natural triplet splitting due to the action of the internal magnetic field on the core–inner electron system.

Heisenberg described the above theory of the anomalous Zeeman effects in a letter to Pauli, dated 19 November 1921, by first stating his guiding principle: '*Der Erfolg heiligt die Mittel*' ('Success sanctifies the means'). He was willing to use every theoretical idea, even if it was not generally accepted by others, in order to arrive at formulae fitting the data. One of the unconventional concepts that he applied was the notion of half-integral quantum numbers about which he had debated a year previously with Sommerfeld. Although Sommerfeld did not really like half quantum numbers, he conceded that they might play some role in the theory of complex spectra, especially since Landé also used them. But it appeared that, in addition to half-integral quantum numbers, there were other changes necessary in order to arrive at a proper dynamical description of the anomalous Zeeman effects. Heisenberg discussed some of them in his letters to Landé, in particular the possible violation of Rubinowicz's selection rules. These rules had been invented in Munich and, in the beginning, Sommerfeld would criticize anybody who departed from them. They were based on the assumption that radiation from atoms is emitted in the form of spherical waves and that the total angular momentum of the system consisting of the atom *and* radiation is always conserved in every process of emission and absorption of a spectral line. However, dear as Rubinowicz's rules were to Sommerfeld, he also knew that they somehow had to be broken in order to explain the anomalous Zeeman effects. In his model of the doublet atom Heisenberg did not have to introduce explicit violations, although he suspected some difficulty in explaining the 'quantum condition,' Eq. (4); but he knew that the middle line of the triplet states was not in agreement with the physical principles applied by Rubinowicz. He argued on that point with Landé, who did not wish to give up the theoretical basis of the selection rules. Heisenberg then talked to Sommerfeld, who agreed that slight violations of Rubinowicz's rules might be possible; in particular, he now tended to believe that the difficulties of explaining the photoelectric effect on the basis of the classical wave theory might be removed if certain conservation laws were valid only on a statistical basis.[52] Heisenberg informed Landé about the conclusion they had reached: 'Thus one cannot regard Rubinowicz's result as an absolute criterion for the correctness of a theory. Selection rules and (what amounts to the same) intensities must be calculated by using the correspondence principle' (Heisenberg to Landé, 28 November 1921). At the same time, he had decided that neither the conservation of energy nor that of angular momentum must be given up. Heisenberg expressed the only violation that was necessary in the following words: 'In order not to contradict experience, we may therefore assume the validity of Rubinowicz' principle only for the totality of atoms' (Heisenberg, 1922a, p. 281). With this statistical formulation of Rubinowicz's principle, however, he succeeded in deriving the sum-rules for the magnetic energies of doublet and triplet term systems; for the triplet, say, he obtained the equation

$$\Delta W_1 + \Delta W_2 + \Delta W_3 = h\left(3m - \frac{3}{4}\,\frac{v}{n-\frac{1}{2}}\right)\Delta\nu_{\text{normal}}. \qquad (10)$$

The sum-rules served Heisenberg to prove the most important point in his theoretical model: namely, that in excited doublet atoms the atomic core can assume exactly two positions (i.e., the 'quantum conditions,' Eqs. (4), are satisfied), and that in triplet atoms the core (or the core–inner electron system) can assume exactly three positions (i.e., the 'quantum conditions,' Eqs. (6), are valid). For this reason he defended the violation of Rubinowicz's principle against Landé's criticism quite forcefully; ultimately, after a debate of two months, Landé agreed with Heisenberg, whereupon the latter submitted his results in a paper to *Zeitschrift für Physik* (Heisenberg, 1922a).

Heisenberg's first scientific paper, which was published in February 1922 right after Sommerfeld's reformulation of Voigt's theory (Sommerfeld, 1922a), contained a full model-dependent explanation of the then known observations on the anomalous Zeeman effects of doublet and triplet spectra. However, the styles of the two papers were quite different, as were the mathematical tools applied. Although they emerged from the same background of changing ideas in physics, Heisenberg went deeply into the mechanical details of atomic models, whereas

[52]The possibility of statistical conservation laws was also mentioned by Sommerfeld in the third edition of *Atombau und Spektrallinien*, whose preface he wrote in January 1922. (See pp. 253–255 of the English edition.)

Sommerfeld confined himself to applying only general principles. In applying the laws of dynamics, Heisenberg allowed himself certain liberties, which Sommerfeld would have avoided. He employed selected parts of mechanics, electrodynamics, quantum theory and whatever else he needed in order to achieve his goal —in agreement with his motto, 'success sanctifies the means'—and he was less concerned about the internal consistency of these theoretical parts and the basic assumptions of his model.[53] Still he was not unaware, as he wrote to Pauli on 19 November 1921, of the dark spots ('*Schattenseiten*') of his theory. He admitted that the mechanical stability of the atomic-core model was a crucial problem to be explained, but he argued that the stability would follow from the 'quantum condition' (Heisenberg to Pauli, 25 November 1921). Another touchy mechanical problem was connected with the various precessional motions of the core's angular momentum, with respect to which Heisenberg had made peculiar assumptions.[54] And finally, he did not know how to relate his atomic core–valence electron model to the Einstein–de Haas effect, i.e., the observed ratio of the magnetic moment over the mechanical (angular) momentum of electrons in metals (Heisenberg to Pauli, 17 December 1921). There was no doubt that the experienced Sommerfeld would not have cared to write a paper quite like Heisenberg's, but he felt that his student, who had done competent work on the anomalous Zeeman effects without having his name connected as yet with a

[53] Heisenberg explained later on that in his first paper on the anomalous Zeeman effect he had really mixed together many different things. 'It was also a proof,' he said, 'that my knowledge of classical mechanics was so weak that I didn't even realize where I departed from decent mechanics and where I did something that was still in agreement with classical mechanics' (Heisenberg, Conversations, p. 77). The uneven physical and mathematical treatment of his first paper was probably due to his uneven scientific preparation at that time: on the one hand, he attended Sommerfeld's lectures on theoretical physics and did research; on the other, he had not yet followed a regular curriculum leading up to a full preparation for starting research work.

[54] Heisenberg summarized the dynamical assumptions of the '*Rumpf*' (atomic-core) model in his letter to Pauli, written after the submission of his paper, as follows:

> The essential mechanical and physical meaning of my theory seems to me (so far as I understand it now) to be the following. The stationary states of the atom are always given by the spatial quantization of the *outer* electron: if no external field is present, [the orbit of the electron is quantized] relatively to the magnetic field of the core; if an external field is present, [it is quantized] *in addition* relatively to the latter. This "in addition" is the important point. For small [magnetic] fields **H**, of course, the two quantizations [i.e., those with respect to the field of the core and the external field, respectively] harmonize very well with each other, because two degrees of freedom have to be quantized: 1. The position of the [orbit of the] electron relative to the core; 2. the position of the atom relative to the [external] field, given by the *position of the [orbit of the] outer electron* relative to the field. Now, what happens when **H** increases, I shall briefly describe as follows: The degree of freedom 1 dies away gradually, and its place is taken by the position of the core relative to the field, which must now be quantized. We may picture this in the following manner. For small [external field] **H** the orbits of the core and the [outer] electron are parallel. With increasing field strength [**H**] the [orbit of the] core will be rotated adiabatically until it assumes, in the case of total Paschen–Back effect, a position perpendicular to [the direction of] **H**. The inner field $\mathbf{H}_i$, then, is already very weak [compared to **H**], and its average value [over a full period of the atom] will have the same direction as **H**; in this case only the normal Larmor precession exists, whose magnitude [compared to the normal situation, in which only the strength of the external field enters] is modified just a little by $\mathbf{H}_i$. The above-mentioned adiabatic rotation of the core comes about exactly by the change of direction [of the core's angular momentum] towards the direction of the vector sum [i.e., the resultant of the field vectors, $\mathbf{H} + \mathbf{H}_i$]. (Heisenberg to Pauli, 17 December 1921)

publication on that subject, should publish a paper even if it was not a perfectly polished one.[55]

Heisenberg was quite happy. He had now joined the club of those people who worked successfully on problems of atomic theory. He had shown that his first exercise in number theory, leading to the assignment of half-integral quantum numbers to explain the anomalous Zeeman effect of doublet spectra, could be turned into a successful model describing the atoms emitting doublet and triplet spectra. He had learned to use his knowledge of mechanics in this problem and had become acquainted with some of the most powerful tools of quantum theory, especially the correspondence principle. At that time, hardly anyone except Niels Bohr himself or his closest collaborator Hendrik Kramers had succeeded in obtaining any concrete results with the help of the correspondence principle. Sommerfeld felt that Heisenberg had achieved important results with the latter, comparable to the ones obtained earlier in Copenhagen. He did not hesitate to incorporate them immediately into the third edition of his *Atombau und Spektrallinien*.[56] Heisenberg became his expert collaborator on the anomalous Zeeman effects. But he also learned about other aspects of atomic theory in the Seminar, which he attended regularly. For example, towards the end of his first semester, Heisenberg gave a report on a paper of Kramers on the quadratic Stark effect that interested Sommerfeld (Kramers, 1920b).[57] At that time Sommerfeld was not satisfied with Heisenberg's presentation and told him afterwards, 'Maybe you have understood it yourself, but you have not explained it to others' (Heisenberg, Conversations, p. 40). But in the following one and a half years Heisenberg made so much progress in understanding the problems of atomic physics that Sommerfeld offered to take him to Göttingen in June 1922 to attend Bohr's lectures there. Heisenberg's familiarity with the quadratic Stark effect and Kramers' treatment of it led to the most exciting experience of his student years. When

[55] Heisenberg recalled that Sommerfeld told him: 'Now you have done so much with these things. Phenomenologically the things are now being written by Landé and by myself, but wouldn't you try to do the model on it?' (Heisenberg, Conversations, p. 87; also AHQP Interview). Sommerfeld also hoped that by trying out the model, Heisenberg would learn how to work in theoretical physics. He certainly did to some extent, and Sommerfeld wrote about his paper to Bohr: 'The presentation in Heisenberg's paper is not quite fitting and should have been more polished. Heisenberg is a student in the third semester and is enormously talented. I could not hold back his desire to publish, and I find his results to be so important that I agreed to publish them, although the formulation of the derivations might not yet be the final one' (Sommerfeld to Bohr, 25 March 1922).

[56] In the 'Supplement' to Chapter Six, Sommerfeld noted: 'In as much as the (Sommerfeld–Voigt) formula, which has been obtained empirically or semi-empirically, may be deduced accurately from our model of doublet atoms, our explanation of the doublets in a free field seems quite assured' (Sommerfeld, 1922d, English translation, p. 409). Similarly, he talked about the triplets: 'We may also assert that in the case of atoms which generate triplet lines, the objects of the theory of atomic models are fully realized' (Sommerfeld, 1922d, English translation, p. 412). He acknowledged Heisenberg's help in writing this Supplement.

[57] Kramers submitted his paper on the influence of the electromagnetic field on the fine structure of hydrogen lines in late September 1920. Some time after its publication Sommerfeld submitted a note on the second-order Stark effect (Sommerfeld, 1921b). The latter was received by *Annalen der Physik* on 27 January 1921. Since Sommerfeld did not quote Kramers' paper, it is certain that he had not seen it. As soon as he came across it, he gave it to Heisenberg to report on it in the Seminar.

Bohr discussed this topic during his lectures, Heisenberg raised a crucial objection. Bohr was so impressed that he invited Heisenberg to go for a walk with him on the Hainberg in Göttingen to deal with the question in detail. This encounter with Niels Bohr would have a decisive influence on his scientific development.[58]

On their return to Munich from Göttingen, Sommerfeld and Heisenberg immediately became involved in a problem which Bohr had raised in his lectures. Bohr had pointed out that his theory of stable states of atoms was limited insofar as radiation damping had not been taken into account. On the other hand, he had indicated that the energies of the quantum states, and therefore also the transition frequencies, were not infinitely sharp. The accuracy of the frequencies was related to the duration of the emission process, about which only an upper limit was known from Wilhelm Wien's experiments on the decay time of the luminescence of canal ray particles (Wien, 1919, 1921). Bohr had remarked that the fact that Wien's data agreed in the order of magnitude with the classical calculations of radiation damping 'might be suited to throw light on various problems of quantum theory' (Bohr, 1977, p. 361). He had also made it clear in discussions and conversations that 'a quantum-theoretical calculation, which takes into account terms that are smaller than the neglected energy loss by classical emission of radiation, is not allowed in principle' (Sommerfeld and Heisenberg, 1922a, p. 393). Towards the end of his lectures, Bohr had then raised a cautious criticism with respect to Sommerfeld's relativistic theory of X-rays (Sommerfeld, 1920b, 1921a). Sommerfeld understood that his explanation of the fine structure of X-ray lines, which seemed to fit the observations so well, was in danger. With Heisenberg's help he undertook to show that Bohr's verdict did not really hold (Sommerfeld and Heisenberg, 1922a).[59] In particular, they studied the question of whether the higher-order relativistic corrections in the case of heavy atoms were larger or smaller than the corrections arising from radiation damping. To solve this problem, they first calculated the classical radiation damping term, $\frac{2}{3}(e^2/c^3)\ddot{v}$ (where $-e$ is the charge of the electron, c the velocity of light, and $\ddot{v}$ the second time derivative of the velocity of the electron), for a circular orbit. They compared this quantity to the relativistic acceleration of the electron, i.e., the expansion of the time derivative of the electron's momentum, $m_e v(1 + \frac{1}{2}\beta^2 + \frac{3}{8}\beta^4 + \dots)$, in terms of $\beta = v/c$, m_e being the mass of the electron. Since each of the terms, β^{2s}, in the latter expansion contributes a term of the order α^{2s} to the relativistic fine structure of the atomic energy levels ($\alpha = 2\pi e^2/hc$), the ratio of the radiation loss to the term of order α^{2s} is given by the equation

$$\frac{e^2\omega}{m_e c^3 \beta^{2s}} = \frac{\alpha}{n}\left(\frac{\alpha Z}{n}\right)^{2-2s}, \qquad s = 1, 2, \dots, \tag{11}$$

[58] We shall discuss Heisenberg's encounter with Niels Bohr in Section III.1.

[59] The X-ray problem was a standard one discussed in the Seminar. Gregor Wentzel was Sommerfeld's main collaborator on this subject. Heisenberg never got really involved in it, except in the short note he wrote with Sommerfeld (Sommerfeld and Heisenberg, 1922a)—but this note did not go into details.

up to factors of the order of unity. (In Eq. (11), ω is 2π times the frequency of the electron's motion, n is the principal quantum number of the orbit, and Z is the charge of the atomic nucleus or core under consideration.) This ratio must be less than 1 in order that the corresponding relativistic limit still make sense according to Bohr. Sommerfeld and Heisenberg found that in the case of hydrogen ($Z = 1$), one is allowed to consider only the corrections of order α^2; for uranium, however, also corrections up to the order α^6 could be admitted, as was required by the comparison of Sommerfeld's theory with the L-doublet lines, where empirically Z was found to be equal to 92–3.5 (for uranium the nuclear charge being 92).

Having answered Bohr's criticism of the relativistic interpretation of X-ray doublets, Sommerfeld and Heisenberg proceeded to develop the quantum theory of the linewidth of transition radiation based on Bohr's assumption that radiation damping or losses were responsible for it. For this, they considered U/W, the ratio of the energy losses, U, to the energy of an atomic state, W, which is given to a good approximation by the right-hand side of Eq. (11) with $s = 0$. On substituting the absolute value of the energy of a state with the principal quantum number n of a hydrogen-like atom, i.e., $W = hRZ^2/n^2$, R being Rydberg's constant, they obtained the radiation loss connected with this state, and from it $\Delta\nu_{\text{qu}}$, the quantum-theoretical linewidth of transition radiation, say, between the states denoted by the quantum numbers n_1 and n_2. The result was

$$\Delta\nu_{\text{qu}} = \alpha^3 RZ^4\left(\frac{1}{n_1^5} - \frac{1}{n_2^5}\right) \times \text{factor of order of 1.} \tag{12}$$

In order to determine the factor in Eq. (12), they equated $\Delta\nu_{\text{qu}}$ with the classical linewidth, $\Delta\nu_{\text{cl}} = (4\pi/3)(e^2/m_e c^3)\nu^2$, in the correspondence limit (i.e., $n_2 = n_1 + 1 \rightarrow \infty$); thus they obtained the factor as 4/15. Finally, they generalized Eq. (12) to elliptic orbits and compared the results with Wien's observations of the Balmer lines of hydrogen, noting a satisfactory agreement between experiment and the new theory. They concluded that their considerations threw light on the nature of the damping mechanism in atoms without, however, giving a causal description. 'Indeed,' they wrote, 'this is not in the spirit of the correspondence principle, which renounces any model-dependent interpretation . . . Nevertheless our study seems to allow [us to draw] useful conclusions about the process of damping. By its agreement with experiment it confirms, on the other hand, the basis of our considerations: the correspondence-like treatment of radiation resistance, as proposed by Bohr' (Sommerfeld and Heisenberg, 1922a, p. 398).

While the paper on the linewidth, which was received by *Zeitschrift für Physik* on 3 August 1922, grew out of the meeting with Bohr in Göttingen, Sommerfeld and Heisenberg had begun to work on their joint paper on the intensities of multiplets in the fall of the previous year (Sommerfeld and Heisenberg, 1922b). It originated in connection with their investigations on the anomalous Zeeman effects, which had led to Sommerfeld's paper on the reformulation of Voigt's classical theory (Sommerfeld, 1922a) and to Heisenberg's development of the core model of doublet and triplet atoms (Heisenberg, 1922a).[60] After both of them had completed their respective papers on the anomalous Zeeman effects, Heisenberg had written to Pauli: 'Very soon I will publish [my work on] the Zeeman effects in *Zeitschrift für Physik* [referring to his paper, Heisenberg, 1922a, which he had already submitted] and then I shall write a paper with the Professor [Sommerfeld] on the problem of intensities, which come out very beautifully with the help of the correspondence principle' (Heisenberg to Pauli, 17 December 1921). Still it took another eight months until the paper was submitted by the authors to *Zeitschrift für Physik* (and received there on 26 August 1922). The main reason for the delay could be inferred from the final remarks in the paper:

> Altogether we may say that the correspondence principle has proved its value excellently in its application to the intensity questions studied above, both in the case of spontaneous, [external] field-free term splittings, and in the case of real Zeeman splittings. The fact that these problems are in general of a qualitative nature implies that we do not need detailed considerations on atomic structure; hence the certainty of our conclusions is enhanced. (Sommerfeld and Heisenberg, 1922b, p. 154)

That is, during the period between December 1921 and August 1922 they had abandoned all detailed model-dependent calculations and stressed those purely qualitative conclusions that could be derived from applying the correspondence principle *alone*.

As we have noted, since the beginning of 1921 Sommerfeld had become increasingly more appreciative of Bohr's correspondence principle and his approach to the problems of atomic structure. It was therefore natural that he should send the two papers on the anomalous Zeeman effects of himself (Sommerfeld, 1922a) and Heisenberg (1922a), where use had been made of Bohr's ideas, to Copenhagen. Bohr was grateful for the understanding which he had received from Munich. 'I must confess, however,' he wrote, 'that several of the assumptions employed by you and your collaborators in the very promising theory of the anomalous Zeeman effect hardly appear to me to be consistent with a unified picture of quantum theory' (Bohr to Sommerfeld, 30 April 1922). In particular, Bohr spoke against the use of half-integral quantum numbers and the atomic-core model, declaring that the latter pointed to the necessity of basic changes in the classical conception of the magnetic properties of a system of moving charged particles.[61] During his Göttingen lectures he was even more

[60] Heisenberg mentioned the work on the intensities of Zeeman lines in his letters to Landé. Some results were already available in October 1921 (Heisenberg to Landé, 16 October 1921); in November, Heisenberg wrote that they had decided to publish the intensity calculations later (Heisenberg to Landé, 28 November 1921).

[61] Bohr made this remark in the note added-in-proof of *Drei Aufsätze über Spektren und Atombau* (Bohr, 1922e, pp. 93–94). He first wrote to Landé on 15 May 1922 about the unacceptability of half-integral quantum numbers.

explicit. He mentioned Landé's rules for the anomalous Zeeman effect and continued: 'How to explain this [especially the half-integral quantum numbers] is a very difficult question. We must conclude from the anomalous Zeeman effect, that the classical theory is inadequate. On the basis of peculiar assumptions about the types of occurring orbits, Heisenberg, in a very interesting paper, has attempted to arrive at Landé's results . . . It is difficult to justify Heisenberg's assumptions' (Bohr, 1977, p. 391). It appears that because of this criticism Sommerfeld and Heisenberg scrupulously avoided any real use of specific models of the kind Heisenberg had employed in his paper on the anomalous Zeeman effects. They preferred to maintain the following point of view:

> Concerning the model-dependent interpretation of the atomic orbits also we need not develop more accurate views; in what follows the general kinematic description given above will suffice. We shall be able to treat doublet and triplet systems (more generally, even- and odd-numbered spectra) according to the same theoretical scheme, while the accurate model theory must employ very different assumptions in both cases. Since the intensity questions are essentially qualitative questions, in many cases we would not have to consider quantitative details. For example, it is unimportant whether we put the angular momentum of the series spectrum [i.e., of the series electron responsible for the series spectrum] directly equal to the azimuthal quantum number k or equal to $k^* = k - \frac{1}{2}$, as the magneto-optical model proposed by Heisenberg demands in the case of doublet systems. (Sommerfeld and Heisenberg, 1922b, pp. 132–133)[62]

Thus, the reason why they delayed the publication of their results on the intensities of multiplet lines and the related anomalous Zeeman effects was clearly this: they wanted to determine exactly what Bohr's objections were when they would meet him in Göttingen. Afterwards, they went ahead and submitted their paper.

Instead of employing any specific models of atomic structure, Sommerfeld and Heisenberg started from the assumption—which seemed to represent all the known data on spectral lines—that the atomic states were characterized by certain quantum numbers, i.e., the principal quantum number n, the azimuthal quantum number k (where $k = 1$ denotes the s-term of a series, etc.), the inner quantum number j (in case of multiplet terms), and the magnetic quantum number m (in case of an external magnetic field). These quantum numbers determined the kinematical character of the orbits completely; n denoted the shape of the orbit ($n = k$, a circular orbit; $n > k$, an elliptic orbit); k was associated with the full orbit of the perihelion of the motion; and j was always the total angular momentum of the atom in a given state.[63] There existed two

[62] There was only one further reference to Heisenberg's '*Rumpf*' (atomic-core) models towards the end of the paper on the intensities (Sommerfeld and Heisenberg, 1922b, p. 152). And this point was not a very important one, as it dealt with a correction to an equation. (See below in the text.)

[63] Note that this is in slight contradiction to what Heisenberg had found in his '*Rumpf*' (atomic-core) model: 'The "inner" quantum number n_j [i.e., our j in the above text] . . . coincides for doublet systems with the total angular momentum of the atom . . . For triplets the connection between n_j and the total angular momentum is more complicated' (Heisenberg, 1922a, p. 297).

precessional motions of the orbits of a valence electron: one precession occurred around the j-axis, another around the axis of the external magnetic field if the latter was present. Altogether the multiply periodic motion of any atom exhibited four different circular frequencies: ω_n $(=2\pi\nu_n)$, ω_k, ω_j, and ω_m. In the calculations one could make use of the fact that $\omega_n \gg \omega_k$, $\omega_k \gg \omega_j$, and for weak magnetic fields also $\omega_j \gg \omega_m$. The main tool for analyzing the kinematics of multiply periodic systems was the expansion of their orbits into Fourier series involving integral multiples of the above frequencies. Niels Bohr had employed this tool in his fundamental work on the correspondence principle (Bohr, 1918a, b; 1922d), and Sommerfeld and Heisenberg now made use of it because it seemed to offer the selection rules and the intensities, both connected with the numerical coefficients occurring in the Fourier series. They expanded the coordinates of the motion of series electrons in arbitrary atoms in the following way. In the absence of a field of force (be it an external or an internal magnetic field), the rectangular coordinates in the orbital plane, ξ and η, yield an adequate description of motion; they are expanded as

$$\xi + i\eta = \exp\{i\omega_k t\} \sum_{s=-\infty}^{+\infty} a_s \exp\{is\omega_n t\}, \tag{13a}$$

with a_s $(s = 0, \pm 1, \dots)$ the Fourier coefficients or amplitudes and t the time coordinate. If the direction of j, the total angular momentum of the atom, is different by an angle ϑ from that of the normal to the plane of the orbit (the ζ-direction), then the orbit performs a precession around the j-direction. In this case, which occurs when an internal field is present (i.e., in the case of doublet and triplet states), the motion is most suitably described in the coordinate system, x, y, z, with z in the direction of j and $y = \eta$. The corresponding Fourier expansion then reads,

$$x + iy = \exp\{i\omega_j t\}(\xi \cos\vartheta + i\eta). \tag{13b}$$

Finally, in the presence of an external magnetic field, they chose the rectangular coordinates X, Y, Z, with Z in the direction of the magnetic field, which makes an angle θ with the j-direction, and $Y = y$. The Fourier expansions in this case become

$$\begin{aligned} X + iY &= \exp\{i\omega_m t\}(x\cos\theta + iy - z\sin\theta) \\ &= \sum_{s=-\infty}^{+\infty} \sum_{r=-1}^{+1}{}' \sum_{q=-1}^{+1} C_{s,r,q} \exp\{i(s\omega_n + r\omega_k + q\omega_j + \omega_m)\}, \\ Z &= x\sin\theta + z\cos\theta \\ &= \sum_{s=-\infty}^{+\infty} \sum_{r=-1}^{+1}{}' \sum_{q=-1}^{+1} D_{s,r,q} \exp\{i(s\omega_n + r\omega_k + q\omega_j)\} \end{aligned} \tag{13c}$$

(with the r-sum, $\sum'$, over $r = -1$ and $+1$, and the q-sum over $q = -1, 0, +1$). The coefficients $C_{s,r,q}$ and $D_{s,r,q}$ are functions of the Fourier coefficients a_s of the expansion (13a) and of the angles ϑ and θ. These angles are different for different transitions in the quantum numbers n and k, but for a given transition of such kind they are identical, i.e., the same for all components of a natural or magnetic multiplet. The C's and D's contain all the information necessary for calculating the intensities of lines and to derive the selection rules.

It is important to note that Eqs. (13a)–(13c) express the correspondence principle completely; from them the selection rules for the various quantum numbers describing the atoms could be easily obtained. Bohr had already considered the rules for the change of the principal quantum number n (i.e., $\Delta n = s$ arbitrary) and the azimuthal quantum number k (i.e., $\Delta k = \pm 1$); these are expressed in the s- and r-summations of Eq. (13c). From this equation, Sommerfeld and Heisenberg found that the selection rule for the inner quantum number was

$$\Delta j = 0, \pm 1. \tag{14}$$

Hence they concluded proudly: 'The selection rule for the inner quantum number, especially its difference from that of the azimuthal quantum number, which had been deduced originally in a purely empirical manner from the line structures of the subordinate series, now follows as a natural kinematic consequence' (Sommerfeld and Heisenberg, 1922b, p. 137). For strong magnetic fields, however, the selection rule (14) may be violated, in agreement with the correspondence principle, because then the precession frequency ω_j is *not small* compared to the magnetic precession frequency ω_m, and the transitions $\Delta j = 2$, $3, \ldots,$ may occur, in agreement with observations. (See Paschen and Back, 1921.) In the presence of any magnetic field, the corresponding selection rule is $\Delta m = 0, \pm 1$. Sommerfeld and Heisenberg were also able to prove that the empirically nonexistent transitions, $j = 0 \to 0$ and $m = 0 \to 0$, were forbidden by the correspondence principle.[64]

Sommerfeld and Heisenberg were not interested in calculating the absolute values of the intensities, which had to be obtained by the methods which F. W. Bessel had applied to the analysis of planetary orbits and H. A. Kramers had transferred to the hydrogen orbits in his thesis (Kramers, 1919); instead they restricted themselves to relative intensities. By this restriction they avoided very complicated calculations (which probably could have been performed only by invoking many model-dependent assumptions); on the other hand, they computed those results which could be compared with reliable experimental data, for the relative intensities of multiplet components did not depend on the excitation

[64] In the case of the j-rule (i.e., $j \nrightarrow 0$) they argued: 'We may say with Landé that because of $\Delta j = 0$, the emitted light should be parallel to the j-axis. However, [in the case of $j = 0$] this axis possesses no physical meaning. Hence the emission of light [in this direction] should also not have any physical meaning' (Sommerfeld and Heisenberg, 1922b, p. 139). They proved the other rule (for Δm) directly from intensity considerations. (See below in the text.)

conditions of the emission lines. They first turned to the intensities of the natural multiplet components, i.e., of doublet and triplet atoms in the absence of an external magnetic field. Hence, they assumed that in a transition, Δn ($= s$ in Eqs. (13a, c)) and Δk ($= r$) were given numbers, while Δj ($= q$) took the values $0, \pm 1$ and j, the total angular momentum of the states involved was left free to some extent—as corresponded, for example, to the two different situations for any state of a doublet atom in Heisenberg's previous model. In the case of $\Delta k = +1$ transitions they obtained from Eq. (13c)—with θ and ω_m put equal to zero—the Fourier coefficients

$$
\begin{aligned}
\Delta j = +1: \quad & C_{s,1} = \frac{1 + \cos\vartheta}{2} a_s, \\
\Delta j = 0 \;\;: \quad & D_{s,1} = \frac{\sin\vartheta}{2} a_s, \\
\Delta j = -1: \quad & C_{s,-1} = \frac{1 - \cos\vartheta}{2} a_s.
\end{aligned}
\tag{15}
$$

Hence, since the intensities of the corresponding multiplet lines, I_{+1}, I_0, I_{-1}, are proportional to $\frac{2}{3}|C|^2$ or $|D|^2$, respectively, they obtained the result

$$
I_1 : I_0 : I_{-1} = \cos^4\frac{\vartheta}{2} : 2\sin^2\frac{\vartheta}{2}\cos^2\frac{\vartheta}{2} : \sin^4\frac{\vartheta}{2}. \tag{16}
$$

A similar result held in the case of transitions with $\Delta k = 1$ (where the roles of I_{+1}, and I_{-1} were simply exchanged). This formula applied to doublets and triplets alike; in the former case, the last line-component, denoted by the subscript -1 (i.e., $\Delta j = -1$) did not exist. Sommerfeld and Heisenberg compared Eq. (16) with the experimentally observed intensities and found a reasonable agreement; e.g., in the case of Ca($2p4d$)–triplet the observed intensity ratio was 10:8:4 as compared to $1 : \frac{1}{2}\vartheta^2 : \frac{1}{16}\vartheta^4$ with small ϑ. They also discussed the situation that the $\Delta k \neq 0$ rule was broken, which had been noticed empirically in some (pp') to (dd') combinations of the Ca and Sr atoms. They assumed that in this case the main contribution to the intensity was due to a nonzero ζ-coordinate and derived from this an intensity ratio

$$
I_0 : I_{+1} : I_{-1} = 2\cos^2\vartheta : \sin^2\vartheta : \sin^2\vartheta, \tag{17}
$$

again in agreement with recent observations (Götze, 1921).

The generalization of the intensity to magnetic multiplets, i.e., to components obtained by the splitting of each line of the natural multiplets by an external magnetic field, was now straightforward. Sommerfeld and Heisenberg had just to consider transitions given by fixed Δn, Δk *and* Δj. Then they calculated the intensity ratio for each magnetic triplet, $I_{+1} : I_0 : I_{-1}$ (where the suffix denotes the corresponding Δm), that is,

$$I_{+1}:I_0:I_{-1} = \begin{cases} \frac{1}{4}(1+\cos\theta)^2 : \sin^2\theta : \frac{1}{4}(1-\cos\theta)^2 & \text{for } \Delta j = +1 \\ \frac{1}{4}\sin^2\theta : \cos^2\theta : \frac{1}{4}\sin^2\theta, & \text{for } \Delta j = 0 \\ \frac{1}{4}(1-\cos\theta)^2 : \sin^2\theta : \frac{1}{4}(1+\cos\theta)^2 & \text{for } \Delta j = -1. \end{cases} \tag{18}$$

In order to evaluate the formulae (18), one had to know the value of the angle θ. Sommerfeld and Heisenberg took the equation

$$\cos\theta = \frac{m}{j} \tag{19}$$

(which was valid for the hydrogen atom) in the case of atoms having doublet and triplet spectra, noting that for small j the formula might not be reliable.[65] With this assumption, they then described the observed data on the intensities of Zeeman multiplets quite well. In addition, they proved that for the $\Delta j = 0$ transitions the transverse polarized magnetic component $m = 0 \rightarrow 0$ had the intensity zero since it was proportional to sin θ with $\theta = 0/j$.

The work on the intensities of the multiplet components satisfied both authors very much. For Sommerfeld, the centre of interest was not the detailed model. As Heisenberg recalled later:

> The centre of interest for Sommerfeld was that one had to have integral numbers as ratios between intensities. The intensity ratio of the two D-lines was two-to-one, and that appealed to him. He was happy that one could, in this case, use the correspondence principle to guess such integral relationships. On the other hand, he saw—that came [after] many discussions—that this idea of combining angular momenta was a tool to derive such formulas just by guesswork, by always guessing from a classical formula to a quantum formula. He probably did not put much emphasis on what this inner quantum number meant. Of course, that vector, which remained constant in time and could only be changed by means of an outer field, could only be the total angular momentum if [one believed] in the ordinary classical conservation laws. But since even these conservation laws and everything else was in doubt [at that time], there was always a state of vagueness about such concepts. For Sommerfeld, these pictures of the total angular momentum were just pictures by means of which one could derive empirical laws about the integral relations between intensities. (Heisenberg, Conversations, pp. 95–96; also AHQP Interview)

Heisenberg, on the other hand, liked building models and thus complemented Sommerfeld's work. In addition, he shared Sommerfeld's optimism about the usefulness of the correspondence principle. He had reason to be quite content with his progress in research work. By the time he finished his fourth semester, he

[65] At this place they referred to Heisenberg's earlier models by saying: 'Only for small values of j . . . shall we correct Eq. [(19)] in the sense of the magneto-optical models proposed by Heisenberg' (Sommerfeld and Heisenberg, 1922b, p. 152).

had completed three papers—one alone and two with Sommerfeld—on atomic theory; apart from some criticism, they were well regarded by the experts.[66]

I.7 Success in a Problem of Classical Physics

Active research on current problems in atomic theory provided Heisenberg with an entrance into the heart of modern theoretical physics. He entered the field boldly and without delay, however, before he had digested the classical theory. At about the same time as he thought of mastering quantum theory by means of its application to the intricate problem of the anomalous Zeeman effects, he had also to become familiar with the full range of classical mechanics and electrodynamics. At school he had had very little physics, only the bare elements of Newton's laws, and he had tried on his own to solve the differential equations for the Kepler problem. At the university he immediately enrolled in Sommerfeld's lectures on theoretical physics, beginning with the course on classical mechanics in his first semester, continuing with the mechanics of continuous media and hydrodynamics in summer 1921.[67] From these courses he acquired enough competence to be able to work on original problems in classical physics, too. He became especially interested in one of the most important problems of classical hydrodynamics: the problem of turbulence.

Heisenberg became involved in this problem very early by following Sommerfeld's advice. When he had first visited Sommerfeld in fall 1920 and had told him about having read and 'understood' Hermann Weyl's *Space, Time, Matter,* Sommerfeld had remarked: 'It is a lucky chance that I am going to give a course in elementary mechanics this semester [i.e., winter semester, 1920–1921]. Just do the exercises diligently; then you will find out what you have understood and what you have not' (Sommerfeld, 1949, p. 316). This advice appealed to Heisenberg and he followed it carefully. And it gave rise to his first original work in classical theory, about which Sommerfeld recalled:

> But, during his second semester, when I gave a course in hydrodynamics, I agreed to his publishing a note on vortices in the *Physikalische Zeitschrift*. I said to my colleague [August] Heisenberg: "You belong to an irreproachable family of philologists, you, yourself being a great expert on the late Greek period, your father-in-law [Nikolaus Wecklein] a famous expert on Homer, and now you have the misfortune of seeing the sudden appearance of a mathematical–physical genius in your family." (Sommerfeld, 1949, p. 316)

[66] For example, Niels Bohr, in his lectures at Göttingen in June 1922, remarked that 'Heisenberg's paper [1922a] is very promising and has contributed much to the formal interpretation of the experimental results' (Bohr, 1977, p. 391).

[67] In the winter semester of 1921–1922 Heisenberg attended Sommerfeld's lectures on electrodynamics; at the same time he applied Biot and Savart's law, which he had just learned, to calculate the forces in his '*Rumpf*' (atomic-core) model. (See Heisenberg to Pauli, 17 December 1921.)

The 'note,' to which Sommerfeld referred and which was published finally in the issue of 15 September 1922 of *Physikalische Zeitschrift*, was entitled '*Die absoluten Dimensionen der Kármánschen Wirbelbewegung*' ('The Absolute Dimensions of Kármán's Vortex Motion,' Heisenberg, 1922b). It came about in the following way. Sommerfeld, in one of his lectures on continuum mechanics, treated the vortex motion in hydrodynamics and Helmholtz's laws governing it. The discussion on this subject was continued in Sommerfeld's office after the lecture, and he told Heisenberg that a certain problem was involved. It had been proved that in a frictionless (nonviscous) liquid no vortices can be created or made to disappear. However, said Sommerfeld, 'Isn't it funny that if you move a spoon in a cup of tea or an oar through the water, you will see that there remain two vortices in the liquid when the spoon or the oar is removed. That is against the conservation law, and you cannot claim that this is due to the very small amount of friction existing in the water' (Heisenberg, Conversations, p. 49). This contradiction aroused Heisenberg's interest in the problem of vortex motion.

Sommerfeld also told Heisenberg about a recent paper of his Leipzig colleague George Jaffé, in which it had been suggested that, in general, vortex motion in a fluid was created, not by viscosity, but by discontinuities in the velocity of the fluid or in the external forces acting upon it (Jaffé, 1920).[68] Now such a discontinuity does indeed exist when one withdraws the oar, and Heisenberg had to follow up this suggestion in some detail. Thus, he was able to *calculate* two quantities which Theodore von Kármán, in his pioneering work, had had to derive from experiment.

In discussing the force of resistance exerted on a plate of diameter d, which is carried along with a velocity U through a fluid, von Kármán (1911) had introduced the concept of a *vortex street*. The vortices then occur at some distance behind the plate on both sides of a central line parallel to the motion, the distance between the centers of two consecutive vortices having the same direction of rotation being l, and the width of the vortex being h. In order to calculate the resistance, von Kármán needed the ratio of the vortex distance, l, to the plate diameter, d, as well as the ratio of the vortex velocity, u, to the velocity of the plate, U. For these ratios he took the values obtained experimentally (von Kármán and Rubach, 1912, p. 58), namely,

$$\frac{u}{U} = 0.20 \quad \text{and} \quad \frac{l}{d} = 5.5. \tag{20}$$

[68]George Jaffé, born on 16 January 1880 in Moscow, became a *Privatdozent* at the University of Leipzig in 1908; he was promoted there to an '*Außerplanmäßiger Professor*' of Theoretical Physics in 1916 and to an *Extraordinarius* (associate professor) in 1923. He left Leipzig in 1926 to become a full professor at the University of Gießen, where he stayed until 1933. Afterwards, Jaffé emigrated to the United States and made his second career at Louisiana State University, Baton Rouge, Louisiana (visiting lecturer, 1939; associate professor, 1942; professor, 1946; emeritus, 1950). Jaffé worked on the electrical conductivity of polarizable substances, on the theory of fluids, and later, on semiconductor and neutron transport theory. When he left Leipzig in 1926, his position was offered to Heisenberg and Pauli, but both declined.

Heisenberg now tried to obtain the desired ratios by formulating Jaffé's idea about the origin of vortex motion quantitatively. He assumed that in the vicinity of a plate, which is carried through a nonviscous fluid, an unsteady potential force arises, and the instable motion of the fluid connected with it then develops —at some distance from the plate—into von Kármán's vortex street. He started by calculating ζ, the momentum of the vortex, from the velocity difference between the plate and the 'dead water' behind the plate, obtaining the equation

$$\frac{\zeta}{l}(U-u)=\frac{1}{2}U^2, \tag{21}$$

under the assumption that the vortex momentum is a conserved quantity. He obtained a second equation for ζ by considering the continuous conserved flow of the fluid in the vortex street and putting it equal to the flow dragged away by the plate, that is,

$$\frac{\zeta h}{l}=U\cdot d. \tag{22}$$

These equations, together with von Kármán's previous theoretical values on the ratios h/l $(=0.283)$ and ζ/l $(=\sqrt{8}\,u)$(see von Kármán and Rubach, 1912, pp. 52–53), yielded the results

$$\frac{u}{U}=0.229 \qquad \text{and} \qquad \frac{l}{d}=5.45, \tag{23}$$

in agreement with the empirical values (20).

Though this calculation was done already in summer 1921, the paper dealing with it was received by the *Physikalische Zeitschrift* only on 18 July 1922 (Heisenberg, 1922b). The reason for the delay was that Heisenberg wanted to discuss his results with the hydrodynamicists before publishing them. He obtained the opportunity of discussing them with Ludwig Prandtl in Göttingen during the *Bohr Festival* (June 1922).[69] Prandtl appended a note (Prandtl, 1922b) to Heisenberg's paper, in which he raised an objection to the assumption leading

[69] Ludwig Prandtl, born at Freising in Bavaria on 4 February 1875, was one of the leading experts of hydrodynamics in his day. He graduated from the *Technische Hochschule*, Munich, with a thesis on elastic stability in 1900, and became Professor of Applied Mathematics at the University of Göttingen four years later. There, in 1925, he was also appointed Director of the *Kaiser-Wilhelm-Institut für Strömungsforschung*. He died in Göttingen on 15 August 1953.

In 1904 Prandtl discovered the so-called 'boundary layer,' which adjoins the surface of a body moving in air or water, a concept that contributed fundamentally to the understanding of phenomena connected with the operation of an air-wing or the flows of fluids around abstacles. Prandtl was one of the founders of the science of aerodynamics; he made important contributions to subsonic and supersonic flow, as well as to the understanding of the problem of turbulence. In his Institute he developed the basic tools of aerodynamical research, such as wind tunnels.

to Eq. (21); he thought that the latter should be replaced by

$$\alpha(1-\beta)\frac{U^2}{2}=\frac{\zeta}{l}(U-u), \tag{24}$$

where $\alpha \sim 2$ and $\beta < 1$.[70]

Prandtl's objection corresponded to the information available at that time. However, more general treatments, which were given later on by A. W. Maue (1940) and Sommerfeld (1945a, p. 231), confirmed Heisenberg's results.[71] Heisenberg was encouraged by the fact that his physical intuition about the problem had been right, that his calculation was in complete agreement with experiment, and that an expert like Prandtl took his work seriously.[72]

Heisenberg enjoyed working on problems in hydrodynamics because it was a subject in which he 'could *see* what was happening in mathematics'; the correspondence between the apparently complicated calculations and the physical picture, such as the one in a teacup or in a lake while rowing, seemed to be direct. It seemed to be a kind of game: to represent mathematically what one observes in nature. 'This kind of "unclear" mathematics,' he confessed later on, 'appealed to me very much. This way of representing something in nature by means of a nice mathematical scheme and then finding out something which you could actually see, that was quite a bit like my own nature' (Heisenberg, Conversations, pp. 49–50).

The situation in hydrodynamics differed from the other subject of Heisenberg's early research at an essential point. In atomic theory, especially in the problem of the anomalous Zeeman effects, the physical picture was comparatively very unclear and very difficult to derive from the observed phenomena. There also Heisenberg tried to start from a physical picture, i.e., his atomic-core model; however, it did not escape him that this model was only an incomplete attempt to account for what really happened in the atom. Still, he would stick to the model and try to improve on it for many years because no clearer physical picture was available and because he needed a *definite* picture in order to formulate his mathematical equations. With respect to the mathematical tools, on the other hand, Heisenberg did not mind using what he called 'unclear' mathe-

[70] Prandtl argued that Heisenberg's *Ansatz* leading to Eq. (21) could not be correct in general. He thought that the pressure difference between the rim and the plate was about U^2, rather than half of this quantity, and that only a part of the vortex momentum generated, $1-\beta$, was available to the vortex street far behind the plate. He concluded: 'In my opinion Heisenberg's calculation, though quite instructive in itself, can provide information about the above-mentioned quantities only in connection with the experimental results' (Prandtl, 1922b, p. 366).

[71] By generalizing von Kármán's theory of the vortex street, Maue obtained the results, $u/U = 0.20$ and $l/d = 5.5$, in exact agreement with von Kármán and Rubach's observations, Eqs. (20).

[72] It is of interest to note that Niels Bohr also started out his scientific career with research on problems of hydrodynamics. In this early work he extended Lord Rayleigh's theory of liquid jets (of 1879) by including the viscosity of the fluid as well as amplitudes that were not infinitely small (Bohr, 1906, 1909, 1910). In contrast to Heisenberg, Bohr also carried out experimental investigations in his early work; however, this was the only experimental work he ever did.

matics. By that he meant an approach involving a certain freedom to play with the mathematical rules. Thus, he would try to derive from his physical picture of a given problem a mathematical description in terms of working rules, which allowed him to obtain results, but he would care less about how these rules were embedded in rigorous mathematical formalism or could be proved from it. Heisenberg had already applied this manner of doing physics, this playing with mathematics to understand the phenomena of nature, in his early research; he would pursue it time and again in his scientific career, including the work on his doctoral dissertation, for which Sommerfeld selected the problem of turbulence in hydrodynamics.

Osborne Reynolds had treated the classical problem of the transition from laminar to turbulent flow on the basis of energy considerations (Reynolds, 1883, 1895).[72a] The problem in which Sommerfeld became interested, one which he would describe later on as 'the most difficult problem of hydrodynamics,' was more ambitious: namely, whether the fundamental equations of hydrodynamics were sufficient to explain the observed phenomena connected with turbulence and, in particular, whether they could describe the fact that laminar flow becomes instable once the critical Reynolds number is exceeded.[73] In a paper presented in 1908 at the fourth International Congress of Mathematicians in Rome, Sommerfeld had made an attempt to understand this problem (Sommerfeld, 1909a). He applied the method of small perturbations to study the stability of the so-called Couette flow, that is, the flow of a fluid between two plane walls, one of which moves with a constant velocity u. The question was whether, by considering small oscillations around the original laminar flow, the latter could be seen to remain stable or not; if the oscillations would increase with time the original flow would become instable, but if they decreased it would remain stable.[74]

Sommerfeld did not carry through the calculations himself, but he encouraged his student Ludwig Hopf to investigate experimentally the problem of the flow in an open pipe under the action of gravity (Hopf, 1910). Hopf found that the dimensionless 'Reynolds number' governing this flow was proportional to the density, ρ, and inversely proportional to the viscosity, μ—just as Reynolds had

[72a]Osborne Reynolds was born on 23 August 1842 in Belfast. He was educated at Cambridge University and in 1868 became professor of engineering at Owens College, Manchester. Reynolds was a pioneer in fluid dynamics and its applications. He was elected a Fellow of the Royal Society of London in 1877. He retired in 1905 and died at Somerset on 21 February 1912.

[73]In the introduction to Section 38 on turbulence of his *Mechanics of Deformable Bodies*, Sommerfeld wrote: 'Now we turn to the most difficult problem of all hydrodynamics . . . *Do these equations* [i.e., the Navier–Stokes equations] *suffice to explain the observed facts?* In particular, do they represent the fact that the laminar flow becomes instable for a certain critical value of the Reynolds number?' (Sommerfeld, 1945a, pp. 260–261).

[74]The method of applying free small oscillations or perturbations to a laminar flow profile and checking whether they increase or decrease—in the former case the flow might become instable—was first suggested by Lord Rayleigh (1892). William McFadden Orr took it up again in order to investigate the problem of the stability of Couette flow (Orr, 1907). Sommerfeld did not know about Orr's work when he treated the same problem.

suggested—and that it became instable for Reynolds numbers of about 300. A problem arose, however, when Richard von Mises (1912) and Hopf (1914), using Sommerfeld's *Ansatz* (1909a), discovered that the Couette flow remains stable (and laminar) for *all* Reynolds numbers, in contradiction to M. Couette's experiments (1890).

At the German Physical Society's meeting in Jena (18–24 September 1921), Ludwig Prandtl, again employing the method of small perturbations which Sommerfeld had used, discussed certain cases in which a laminar flow profile becomes turbulent (Prandtl, 1922a). Sommerfeld, who was present at the meeting, immediately inquired for the reason why the treatment of Hopf and von Mises had not yielded an instability in the case of Couette flow. Prandtl replied that the Couette flow profile might well become instable with respect to oscillations which were *large* enough.[75]

Sommerfeld was not satisfied with this state of affairs in the theory of turbulence. On his return to Munich from Jena, Sommerfeld discussed the matter with Heisenberg since he knew about his interest in hydrodynamics. He told Heisenberg about various papers dealing with the problem of the stability of laminar flow, especially Poiseuille motion, and asked him to investigate whether an instability would arise at the appropriate Reynolds number.[76] Sommerfeld thought that since Heisenberg was always too occupied with atomic theory it would be good for him once to do some really classical physics.

What appealed to Heisenberg, but also astonished him, was that a fundamental question in hydrodynamics, such as the criterion for the onset of turbulence, had not been answered up to that time. After all, one 'saw' turbulent motions wherever one looked. This was very much the kind of problem which challenged his attitude as a sportsman and his ambition as a scholar. Moreover, it was a problem in a field which was one of Sommerfeld's favourites, and Heisenberg, like his friend Pauli, was always willing to undertake tasks that would please his teacher. Besides, the problem for his doctoral thesis was thus automatically chosen.

Heisenberg started to work on this problem immediately, but it was only one of several things in which he was engaged at the time. He was still deeply involved in the question of the anomalous Zeeman effect. Then came the *Bohr Festival* at Göttingen in June 1922, following which he completed two joint

[75] Prandtl said in detail:

> The case of Couette [flow] is, as must be presumed to be certain, instable with respect to finite perturbations, and they may well have been present in the experiments [of Couette and others]. The exceptional role of the Couette flow with respect to small perturbations I see, as I mentioned in my talk, in the fact that a frictionless motion of this type is not capable of oscillations, while other velocity distributions can perform oscillations. (Prandtl, 1922a, p. 23)

[76] Fritz Noether (1921) wrote an extensive review of the turbulence problem. He discussed all the mathematical methods that had hitherto been applied to this problem: first, the method of small periodic perturbations (used by Rayleigh, Orr and Sommerfeld); second, the energetic considerations of Reynolds, who studied laminar flow and the specific changes in it; third, the method of finite perturbations, which had been introduced by Lord Kelvin (1887). Heisenberg, who referred to Noether's review in the first reference of his published thesis (Heisenberg, 1924c, footnote 1, p. 577), used the first and the third methods.

papers with Sommerfeld on the relativistic linewidths in the calculation of relativistic Röntgen doublets and the intensity of the multiplets and their Zeeman components, respectively (Sommerfeld and Heisenberg, 1922a, b). Yet Heisenberg had been able to devote some time to the subject which Sommerfeld had given him for the thesis, and in September 1922 he gave a talk at Innsbruck on his method of deriving the critical Reynolds number for the Couette flow (Heisenberg, 1924f).[77]

On his return from Innsbruck Heisenberg worked on improving the mathematical technique for dealing with the flow problem at hand.[78] In order to do so, he sought to generalize his previous calculation on a kind of Couette flow and looked for the (laminar) velocity profile which would admit induced oscillations. He found that, 'for large Reynolds numbers, R, a [velocity] profile admits undamped oscillations *only if* its second [spatial] derivative is greater than (or equal to) R, at least in a given [flow] region' (Heisenberg to Sommerfeld, 17 October 1922). 'Largely,' Heisenberg continued, 'I have arrived at results [which have been] obtained previously. However, to a certain extent, the calculations are still too involved to permit definite conclusions.' In any case, he regarded it as certain that the kind of Couette flow which he had in mind was instable for high Reynolds numbers: it consisted of a middle region having close to linear profile and two boundary layers of thickness $b/\sqrt{R}$, where b is the distance between the two walls (i.e., the characteristic distance of the flow problem).[79]

[77]The '*Hydro-aerodynamische Konferenz*' took place in Innsbruck, Austria, 10–13 September 1922. It was organized by T. von Kármán (Aachen), T. Levi-Civita (Rome), C. W. Oseen (Uppsala) and L. Prandtl (Göttingen), in order to bring together the leading experts in fluid dynamics from different countries. Among the talks announced were those of V. W. Ekman (Lund) on sea-currents, L. Hopf (Aachen) on the aerodynamics of airplanes, T. von Kármán on surface tension of moving fluids, T. Levi-Civita on the transport velocity of stationary wave motion, W. C. Oseen on the analytic theory of the equations of motion of a viscous, incompressible fluid, and L. Prandtl on the origin of vortices in ideal fluids and their application to the theory of wings. (See *Z. angew. Math. u. Mechanik 2* (1922), p. 322.)

Heisenberg attended this conference because Innsbruck, where it was held, was not far from Munich, and he was naturally interested in learning what the experts had to say about the turbulence problem. In the case of the Couette flow he had obtained, by using the Rayleigh–Orr–Sommerfeld method of small perturbations, a critical Reynolds number of about 1600 below which the laminar flow was stable. As Heisenberg reported to Sommerfeld:

> I found this result to be so encouraging that I decided to give a short talk at Innsbruck, although I already knew at that time the limits of its validity. At Innsbruck—where, by the way, it was very interesting—mine was the shortest talk given there, and I discussed the main method and stated the preliminary results of my calculations. At the same I added, however, that the accuracy was in no way sufficient to demonstrate rigorously the existence of the solution; hence, in any case, I have not claimed more than I can safely justify. It seemed to me that most participants did agree with me, and I hope that you would agree also. (Heisenberg, 17 October 1922)

[78]Heisenberg's work on the turbulence problem in fall 1922 was interrupted by his attending the 87th *Naturforscherversammlung* at Leipzig, which took place from 17 to 24 September 1922; from there he returned to Munich several days after the meeting.

[79]This 'Couette flow' was somewhat different from the one considered by Hopf and others earlier, insofar as the profile possessed two points at which the first spatial derivative changed by a large amount, while in the usual case—without a boundary layer—the change in the profile was rather smooth, the velocity profile exhibiting a linear increase from zero (filled wall) to a maximum value (moving wall).

Soon after reporting the status of his work on turbulence to Sommerfeld, who was then visiting the United States, Heisenberg turned his attention to the problem of helium in quantum theory, on which he worked with Max Born during the winter semester, 1922–1923.[80] He returned to Munich in spring 1923, plunging directly into the work on his thesis, and laboured strenuously to complete it. His friend and fellow student Otto Laporte watched his 'rather wild and frenetic blasts of research' and recalled: 'While we all said that there was a talented man, it was not so sure that it would lead anywhere. It was getting more and more mystical' (Laporte, AHQP Interview, Second Session, p. 10). However, this time Heisenberg's endeavours led to his thesis, which was completed in summer 1923, though it appeared in *Annalen der Physik* about a year later (Heisenberg, 1924c).[81]

Heisenberg's thesis consisted of two parts. In the first part, he treated the question: under what conditions will a given laminar flow become instable? Rayleigh, Orr, Sommerfeld, Hopf and many others had discussed this problem without achieving agreement with the experimental observations. Heisenberg succeeded in putting together his ideas of summer 1922 into a theoretical scheme that yielded, for the flow between parallel walls having originally a parabolic velocity profile, a critical Reynolds number of the right order of magnitude. In the second part (*Teil* II of the published paper) he used a different approach to describe the turbulent motion of a fluid, based on semiempirical arguments of similarity. Again he succeeded in deriving results from his theory that agreed with the experimental ones.

To investigate the stability of a laminar flow between two parallel plates, which is described by $w(y)$, the velocity in the x-direction as a function of the y-coordinate—the third coordinate being irrelevant—Heisenberg used the method of small oscillations of Rayleigh, Orr, and Sommerfeld. He assumed that the velocity profile and its perturbations can be derived from a potential ψ, that is,

$$\psi = \Phi(y) + \phi(y)\exp\{i(\beta t - \alpha x)\}, \tag{25}$$

such that $w = \partial\Phi(y)/\partial y$ and $\phi(y)$, the amplitude of the second (periodic) term, is small. Inserting ψ into Stokes' equation for an incompressible fluid, he obtained a

[80] In his letter to Sommerfeld, dated 17 October 1922, Heisenberg reported that he had been thinking about the helium problem; but he also mentioned that since he had been so occupied with the turbulence problem, he had not had enough time to work on it. However, the helium problem interested him so much that he reported about it again to Sommerfeld from Munich on 28 October 1922.

[81] Heisenberg's doctoral oral examination took place on 23 July 1923; hence his thesis would have had to be ready latest by June of that year. The published paper on the results of the thesis, entitled '*Über die Stablität und Turbulenz von Flüssigkeitsströmen*' ('On the Stability and Turbulence of Flows'), was signed from Munich and received on 20 February 1924; he therefore had taken the opportunity of polishing his work for publication.

well-known fourth-order equation for ϕ $(= \phi(y))$,

$$(\phi'' - \alpha^2\phi)(w - c) - \phi w'' = \frac{i}{\alpha R}(\phi'''' - 2\alpha^2\phi'' + \alpha^4\phi), \tag{26}$$

involving the constants R, the Reynolds number, $c(= \beta/\alpha)$, the wave velocity, and α, the wave number; the primes denote *ordinary* differentiation with respect to y. All the quantities in Eq. (26) may be assumed to be dimensionless and normalized, i.e., $x \to x/h$, $y \to y/h$, $v \to v/U$, where h is a characteristic length (say, half the distance between the two parallel walls) and U a characteristic velocity of the (undisturbed) flow profile. The main task was now to solve Eq. (26) with the boundary conditions

$$\phi = 0 \quad \text{and} \quad \phi' = 0, \tag{27}$$

at the walls, where $y = 1$ and $y = -1$, respectively.

Heisenberg simplified his task by making numerous approximations, some of which had been applied by other authors in hydrodynamics (e.g., Hopf), while others arose from the translation of his intuitive picture of the physical situation into mathematical equations. For the problem he had in mind, namely, to find a possible instability of the laminar flow, he knew that the Reynolds number R $(= U \cdot h \cdot \rho/\mu$, where ρ is the density of the fluid and μ its viscosity) was a number of the order of 1000. He assumed that α, the wave number of the perturbation, was small, but such that the product $(\alpha R)^{1/2}$ was still a large quantity. Hence he searched for approximate solutions of Eq. (26), keeping only the terms of order α^2 and $(\alpha R)^{-1/2}$. Putting $\phi = \exp\{\int g\, dy\}$ and determining g (in the above approximation), he then obtained the particular solutions

$$\phi_{1,2} = (w - c)^{-5/4} \exp\left\{ \pm \int_{y_0}^{y} \sqrt{-i\alpha R(w - c)}\, dy \right\}, \tag{28}$$

where y_0 is a coordinate point at which the velocities of the perturbed and the unperturbed flows are identical. The other two integrals, ϕ_3 and ϕ_4, followed by solving the second-order differential equation, which is derived from Eq. (26) in the case of no friction (i.e., the right-hand side equal to zero), and by correcting these solutions to the order $(\alpha R)^{-1/2}$. Hence only the solutions ϕ_1 and ϕ_2 depend strongly on the change of the Reynolds number, and they determine the perturbed profile of the flow close to the walls. The profile in the interior, on the other hand, is close to the one found from the frictionless differential equations, i.e., determined by ϕ_3 and ϕ_4. Heisenberg also noted that the solutions for ϕ_1, ϕ_2, ϕ_3 and ϕ_4 thus obtained did not hold at points y_0, with $w = c$; hence he had to use different expansions in the vicinity of these points.

Having performed this task, he turned to select proper combinations of the ϕ's that satisfied the boundary conditions, Eqs. (27). He found that these conditions required one of the following equations to hold: that is, either

$$\exp\left\{2\int_{y_0}^{1}\sqrt{-i\alpha R(w-c)}\right\} = -i \tag{29a}$$

or

$$\int_{-1}^{+1}\frac{dy}{(w-c)^2} = 0 \qquad \text{(in the approximation } \alpha = 0\text{)}. \tag{29b}$$

These equations could be considered as the equations for the possible values of the perturbation velocity c, and these values determined whether the original flow profile was stable or instable under small perturbations. For example, from Eq. (29a) it follows that β $(= \alpha c)$ always possessed a positive imaginary part. As a result, the perturbations, Eq. (25), are damped, and this fact implies that any laminar profile remains stable for arbitrary values of the Reynolds number. Since Ludwig Hopf (1914) had considered only the solutions satisfying Eq. (29a), his conclusions could be easily explained.

On the other hand, Heisenberg found that the solutions satisfying Eq. (29b) behaved differently, especially since they admitted real and complex values for the perturbation velocity c. Heisenberg called the velocity profiles in frictionless fluids $(R = \infty)$, which lead to real c, '*schwingungsfähig*' [i.e., 'capable of performing oscillations' (see Heisenberg, 1924c, p. 596)], in agreement with Prandtl (1922a). He found that such profiles were characterized by the following properties: either, at some point, where $w = c$, the second derivative w'' was zero; or, the unperturbed velocities at the boundaries were equal (i.e., $w(+1) = w(-1)$). If one included the viscocity of the fluid (i.e., $R < \infty$), the value of c obtained an imaginary part, and the flows turned out to be stable as long as the value of the Reynolds number was small, and became instable when it was very large. The reason was that in between, at a critical value R_c, the imaginary part of c changed its sign.

Heisenberg showed how to determine the critical value for the Reynolds number in the two cases. In the case of a flow profile with $w'' = 0$, he proceeded in two steps. He first found that in the solution ϕ $(= \phi_4)$, that is,

$$\phi = (w-c)\int_{y_0}^{y}\left\{1 + \alpha^2\int_{y_0}^{y}(w-c)^2 dy\int_{y_0}^{y}\frac{dy}{(w-c)^2} + \cdots\right\}\frac{dy}{(w-c)^2}, \tag{30}$$

which is valid in the interior of the flow, α^2 assumes only values between certain boundaries. Taking the upper bound, α_{max}, he then derived the minimal value for R_c from the estimate

$$\left(\alpha_{max}R_c^{min}\right)^{1/3} \sim \frac{\left[w'(+1)\right]^{2/3}}{w(+1) - c_0}, \tag{31}$$

with c_0 being the real part of c.[82] Since $w(+1)$ is, in general, rather close to c_0, the product αR and, therefore, also the minimal critical Reynolds number becomes rather large. In the case of a flow profile with $w(+1) = w(-1)$ this procedure cannot be used, for α^2 in Eq. (30) must then be zero. Hence, Heisenberg had to derive another equation, replacing Eq. (29b), that would determine c. He obtained it by selecting a solution different from ϕ_4, Eq. (30). By taking the specific case of parabolic flow, $w = 1 - y^2$, Heisenberg derived an approximate equation expressing R_c as a function of the wave number α. This equation had also the property of yielding a maximum α and a minimal R_c. He estimated a value of the order of 1600 for the latter.[83] Thus, he could give the first quantitative affirmative answer to Sommerfeld's question of whether a laminar flow between two ideal plane parallel walls can become instable.[84]

In the second part of his thesis Heisenberg turned to a different aspect of the turbulence problem, which was closer to the one considered originally by Reynolds. He started from the observation that 'the Reynolds number, which is usually called the "critical" one . . . and which determines the onset of turbulence for sufficiently large perturbations, has nothing to do with stability problems and with laminar flow; it represents entirely a characteristic constant of the *turbulent* motion' (Heisenberg, 1924c, p. 608). That is, for viscous flows there exist two different kinds of motion: the laminar motion, which exists for Reynolds numbers from 0 to ∞, but may become dynamically instable for values greater than a critical one; and the turbulent motion, which exists only above another critical value, but is then *energetically* always more stable than the laminar flow. The two critical Reynolds numbers may be different; thus, within a certain range, the two flows may coexist as long as there are no perturbations which are too large.[85] Since the exact form of the turbulent motion was not known, and it was too difficult to develop a hydrodynamical theory from first principles, Heisenberg applied a semiempirical treatment; that is, he tried to

[82] Equation (31) follows by considering the situation in which the imaginary part of α changes considerably, which is the case if the exponent in Eq. (28), for $y = +1$, assumes values of the order of 1.

[83] Heisenberg did not give all the details of his estimate, for he believed that his approximations only showed the order of magnitude; thus he restricted himself to presenting a curve (Heisenberg, 1924c, p. 604, Fig. 2). He concluded:

> 1. There exists a maximum value of α and a minimum value of R; when the former is exceeded or the latter is not achieved, the instability ceases. 2. For a *given* value of R there exist both a maximum and a minimum value of α; if α assumes values in between these, *instability* occurs, and if α assumes values beyond them then *stability* occurs. 3. The maximum value of α lies close to $\alpha = 0.7$ ($\alpha^2 = \frac{1}{2}$). The minimum value of R has the order of magnitude of 10^3. An estimate of this minimum value, which would be accurate to some extent, cannot be obtained from the figure. (Heisenberg, 1924c, p. 605)

[84] The flow profiles, which O. Tietjens had proved to become instable for higher Reynolds numbers (Prandtl, 1922a), had discontinuous derivatives w'; hence to Sommerfeld and Heisenberg they appeared to be rather singular examples.

[85] Most of this was common knowledge before Heisenberg's work (see e.g., Noether, 1921); Heisenberg only added the criterion that a critical Reynolds number $R_c < \infty$ existed, beyond which laminar flow would become dynamically instable.

'expose the undefined character of the turbulent motion and to idealize it to the extent that it can be treated mathematically with the help of Stokes' equations' (Heisenberg, 1924c, p. 608). This procedure was very much like the kind he knew from atomic theory. For example, in the case of the anomalous Zeeman effects Sommerfeld had also made progress by idealizing the situation in terms of Voigt's classical theory until he could apply the quantum theory of atomic structure.

To proceed with his programme, Heisenberg again considered the flow between two parallel walls. But now he assumed that ψ, the potential of its velocity distribution, could be expanded in a Fourier series,

$$\psi = \phi_0(y) + \phi_1(y)\exp\{i(\beta t - \alpha x)\} + \phi_1^*(y)\exp\{-i(\beta t - \alpha x)\}$$
$$+ \phi_2(y)\exp\{2i(\beta t - \alpha x)\} + \cdots, \tag{32}$$

where ϕ_0, ϕ_1, etc., are the Fourier amplitudes, and β and α are, respectively, the velocity and wave number of the fundamental wave; ϕ_1^* denotes the complex conjugate of ϕ_1. By substituting this expression for ψ in the Navier–Stokes equation, he obtained three differential equations. If only the terms involving $\phi_0(y)$ are retained, there arise the solutions corresponding to laminar flow; if one keeps the first term one obtains, of course, solutions identical to those discussed in the first part of the thesis. By going beyond these approximations, however, Heisenberg obtained some of the known semiempirical formulae of the turbulent flow. In particular, he derived the so-called theorem of Blasius, according to which w ($= \partial\phi_0/\partial x$), the velocity of a point in the vicinity of a smooth wall, varies with the seventh root, $\eta^{1/7}$, of the distance η of this point from the wall (Blasius, 1913). He then treated the situation at a greater distance from the wall, arriving at a linear profile for the turbulent flow of a fluid between two parallel walls with a sharp kink in the middle; this profile corresponded to the parabolic profile in the laminar case. Finally, he estimated by arguments, similar to the ones given in Part One, the critical Reynolds number of the turbulent motion, obtaining again by very rough estimates a value of the order of 1000.[86]

Earlier developments had made one wonder whether the transition from laminar to turbulent motion could at all be described by the equations of hydrodynamics—notably, the Navier–Stokes equations—or whether some new conception beyond the known principles was called for. Heisenberg's thesis gave an answer to this question of principle. By a clever handling of the available equations and the mathematical tools—which had already been applied in earlier treatments of the subject—he was able to present 'the proof that all results obtained thus far, which partially seem to contradict one another, can be

[86] Heisenberg remarked: 'The actual value of R will surely depend on the way how the [velocity] profile had been established; hence we cannot compare [our results] with experience. For this reason we have not yet carried out such a computation of R' (Heisenberg, 1924c, p. 626).

described with the help of simple basic assumptions in a unified mathematical way' (Heisenberg, 1924c, p. 627). This result represented a great success, which was fully acknowledged by Sommerfeld: 'In dealing with the present problem, he [Heisenberg] has again exhibited his extraordinary abilities: complete mastery of the mathematical methods and bold physical intuition. I could not have proposed such a difficult problem to any of my other students' (from the *Protokoll* of Heisenberg's *Examen Rigorosum* of 23 July 1923, quoted by Hermann, 1976, p. 24).

In spite of Sommerfeld's enthusiastic appraisal of Heisenberg's achievement, his work on the turbulence problem did not achieve much notoriety. The reason was historical and may be described here briefly. The story began already at the joint annual meeting of the German Mathematical Association (*Deutsche Mathematiker-Vereinigung)* and the German Engineering Association (*Deutsche Ingenieurwissenschaftliche Vereinigung*), held at Marburg, 20–25 September 1923.[87] At this meeting, the mathematical expert Fritz Noether of Breslau gave a talk on the boundary value problem of turbulence ('*Über die Randwertaufgabe des Turbulenzproblems*'). He published an extended version of his Marburg lecture three years later in *Zeitschrift für angewandte Mathematik und Mechanik* (Noether, 1926). In that paper Noether employed the most rigorous mathematical methods available at that time, including theorems on asymptotic expansions of the solutions of Eq. (26) in terms of orthogonal functions and eigenvalue problems.[88] In particular, he arrived at the result that the flow between parallel walls remained stable under all conditions, and concluded:

> It seems to me that this result is contradicted by the conclusions which Mr. Heisenberg has published. He claims, on a purely asymptotic basis which does not go beyond our asymptotic approximations, that, for instance, the parabolic flow profile is "capable of oscillations," i.e., it possesses purely imaginary roots [Heisenberg, 1924c, p. 604]. According to our theorem such a conclusion cannot be derived on this basis. Without going into the details of Heisenberg's treatment, I just wish to mention here that he makes use of a rather doubtful extension of the "transition situations", which has been proposed by Hopf for a specific case, and that he takes over certain assumptions from Rayleigh's studies of frictionless flows, which cannot be considered justifiable even in Rayleigh's case; for, as I have already mentioned, the problem of the frictionless flow profiles, which are capable of oscillations, actually requires a treatment by considering the detailed limiting process of the case of small friction, and Rayleigh has replaced this limiting process by the assumption of an arbitrary transition equation. (Noether, 1926, p. 242)

Noether's opinions concerning the problem of turbulence were so well respected

[87] At the Marburg meeting, some of the experts on hydrodynamics who had attended the Innsbruck meeting in 1922 (e.g., T. von Kármán) spoke again. (See the announcement in *Z. angew. Math. u. Mechanik 3* (1923), p. 328.)

[88] Fritz Noether was the son of the Erlangen mathematician Max Noether, and a brother of Emmy Noether.

that Heisenberg's solution came to be regarded as mathematically suspect, and one turned to other approaches to the same question.[89]

More than twenty years after this event Heisenberg, in a letter to Sommerfeld, thanking his old teacher for sending him a copy of his book *Partial Differential Equations in Physics* (Sommerfeld, 1945b), wrote:

> By the way, in looking at the chapters dealing with asymptotic representations I was reminded of the problem of my old doctoral thesis (turbulence). Recently Tollmien gave me [the results of] certain investigations on the accuracy of asymptotic formulae used in this differential equation of the fourth order [which describes the flow situation]. I was amused to find that evidently the main content of my dissertation was still all right. In particular, the specialists on hydrodynamics now apparently agree that the parabolic flow profile is indeed instable, as I had stated at that time, and that also my calculation of the region of instability was essentially correct. The same has been found in America by the Chinese [scientist] Lin (Quarterly of Applied Mathematics, 1945 and 1946). (Heisenberg to Sommerfeld, 6 October 1947)

In the last remark, Heisenberg referred to the work of von Kármán's student Chia-Chiao Lin, who had studied the problem of the plane Poisseuille or parabolic flow again in 1944 (Lin, 1944, 1945) and found that 'the asymptotic behaviour of the solution of [Eq. (26)] is indeed very complex, but the method given by Heisenberg was found to be a valid approximation' (Lin, 1955, p. 13).

Lin's work stimulated a new discussion. First, Chaim Leib Pekeris of Princeton University used different asymptotic expressions and concluded from them that the parabolic flow was most likely to be stable, in agreement with Noether's

[89] In Germany, Ludwig Prandtl and his students continued to work on the problem of the origin of turbulence. In his paper, Noether also criticized the work of O. Tietjens (Noether, 1926); but Prandtl defended his student vigorously and suggested carrying out new investigations on this subject (Prandtl, 1926a, b), which were eventually performed by another of his students, W. Tollmien (1929). Tollmien took into account Noether's criticism of Heisenberg's approximations and tried to improve on them. For example, in the case of a plate immersed in a uniform flow parallel to the flow direction, he calculated the lowest critical Reynolds number of instability, obtaining the result $R_c = 420$. He also referred to Heisenberg's treatment of the turbulence problem in the following words:

> Heisenberg has sought to investigate the stability of curved [velocity] profiles in a paper [Heisenberg, 1924c]. In general Heisenberg assumes $\alpha = 0$; he does not clarify the problem of convergence of the expansion of the solution in terms of α. The transformation substitution for ϕ_2 (in Heisenberg's treatment, ϕ_4) was derived [in Heisenberg's paper] by another, not entirely compelling method. The numerical value of the phase transition agrees with ours in the special case he considered. Heisenberg has not gone through to the calculation of the profile, but has restricted himself to making conjectures, which—with respect to the existence of an upper limit for α and a lower limit for R—partially lie in a direction similar to the results we found for the special profile of flow along a plate. (Tollmien, 1929, p. 43)

Tollmien later became a leading expert on the turbulence problem. The main emphasis in this problem shifted from considering it as a purely mathematical question—such as Sommerfeld, Hopf, Heisenberg and Noether had done—to a problem in which the initial conditions of the experimental situation played an important, if not decisive, role. This came about from continued, refined experimental investigations of the conditions under which turbulence was obtained.

result (Pekeris, 1948). Two years later Heisenberg was invited to the International Congress of Mathematicians, held 30 August–6 September 1950 at Cambridge, Massachusetts, and he contributed a paper on 'The Stability of Laminar Flow' to the Conference on Applied Mathematics (Heisenberg, 1952). He discussed the entire development of the problem up to that time and argued that 'the paper of Noether, which in his time had made the whole theory of instability suspicious, seems to contain some mistake, but the mistake has not yet been found' (Heisenberg, 1952, p. 295). When in 1952 Llewellyn Hilleth Thomas performed, following an earlier suggestion of John von Neumann, Lin and Pekeris, a computer calculation of the imaginary part of the velocity c in the solutions of Eq. (26) in the case of the parabolic flow, he indeed found that the sign of its imaginary part did change for a Reynolds number of about 5800, thus confirming Heisenberg's and Lin's results (Thomas, 1952).[90] 'It would be highly desirable,' Lin remarked, 'if the instability of the classical problem could be proved without resorting to such heavy calculations, as used in the methods mentioned above. This has been unsuccessful so far; only *sufficient* conditions for stability have been obtained by simple methods' (Lin, 1955, p. 31). After more than twenty-five years Heisenberg found his work on the doctoral dissertation fully rehabilitated, both in its mathematical methods and its physical content, and he had no reason to look back on it with unease.[91]

I.8 Life in Munich and a Near Failure in the Doctoral Examination

Heisenberg studied for three years at the University of Munich, from fall 1920 to summer 1923, with the exception of the winter semester 1922–1923 which he spent at the University of Göttingen. Heisenberg loved Munich, his second

[90] Thomas' calculation was done on IBM Corporation's Selective Electronic Calculator. 'It took about 150 hours of operation time, equivalent to about 100 years of hand computing' (Thomas, 1952, p. 813).

[91] It is of interest to note that Heisenberg, even before learning about Lin's confirmation of his turbulence calculations of 1922–1923, became interested in this subject again. During his internment in England after World War II, from July 1945 to January 1946 (at Farm Hall, near Cambridge, together with Otto Hahn, Max von Laue, Walter Gerlach and others), he discussed the application of statistical ideas to the problem of turbulence with Carl Friedrich von Weizsäcker. The latter, in connection with his theory of the origin of the solar system, had argued that turbulence might be described by two parameters: a length, corresponding to the dimensions of the largest turbulence element, and a velocity, corresponding to the quadratic average of the turbulent velocity. Heisenberg extended this theory of homogeneous turbulence and proposed a mechanism governing the energy transfer (Heisenberg, 1948a, b, c). This theory, which provided a reasonably accurate description of many overall properties of turbulence, was well received by the experts, and people began to refine it immediately afterwards. Heisenberg later applied some of the statistical ideas involved in his treatment of turbulence to the problem of multiparticle production in high energy collisions between elementary particles (Heisenberg, 1949).

home. He lived with his parents in the *Äußere Hohenzollernstrasse* at the western rim of the famous quarter of *Schwabing*. Although the times were past when Munich had been one of the leading centres of cultural progress in the German Empire—one may just think of the outstanding painters like Franz Marc, Wassily Kandinsky and their colleagues of the '*Blaue Reiter*' association, of playwrights like Max Halbe, Carl Sternheim and Frank Wedekind, of men like the cartoonist Thomas Theodor Heine and the writer Ludwig Thoma, who determined the style of the political-satirical journal *Simplicissimus*, the most critical journal in Germany before World War I—the cultural level of the city was still very high. There was the famous Opera, where Richard Wagner's *Tristan* and *Meistersinger* had their premieres in the 1860s. Bruno Walter was now the chief conductor at the Opera; he introduced, among others, the new work of Hans Pfitzner, and after he left for Berlin in 1922 he was succeeded by Hans Knappertsbusch, an excellent conductor, though less interested in modern music. The opera played an important role in Munich's cultural life; going to the opera was not considered as a privilege of the upper class, but everybody went. In addition, the outstanding musicians of the day stopped in Munich to give concerts that were well attended. In the two large theatres, one run by the State of Bavaria and the other by the City, classical and modern plays were performed. The numerous museums of Munich displayed many exceptional treasures of art and civilization; however, the building of the *Deutsches Museum* was completed only in 1925. There were those (and still are) who maintained that the City of Munich itself represented a continual exhibition of art, not to forget the surrounding landscape of lakes and mountains. Heisenberg once recalled the following images of Munich:

> When the name of Munich resounds, who thinks of the sobriety of the natural sciences? Other pictures come to mind . . . The *Ludwigstrasse* bathed in sunlight from the *Siegestor* to the *Feldherrnhalle*; the view from *Monopteros* across the flower-strewn grass of the *Englischer Garten* to the *Frauenkirche*; "*Nozze di Figaro*" at the *Residenztheater*; the Dürers in the *Pinakothek*; the train to *Schliersee* and *Bayrischzell* filled with skiers; and the beer-tent on the *Oktoberfest* meadow, crowned by the Bavarian lion. All that is Munich. (Heisenberg, 1958b, p. 3)

Unlike his friend Wolfgang Pauli, who enjoyed staying up late at night, walking around the streets and boulevards of the City and visiting cafés and pubs, Heisenberg was not a *Schwabing Bohemian*. He preferred the daytime, rose early in the morning and used most of his spare time hiking or bicycling in the Munich countryside. He was frequently accompanied in these excursions by friends and other young men his own age. Heisenberg also participated in meetings by young people after the war (the so-called *Jugendbewegung*, the Youth Movement); though he did not feel competent to join in the political and sociological questions that came up from time to time, he did enjoy the joint hikes over long distances lasting several days and staying together in camps overnight.[92] He relaxed from physics and hard physical exercise by playing chamber music with his friends, Bach and Mozart being his favourite composers. Heisenberg was a good pianist; as a musician he would join the illustrious circle of physicists that included Planck and Einstein.[93]

Altogether, Heisenberg spent an active, healthy and harmonious life in Munich. He did not notice much of the political unrest of those days that would culminate a little later in Munich.[94] In summer 1923, after only six semesters of

[92]Heisenberg also reported about some other activities connected with his membership in the Youth Movement in those days. Thus he recalled having told Bohr:

> Four years ago [in 1920], for instance, we helped run extracurricular classes in Munich, and I myself was rash enough to give a series of lectures on astronomy, pointing out the various constellations to some hundreds of workers and their wives, describing the motions of the planets and their distances from each other, and trying to interest them in the structure of our Milky Way. With a young lady, I also helped give a course of lectures on the German opera. She sang arias and I accompanied her on the piano; and afterward she would give brief summaries of the history and structure of the various operas. The whole thing was amateurish in the extreme, but I do believe that the audience appreciated our good intentions and that they enjoyed our recitals as much as we did. This was also the time when many young people in the Youth Movement turned to elementary school teaching, as a result of which I imagine that many of our elementary schools have much better teachers than quite a few of our so-called high schools. (Heisenberg, 1971, pp. 54–55)

[93]As Alfred Landé reminded Heisenberg many years later: 'As to your music—fifty years ago you played on our grand piano in Tübingen—I just wish to express my amazement that you played the last movement of Schumann's concerto, which is technically so much more difficult than the earlier ones' (Landé to Heisenberg, 30 April 1974). On the occasion of his sixtieth birthday in December 1961 Heisenberg played—accompanied by an orchestra consisting mainly of professional musicians —a slow movement of one of Mozart's piano concertos, which was broadcast by the *Bayerische Rundfunk*.

[94]We refer to the *Hitler–Ludendorff Putsch*, which took place in November 1923 when Heisenberg was already in Göttingen. The only indication of the difficult political situation in post-war Germany, which he later remembered vividly, was an event that happened during the 87th *Naturforscherversammlung*, which took place at Leipzig, 17–24 September 1922. He recalled that:

> The summer of 1922 ended on what, for me, was a rather saddening note. My teacher, Sommerfeld, had suggested that I attend the Congress of German Scientists and Physicians in Leipzig, where Einstein, one of the chief speakers, would lecture on the general theory of relativity. My father had bought me the return-trip ticket from Munich, and I was looking forward greatly to this chance of hearing the discoverer of relativity theory in person. Once in Leipzig, I moved into one of the cheapest inns in the poorest quarter of the city—I could afford nothing better. Then I made for the meeting hall, where I found a number of the younger physicists whose acquaintance I had made in Göttingen during the "Bohr Festival," and asked them about Einstein's lecture, scheduled within a few hours. I noticed a certain tension all around me, which struck me as being rather odd, but then Leipzig was not Göttingen. I filled in the waiting time with a walk to the Memorial (to the great Battle of Leipzig), where, hungry and exhausted by the overnight railway journey, I lay down on the grass and at once fell asleep. I was wakened by a young girl who had decided to pelt me with plums. She sat down beside me, and made her peace with me with generous offerings of fruit from her ample basket.
>
> The lecture theater was a large hall with doors on all sides. As I was about to enter, a young man—I learned that he was an assistant or pupil of a well-known professor of physics in a South German university—pressed a red handbill into my hand, warning me against Einstein and relativity. The whole theory was said to be nothing but wild speculation, blown up by the Jewish press and entirely alien to the German spirit. At first I thought the whole thing was the work of some lunatic, for madmen are wont to turn up at all big meetings. However, when I was told that the author was a man renowned for his experimental work, to whom Sommerfeld had often referred in his lectures, I felt as if part of my world were collapsing. All along, I had been firmly convinced that science at least was above the kind of political strife that had led to the civil war in Munich, and of which I wished to have no further part. And now I made the sad discovery that men of weak or pathological character can inject their twisted political passions even into scientific life. Needless to say, my immediate reaction was to drop any

study and research—this was exceptionally fast, for the normal time in those days was a minimum of eight semesters—he completed his thesis and took the oral examination to obtain his doctorate. In addition to the principal field of research and dissertation (major), every candidate had to choose two subsidiary subjects (minors). In Heisenberg's case the major was physics; for the minors he selected mathematics and astronomy.

There used to be only the final oral examination for the doctorate at the end, that is, after the thesis had been completed; in the interim the professors, in particular the thesis supervisor, judged a student's progress by his performance in the exercises connected with the courses and his work on research problems. The thesis was the primary thing. Only after the professor had approved it, and the comments and criticism of the other members of the faculty in the candidate's major field had been obtained, could the final examination be scheduled.[95] Heisenberg was examined by Arnold Sommerfeld and Wilhelm Wien in physics, by Oskar Perron in mathematics, and by Hugo von Seeliger in astronomy. The minors presented him no difficulty—he obtained a 'very good,' the best grade, in mathematics and a 'good' in astronomy—nor did theoretical physics in which he was examined by Sommerfeld. Heisenberg's difficulties arose in experimental physics with Wilhelm Wien.

In fact, the difficulties with Wien could have been foreseen to some extent.

> reservations I may have had with regard to Einstein's theory, or rather to what I knew about it from Wolfgang's [Pauli's] occasional explanations. For if I had learned one thing from my experiences during the civil war, it was that one must never judge a political movement by the aims it so loudly proclaims and perhaps genuinely strives to attain, but only by the means it uses to achieve them. The choice of bad means simply proves that those responsible have lost faith in the persuasive force of their original arguments. In this instance, the means applied by a leading physicist in his attempt to refute the theory of relativity were so bad and insubstantial that they could signify only one thing: the man had abandoned all hope of ever refuting the theory with scientific arguments.
>
> Still, so upset was I by this spectacle that I failed to pay proper attention to Einstein himself, and, at the end of the lecture, forgot to avail myself of Sommerfeld's offer to introduce me to the speaker. Instead, I returned somberly to my inn, only to discover that all my possessions—rucksack, linen, socks and second suit—had been stolen. Luckily I still had my return ticket. I went to the station and took the next train to Munich. I was in utter despair because I knew that my father would find it extremely hard to make up my loss. And so, upon discovering that my parents were out of the city, I took a job as a woodman in Forstenried Park, south of the town. The pines there had been attacked by bark beetles, and a large number of trees had to be felled and their bark burned. Only when I had earned enough money to replenish my meager wardrobe did I return to my studies. (Heisenberg, 1971, pp. 43–45)

It seems that Heisenberg, in retrospect, mixed up a few things in connection with these events. First, Sommerfeld did not attend the Leipzig meeting; he was in the United States at that time. Sommerfeld's name was neither mentioned in the official reports of the meeting (*Verh. d. Deutsch. Phys. Ges.* 1922, pp. 67–71), nor in the account of talks presented there (which were reprinted in *Phys. Zs.* **23** (1922) and **24** (1923)). Moreover, Einstein, in summer 1922, had retreated from his promise to give a report on general relativity, suggesting that Max von Laue or somebody else might replace him at the Leipzig meeting (Einstein to Planck, 6 July 1922, quoted in Seelig, 1954, p. 213). Einstein was *not* present at the meeting.

[95] On completing the thesis a candidate would give it to his supervisor. The latter would read it and write his judgment upon it, then pass it around to the other members of the faculty. The whole procedure took at least a couple of weeks; the candidate would then be notified that he could arrange the dates for examination sessions with the professors concerned.

The relationship between Sommerfeld and his experimental colleague had worsened in the early twenties, and this meant that his students had to exercise special care in dealing with Wien. Now Wien had required of Heisenberg that before finishing his doctoral thesis and taking the final examination, he should perform some experiments (i.e., a *Praktikum*) in the Institute for Experimental Physics. He had asked Heisenberg to measure the hyperfine structure of lines in the Zeeman effect of mercury. It was an interesting problem, for, as Heisenberg recalled, in certain cases structures appeared that were very different from the ones described by Sommerfeld's rules, which Heisenberg had explained in his theoretical work.[96] Wien had asked him to photograph the lines resolved by an interferometer. Heisenberg did not succeed in doing this work for several reasons. First, he had not received clear instructions about what was expected of him, and he himself did not know how to proceed.[97] Second, difficulties arose because of the administrative arrangements that prevailed in Wien's Institute. Thus, Heisenberg did obtain some equipment, a Fabry–Perot interferometer for instance, but no camera. It did not occur to him to ask an assistant where to find a camera. He was not aware of the fact that he was entitled to use the Institute's workshop in order to bring the apparatus into a shape fit for his experimental work. He tried to do everything by himself, employing the traditional simple means of string and sealing wax. And finally, his experimental skill was far from adequate for the problem which Wien had assigned to him.[98] In any case, Heisenberg soon lost interest in the *Praktikum* and began to engage himself either in theoretical work or in discussions with the experimental physicists on problems of atomic theory during the time he was supposed to be doing the experiments. As a result, he did not even study certain obvious theoretical questions relating to the apparatus which he had to use, and this fact nearly spoilt the chance of his getting the doctorate.

Wien was annoyed when he learned in the examination that Heisenberg had done so little in the experimental exercise given to him. He then began to ask him questions to gauge his familiarity with the experimental setup; for instance, he wanted to know what the resolving power of the Fabry–Perot interferometer was, how it depended on the distance of the plates, etc. Wien had explained all this in one of his lectures on optics; besides, Heisenberg was supposed to study it anyway before beginning the experiment. But he had not done so and now tried

[96] A little more than a year later W. Pauli explained the hyperfine lines as originating from the magnetic structure (i.e., moments) of the atomic nucleus (Pauli, 1924c).

[97] That Heisenberg had not received clear instructions was by itself not exceptional. At German universities the student was allowed much freedom, and he was supposed to find out for himself how to proceed in most cases. One wonders why Heisenberg did not ask one of the members of Wien's Institute, some of whom he knew personally, for help.

[98] The splittings to be observed were of the order of 0.1 cm^{-1}, i.e., about one-third of the fine structure of hydrogen. This may not have been the right problem to assign to a freshman, which Heisenberg was in experimental physics; however, it is not certain that Wien expected him to complete the task entirely successfully.

to figure it out unsuccessfully in the short time available during the examination. Wien concluded that Heisenberg had simply not been interested in the problem which had been assigned to him and it made him angry. He asked about the resolving power of a microscope; Heisenberg did not know that either. Wien questioned him about the resolving power of telescopes, which he also did not know. The examination on experimental physics could legitimately cover the general laws of physics in different fields, such as thermodynamics, electricity and magnetism, optics, etc. So Wien asked him finally about the principle of the storage battery, which again Heisenberg could not answer. The examiner was naturally very dissatisfied with the candidate's knowledge of experimental physics.[99]

Because of this disheartening performance Wien wanted to fail Heisenberg. However, he could not do so simply on his own. He and Sommerfeld had to agree on a common mark for the physics examination, and Sommerfeld claimed that Heisenberg was a theoretical genius. Finally, after a long discussion, Wien allowed Heisenberg to pass the doctoral examination with the lowest possible grade, *rite*, or just sufficient according to the rules.[100] Since his thesis on turbulence had been judged to be an excellent piece of work, Heisenberg obtained his doctorate with the mention of *cum laude*. On the evening of 23 July 1923, after the examination was over, 'Sommerfeld had invited a little party to his house to celebrate the graduation of his favourite pupil. But Heisenberg was so depressed by his [near] failure that he excused himself very early, went home, packed his suitcase and took the night train to Göttingen' (Born, *Recollections*, 1978, p. 213). He was now almost twenty-two and deeply anxious about his

[99] It might seem that Wien's insistence on questions concerning the resolving power of optical instruments was perhaps unfair; for psychological reasons he should have changed the subject after the candidate's negative answer to the first question. On the other hand, Heisenberg's minor was astronomy; hence Wien could reasonably expect some knowledge of the operation of a telescope. But Heisenberg was not interested in experimental questions at all at that time. Finally, with respect to the theory of the storage battery, he could have learned about it in K. Herzfeld's course on physical chemistry; Wien was surprised and angry that he could not even answer this question.

[100] Heisenberg himself thought that Wien was quite right in his negative reaction, and did not see any reason to complain. He had not realized that his work at Wien's Institute had not been successful at all, and that he had never really come into contact with any real experimental physics. Sommerfeld, on the other hand, was unhappy that his favourite student had all these difficulties with experimental physics at Wien's Institute. He was aware that he had a continuous battle on hand with Wien because of their different temperaments and their feelings about the relative importance of their respective fields. Sommerfeld believed that theoretical physics was more important than anything else, and that theoretical physicists did 'high-brow' physics. Wien thought that Sommerfeld had little concern for the 'normal and decent' part of experimental physics which every physicist ought to know. He therefore insisted that all students of physics, including those working in Sommerfeld's Institute, must learn the fundamental things about experimental physics.

The difficulties between Sommerfeld and Wien were known to Heisenberg's father; he was worried about the fact that a celebrity like Wilhelm Wien disliked theoretical physics as a field of study. Heisenberg recalled later that his father sometimes had doubts about his son going 'into a line which was still so much under suspicion among the physicists at the universities' (Heisenberg, Conversations, p. 21).

career. For it was clear that he could not stay on in Munich, and he did not know what awaited him in Göttingen.[101]

However, Heisenberg took one lesson from his nearly disastrous doctoral examination: he did learn about the resolving power of microscopes and other optical instruments. He even enjoyed the interference calculations by means of which one could show how the instrument worked. 'So one might even assume,' he said many years later, 'that in the work on the gamma-ray microscope and the uncertainty relation I used the knowledge which I had acquired by this poor examination' (Heisenberg, Conversations, p. 68).

[101] The question of a career at this time was very important for Heisenberg, and his father was very concerned how he would make a living. August Heisenberg knew that the study of physics could lead to the possibility of working in industry, but an experimental physicist had a better chance than a theoretician at that time. Companies like *Siemens* and *Allgemeine Elektrizitätsgesellschaft (A.E.G.)* employed people trained in physics, as did the optical industry. However, Heisenberg's interests were theoretical, and a theoretician had almost no possibility of finding work in industry; he was thus limited to pursuing a career in a university, and it did not appear that the University of Munich would have a place for him in the years to come.

Chapter II
Towards the Recognition of the Crisis

The first period of time, which Heisenberg spent in Göttingen during the years 1922–1924, played as important a role in his scientific education as the previous Munich period. On the one hand, he began with it his *Wanderjahre*, his journey years; he had already received extensive training in his field, and he now went to a new place to use and extend his knowledge. On the other hand, since the attitude to theoretical physics in Göttingen was entirely different from the one he had experienced in Munich, he also continued his *Lehrjahre*, his period of apprenticeship, by obtaining a thorough training in those methods which Max Born taught and applied in his research work. Heisenberg adapted quickly and with enthusiasm to the new environment; he was eager to learn the new methods and techniques that were presented to him, and he soon became as much a favourite pupil of Born as he had been Sommerfeld's.

In Göttingen, Heisenberg became acquainted with many of the problems which grew out of Born's interests. But, armed with new and more efficient mathematical techniques, he also attacked with fresh insight the old problems which he had already considered in Munich. Thus, he combined the continued interest in his favourite problems of atomic theory, especially the anomalous Zeeman effects, with the search for new challenging tasks. In doing so, Heisenberg became increasingly aware of the general situation in atomic theory prevailing at that time. Through hard and, at times, painful work he discovered that, in spite of the most detailed and skillful applications of the available theoretical ideas, one did not obtain a really satisfactory solution in most cases; again and again certain features showed up in the analysis of atomic problems, which could not be explained on the basis of the accepted quantum principles. From 1922 to 1924 Heisenberg experienced in Göttingen what later on was referred to as the crisis of quantum theory. With his strong natural optimism he did not hesitate to search for a cure of this crisis; he forcefully joined Max Born's early attempts at replacing the hitherto used (classical) atomic mechanics by a new quantum mechanics.

Heisenberg found that the physical ideas which entered into the theoretical work at Born's Institute frequently came from other places. There existed a fruitful exchange of scientific ideas and problems especially with the experimentalist James Franck, a close friend of Born's. Franck and Born also had close relations with Niels Bohr in Copenhagen, and Born kept regular contact with Albert Einstein in Berlin. Born took the suggestions from Copenhagen and Berlin very seriously and, with their help, tried to construct a new, better quantum theory. In particular, as Heisenberg noticed, 'It was the mathematical aspect which appealed to Born's mind. Born felt that one should repair physics by introducing a new mathematical tool, namely, the difference equations. Born would say, "Perhaps some new mathematical tool is decisive in understanding [atomic] physics"' (Heisenberg, Conversations, p. 230).

Born's interests strongly determined the work of his students and collaborators. Heisenberg recalled his efforts in Göttingen as follows: 'I studied classical mechanics very deeply to see how this mathematical scheme works, and to find out whether I could somehow go over to difference equations instead of differential equations' (Heisenberg, Conversations, p. 227). Heisenberg eagerly sought to apply the mathematical methods, which he invented together with Born, to actual problems of atomic theory, such as the anomalous Zeeman effects. He also assisted Born in reformulating the classical scheme of perturbation theory in the language of difference equations, in order to derive definitive results like Kramers' dispersion formula. He thus helped in establishing Göttingen's Institute for Theoretical Physics as a centre where decisive ideas were born that helped in removing the deep-seated difficulties of quantum theory.

II.1 A New Environment

Heisenberg first visited Göttingen for about two weeks in June 1922 to attend Niels Bohr's lectures on atomic structure. In the beginning of November of the same year he returned to spend the winter semester 1922–1923 with Max Born.[102] The main reason for this change of university was that Arnold Sommerfeld was visiting the United States at that time; he wanted his students to continue to work under proper guidance, and Born seemed to him to be the best choice for this purpose.[103] The second reason was that Heisenberg was supposed to replace Wolfgang Pauli as Born's personal assistant. After obtaining his doctorate in Munich, Pauli had served as Born's assistant on Sommerfeld's recommendation; but he remained in Göttingen only during the winter semester 1921–1922 before he went on to Hamburg and Copenhagen. 'When he left me,' Born recalled, 'he recommended to me his friend Heisenberg as his successor. He too came from Sommerfeld's school in Munich and was no less an "infant prodigy" [than Pauli]. He was working at the time on his doctor's thesis on a problem of hydrodynam-

[102] Heisenberg arrived in Göttingen sometime between 3 November and 13 November 1922; this can be inferred from two letters to Alfred Landé, which bear these dates, the first having been posted in Munich, the other from Göttingen.

[103] Heisenberg was not Sommerfeld's only student to be transferred to Göttingen during the winter semester of 1922–1923. In fact, in a letter of 5 January 1923 to Sommerfeld, Born mentioned the names of three other students whom he had taken over from Sommerfeld: Johannes Fischer, H. Ludloff and Walter Wessel.

ics; Sommerfeld advised him to accept my offer in order to breathe another scientific atmosphere' (Born, *Recollections*, 1978, p. 212).[104]

Heisenberg had impressed many people in discussions during Bohr's lectures in summer; now he did very well in his first extended stay at Göttingen. As Born recalled later: 'When he arrived he looked like a simple peasant boy, with short, fair hair, clear bright eyes and a charming expression. He took his duties as an assistant more seriously than Pauli and was a great help to me. His incredible quickness and acuteness of apprehension has always enabled him to do a colossal amount of work without much effort; he finished his hydrodynamical thesis, worked on atomic problems partly alone, partly in collaboration with me and helped me to direct my research students' (Born, *Recollections*, 1978, p. 212). Indeed, Heisenberg did much more than would correspond to being merely a research student. Born got along very well with him, for he reported to Sommerfeld: 'I have grown *very* fond of Heisenberg; he is liked and esteemed by everybody. His talents are extraordinary, but especially pleasing are his friendly, modest attitude, his good spirits, his eagerness and his enthusiasm' (Born to Sommerfeld, 5 January 1923). Heisenberg stayed in Göttingen until the end of the winter semester and returned to Munich in early March 1923.[105]

Before Heisenberg left, Born asked Sommerfeld whether he could have the young man back after the completion of his doctorate.

> I would like to have a *Privatdozent*, for I am too much burdened with lecturing. Paul Hertz [who had been *Privatdozent* for theoretical physics since 1912] does not count because he has changed over to philosophy, and my doctoral candidates—some of whom are quite able—have not proceeded far enough; also they cannot be compared to Heisenberg. You have Wentzel, and I assume that Pauli will return to you after a year. Could you under these circumstances give up Heisenberg and persuade him to get his *Habilitation* in Göttingen? I shall, of course, make sure that he will be well off financially. Do please think about this matter. Naturally I would also welcome Pauli very much; but he cannot stand, as he claims, the life in a small town. (Born to Sommerfeld, 5 January 1923)

Sommerfeld agreed that Heisenberg should return to Göttingen in fall 1923 to accept Born's offer. But he showed up in Göttingen already on the day after his

[104] In fact, Heisenberg nearly did not go to Göttingen at all, for in October 1922, shortly before going there, he received an inquiry from Wilhelm Lenz of the University of Hamburg about the possibility of succeeding Wolfgang Pauli, who, in the meanwhile, had gone to Copenhagen. Eventually this offer did not materialize, because in Hamburg they decided in favour of the more senior Walter Schottky. Thus Heisenberg reported to Sommerfeld:

> I do believe that in some respects I am better off in Göttingen than [I would have been] in Hamburg, but still I feel very sad; for I would have indeed liked—though my father does not agree with this opinion—not to be dependent on my father's earnings anymore. But it is quite all right the way it has happened, for I did not know whether you would have agreed with the plan of [my going to] Hamburg. (Heisenberg to Sommerfeld, 28 October 1922)

[105] In January 1923 Heisenberg received another offer from Hamburg to replace W. Schottky as Lenz's assistant; Schottky had been called to succeed Otto Stern as *Extraordinarius* (associate professor) in Rostock. He turned down this offer on the advice of Sommerfeld.

doctoral examination, for as Born recalled:

> I was a little astonished when, one morning long before the appointed time, he suddenly appeared before me with an expression of embarrassment on his face. "I wonder whether you still want to have me," he said, and then he explained: he had almost failed in the examination and only got the lowest mark, '*rite*' . . . There he was, putting his fate into my hands. I said: "Let's see what these formidable questions were which brought you down, and whether I can answer them." They were certainly rather tricky. So I saw no reason to send him away—in fact, I was perfectly confident about Heisenberg's outstanding abilities. (Born, *Recollections*, 1978, p. 212–213)[106]

It was fortunate that Max Born's opinion of Heisenberg was too high to be affected by his poor performance in the doctoral examination at Munich. He saw no reason to alter the arrangement that Heisenberg would resume the duties as his assistant in the fall.

At this crucial moment when his scientific career seemed to be in danger, Heisenberg was relieved by Born's reaction. He went on his summer vacation, a part of which he spent with a group of boy scouts (from the Youth Movement) and some of his friends on a hiking tour in Finland.[107] In the beginning of October—his first letter to Pauli from the new place was dated 9 October 1923, earlier than agreed upon—he showed up in Göttingen and plunged into his work.

Already in his first visit to Göttingen, Heisenberg had obtained a favourable impression of the University and its professors. During the memorable event of the 'Bohr Festival' in June 1922 he had been deeply impressed by the active interest taken by the famous mathematician David Hilbert in atomic theory; and he had become acquainted not only with Max Born, his future boss, and the experimentalist James Franck, but also with some other members of the Göttingen faculty such as Ludwig Prandtl. He noticed some striking differences between Göttingen and Munich, and later summarized his impressions in the following words:

> I felt that in Göttingen the Mathematics Institute was in some way the centre of the scientific life. Experimental physics was also quite full of interest; my connection with experimental physics was much easier in Göttingen and much nicer [than it

[106] Born's reaction is confirmed by an incident which Heisenberg recalled. His father was so worried in 1923 that he wrote a letter to Born, asking him for his earnest advice whether Heisenberg was on the right track. 'Because after such a poor examination on the doctor's thesis it seemed quite doubtful to my father whether I would have a chance to enter into science. But Born was quite optimistic about it, so that was all right' (Heisenberg, Conversations, pp. 20–21).

[107] Heisenberg was accompanied by his friends Robert Honsell and Kurt Pflügel and other younger comrades. At this time of inflation in Germany, such a trip brought along many problems as well as exciting experiences; but Heisenberg was courageous and daring not only in problems of theoretical physics, he also loved adventures on long trips and hikes. (For an account of the Finland trip, see Hermann, 1977, pp. 62–63.)

had been in Munich]. I did learn from Franck's papers and I had discussions with him. The connection between experimental physics and theory was as good as it could possibly be—though better with Franck's Institute than with Pohl's Institute. Pohl always disliked these abstract things in quantum theory. He certainly disliked the idea that one should not be able to form models [of the physical situation in atoms]. But still he was very helpful by his interest in band spectra and infrared problems in crystals, and so on. Franck naturally was extremely interested in all the developments in quantum theory. (Heisenberg, Conversations, p. 158; also AHQP Interview)

Heisenberg soon fell into the group of young physicists who worked with Max Born on atomic structure. Most of Born's collaborators in the early 1920s were still engaged in the theory of crystal lattices. His principal collaborator in this field, the Hungarian E. Brody, had left in 1922, but Heisenberg met Carl Hermann who later went to Stuttgart to work with Paul Ewald; Ewald and Hermann edited the fundamental crystallographic compendium, *Strukturbericht*, in 1931. Heisenberg especially got to know Gustav Heckmann, who was interested in crystals, but later changed to philosophy. He found that in Göttingen, as in Munich at Sommerfeld's Institute, there was a group of students working on more conventional topics for their doctoral theses. Heisenberg had naturally more contact with those who did research in atomic theory, and they were primarily Friedrich Hund and Pascual Jordan. Hund, the senior of them, succeeded Erich Hückel, Born's former official assistant (i.e., the assistant who was associated with his Chair and paid by the University) in fall 1922. Heisenberg became 'very good friends with Hund' (Heisenberg, Conversations, p. 164). Jordan had arrived in Göttingen at about the same time as Heisenberg; he was one year younger, had studied before at the *Technische Hochschule*, Hanover, and would complete his doctorate under Born in 1924. Heisenberg had many discussions with Jordan, just as he used to have them with Pauli and Wentzel in Munich, and he realized that Jordan was very good. 'He was certainly one of the best students in Göttingen and understood a lot about this quantum-theoretical business' (Heisenberg, Conversations, p. 164). Heisenberg and Hund were frequently together, also outside the Institute. They used to go on walks and hikes on Sundays, and during these they had many discussions about physics. They were occasionally accompanied by the mathematician Béla von Kerékjártó, who was senior to them, or by some members of other Institutes, or by visitors.[107a]

Like the Göttingen mathematicians for many decades, Born and Franck in the 1920s attracted an increasing number of visitors from Germany and other countries. For example, Enrico Fermi spent the winter semester 1923–1924 with

[107a] Béla von Kerékjártó was born in Budapest, Hungary, on 1 October 1898. He taught at the University of Szeged (*Extraordinarius*, 1924–1928, *Ordinarius*, 1929–1938) and at the University of Budapest (1938–1946); he died in Budapest on 27 May 1947. Kerékjártó worked mainly on geometry and topology. He published a book on topology in 1923 (*Vorlesungen über Topologie*, Volume 8 of the series *Die Grundlagen der mathematischen Wissenschaften in Einzeldarstellungen*, J. Springer, Berlin, 1923.)

Max Born in Göttingen on a travelling scholarship from the Italian Ministry of Public Education. Heisenberg saw him only rarely at that time. 'I had some discussions with Fermi. At that time he was always a bit shy and stayed to himself; it was not easy to get into contact with him. Perhaps he had personal problems, and it was not a good period for him. In any case, I liked him as a rather different type of physicist' (Heisenberg, Conversations, p. 165).[108] While Heisenberg did not get much personal contact with Fermi in Göttingen, it was different with the experimentalist Patrick Maynard Stuart Blackett from Cambridge, who came to Franck's Institute in 1924. 'Our contact was not so frequent as with others because Blackett was married, and therefore he had a family life. I remember that I was in Blackett's home quite frequently and we had many nice discussions on the problems that were going on' (Heisenberg, Conversations, p. 164). They discussed, for example, the energy levels of atoms, the Franck–Hertz experiment, the Compton effect, and resonance fluorescence. Although Blackett had some experimental equipment and was involved in experiments at Franck's Institute, he was very interested in the theoretical aspects of quantum theory and became a fine discussion partner for Heisenberg.

The friendly contact with Franck and his collaborators, Wilhelm Hanle among them, allowed Heisenberg to improve on his knowledge of experimental facts.[109] Besides expanding his scientific horizon, this experience helped him considerably in his theoretical work. However, Heisenberg's greatest experience in his new environment was that he absorbed the mathematical atmosphere of

[108] Heisenberg's view is confirmed by the story which Laura Fermi told in *Atoms in the Family*. She reported that her late husband had a difficult time in Göttingen in spite of his financial security. She wrote:

> Born himself was kind and hospitable. But he did not guess that the young man from Rome, for all his apparent self-reliance, was at the very moment going through the stage of his life which most young people cannot avoid. Fermi was groping in uncertainty and seeking reassurance. He was hoping for a pat on the back from Professor Max Born.
>
> Fermi knew himself to be held in good esteem by scientists in Italy. He also knew that in the kingdom of the blind the one-eyed man is king. How could he tell whether he had one eye or two as a physicist? What was the measure of his abilities by absolute standards? Would he be capable of competing with young scholars like those around Professor Born, of whom Werner Heisenberg was one? The seven months in Göttingen dragged on inconclusively, with little profit, adding to the uncertainty. (Laura Fermi, 1954, pp. 31–32)

Fermi received the desired encouragement more than a year later from Paul Ehrenfest during his stay at Leyden. Heisenberg then got to know a different Fermi. He recalled:

> I remember that when quantum mechanics had come in, in spring of 1926 I made my first trip to Italy with some of my Göttingen friends . . . Coming back from Sicily with Drude [the son of the theoretician Paul Drude], I decided, "Now I must see Fermi," who had been in our Seminar and by that time had become already some kind of professor in Rome. So I visited Fermi on the way back from Sicily, and we had a nice discussion, I think it was already on the exclusion principle and Fermi statistics . . . which we both remembered later on. That discussion was the beginning of our real conversation. In Göttingen we never had a real conversation; we met in the Seminar and occasionally in the streets, but we did not really get in touch. (Heisenberg, Conversations, p. 165)

[109] Heisenberg even tried to do some experimental work of his own in Franck's Institute after his unfortunate experience in Wilhelm Wien's examination. The reason was that August Heisenberg, who was worried about his son's career, had asked Franck to provide him some experimental training. Franck allowed Heisenberg to participate in a *Fortgeschrittenenpraktikum*, i.e., a laboratory course for advanced students. However, he soon found that Heisenberg was thoroughly bored and decided that he should not waste his time upon these unnecessary exercises.

the University of Göttingen. As he recalled:

> In some ways mathematics formed the whole spirit of Göttingen. There was, of course, Hilbert; he was a very strong personality in spite of his rather strange habits . . . . He was very strong in giving a direction to [the entire] University. This famous remark by Hilbert, that "Physics is much too difficult for the physicists," was quite typical, because he felt, "After all, only the mathematicians can really carry things through to the end." Hilbert was just a tremendous man; he could always see when something of interest was happening. (Heisenberg, Conversations, p. 158)

Heisenberg saw Hilbert frequently; he attended some of his lectures and was occasionally invited to his house, especially to the parties that took place there. He found that Hilbert was deeply interested in atomic theory, and he would benefit from that interest in his future work and career. On the other hand, the theoretical physicists in Göttingen took an active interest in mathematics, including mathematical axiomatics in which Hilbert was involved at that time. 'Everybody went [to the discussions], because one knew that was something very exciting and very deep' (Heisenberg, Conversations, p. 159).

Max Born had been profoundly influenced by Hilbert, his former teacher, and he carried over many of his mathematical methods and inventions into physics. Thus, Born's approach to the problems of atomic theory differed greatly from Sommerfeld's. Sommerfeld had a great knowledge of the details of atomic physics and an instinct for how things were; but Sommerfeld was rather removed from the mathematicians in Munich, and strongly separated from the Institute for Experimental Physics. Born had the most cordial relations with his friend James Franck, and the mathematical spirit of Hilbert pervaded his own work. As a person, Heisenberg found Born to be very different from Sommerfeld. In Munich, Sommerfeld cared like a patriarch for the welfare of his students and collaborators, and they visited him in his office without formality. Born, on the other hand, would not see his associates as often. In working with him Heisenberg had to make prior appointments. 'Born would give me a special part of the problem and say, "Would you think about it and we shall discuss it next week on Monday morning?" Then I would do the calculations and go to him at the appointed time. He would see what I had done and then criticize it' (Heisenberg, Conversations, p. 183). Born would also present the things he had done, and they would discuss them together. In the beginning, Heisenberg had fewer scientific discussions with Born than he was used to in Munich. However, Born established and fostered personal relations with the young people. He gave many parties at his house to which he invited all his collaborators and students. At these parties would regularly assemble about ten to twelve people, including those working on crystal problems. Often there were excursions to the surroundings of Göttingen.

> The whole style in Göttingen [Heisenberg recalled] was less formal than the traditional style; I mean, Sommerfeld was still "*der Herr Geheimrat*" ["Privy

> Councillor"] and was a bit traditional, while Born tried to be everything but "*der Herr Geheimrat*." We used to play various games in Born's house and garden. Sometimes we would eat together and go to the Hainberg with the Borns. The whole thing was a kind of family life among the young people and the Borns. So in Göttingen it was more in some respects and less in others than in Munich. (Heisenberg, Conversations, p. 161; also AHQP Interview).

Heisenberg had no difficulty in fitting himself into this overall friendly and pleasant atmosphere. Unlike Pauli, he did not mind living in a small provincial town which Göttingen was in the 1920s with its forty thousand inhabitants. Absorbed in his work, he did not miss too much the rich cultural life of Bavaria's capital, nor the high mountains of the South. After all, he could return to his parents during part of every university vacation.[110] On the other hand, he approached Göttingen and its surroundings with an open mind, as he did the new problems of physics. Thus, he discovered certain replacements for the conveniences he had enjoyed previously. For example, in Munich, Sommerfeld would often visit *Café Hofgarten* with his collaborators to discuss things at leisure. In Göttingen there was the *Café Cron & Lanz*, but Heisenberg and his friends seldom went there. They would, however, often take walks on the Hainberg close to the city. 'There was much hiking and walking around. On Sundays there would always be groups together. On these hiking tours, we would occasionally discuss physics, so that was [in place of] the *Hofgarten*' (Heisenberg, Conversations, p. 190). Since people were so involved in the problems of quantum theory, the discussions were often very excited. Heisenberg remembered the following incident: 'At that time we used to take our modest lunch at a private inn opposite the *Collegienhaus*. Once, to my surprise, I was asked by the proprietress after lunch to see her privately in her room. She told me that in future we could not eat at her place, because the constant discussion on physics ("*ewige Fachsimplei*") at our table was unbearable for people at the other tables, and she would lose the other clients if she kept us any longer' (Heisenberg, 1977, p. 49). Heisenberg and his friends also did some skiing during the winter in the Harz mountains. On one of these trips Wilhelm Hanle got lost. 'We searched for him everywhere,' Heisenberg recalled, 'but could not find him, and we were afraid that he might have injured himself or got lost in the woods. Suddenly we heard a rather plaintive voice from a rather distant spot in the forest, calling '*hν*,' and then we knew where to look for him' (Heisenberg, 1977, p. 49). The relaxed and undisturbed atmosphere of Göttingen provided an adequate balance for the strenuous work in which Heisenberg became involved during the years he spent at Max Born's Institute.

[110]Thus Heisenberg reported to Niels Bohr from his spring vacation in Munich: 'With my hometown, however, I am again fully enraptured: the dark blue sky and the people, none of whom cares about the other and almost all of whom one addresses as "thou" ["*Du*" in German]—that really pleases me very much. Yesterday I heard Beethoven's Ninth Symphony, as beautiful as one can ever hear it only here (not because of the quality of the musicians, but because of that of the listeners)' (Heisenberg to Bohr, 6 April 1924).

II.2 Born's Seminar and the Pursuit of New Subjects

From the very beginning of his stay in Göttingen, Heisenberg found the opportunity of learning new things. For example, in the winter semester of 1922–1923 he took the advantage of attending Max Born's lectures on the kinetic theory of matter and Ludwig Prandtl's course on continuum mechanics; they covered topics different from the ones Sommerfeld and Herzfeld had treated in Munich. He even went now and then to Robert Wichard Pohl's lectures on electricity and optics. 'Göttingen did a lot for my general education in physics,' Heisenberg said. '[It] grew not only from the course-work but also from the subjects people talked about when they got together' (Heisenberg, Conversations, p. 190). Thereby he also obtained inspiration for his current and later work. For instance, his future interest in ions and solid state physics was inspired to some extent by Pohl's involvement in optical problems of the solid state; the papers on ions with Max Born grew from it (Born and Heisenberg, 1924b; Heisenberg, 1924a, d). 'The later work on magnetism actually had its origin in Born's lectures on magnetism which I attended during these years' (Heisenberg, Conversations, p. 189).[111] However, of greater importance for his personal development was that during his stay at Göttingen, Heisenberg became more thoroughly acquainted with mathematics and mathematical techniques. Already during his first stay in Göttingen, in the winter semester of 1922–1923, he began attending some lectures in mathematics there, especially David Hilbert's.[112] From other courses he sought to learn the particular mathematical methods that he needed for his current work in physics. Thus he attended the Seminar on 'Difference Equations,' organized by Richard Courant and Carl Ludwig Siegel during the winter semester of 1923–1924, because he was interested in this subject in connection with the theory of the anomalous Zeeman effects.[113] So it was only in Göttingen

[111] This was Born's course on '*Theorie der Elektrizität und des Magnetismus*' ('Theory of Electricity and Magnetism'), which he held in summer semester of 1924. (See *Verzeichnis der Vorlesungen, Sommerhalbjahr* 1924, *Dieterichsche Universitäts-Buchdruckerei*, Göttingen, 1924.)

[112] For the winter semester of 1923–1924, Hilbert announced three courses: one on '*Wissen und mathematisches Denken*' ('Knowledge and Mathematical Thinking') (one hour per week); another, with Paul Bernays, on '*Grundlagen der Arithmetik*' ('Foundations of Arithmetic') (two hours per week); and the third on '*Mathematische Grundlagen der Quantentheorie*' ('Mathematical Foundations of Quantum Theory') (two hours per week). (See *Phys. Zs.* **23** (1922), p. 532.) In his lectures on the mathematical foundations of quantum theory, of which notes exist in the *Staats- und Universitätsbibliothek* at the University of Göttingen, he discussed the Bohr–Sommerfeld theory of atomic structure, making use especially of the Hamilton–Jacobi methods of mechanics and the variational principles of mathematics. Heisenberg attended some of Hilbert's lectures on quantum theory as well as his general lectures on knowledge and mathematical thinking.

[113] This Seminar, entitled '*Seminar über Differenzengleichungen*' ('Seminar on Difference Equations') was announced by Courant and Siegel already for the winter semester of 1922–1923 (see *Phys. Zs.* **23** (1922), p. 533). However, it is unlikely that Heisenberg attended it at that time. He attended it later on when the idea of changing the differential quotients in atomic mechanics into difference quotients came up in Born's circle; this was only in the second half of the year 1923, and the Seminar on Difference Equations was also repeated at that time. This is also confirmed by Heisenberg's recollection: he mentioned that the idea of discretization of the mathematical description came up in Born's Seminar, in which Enrico Fermi was also participating (Heisenberg, Conversations, p. 157). That happened in the winter semester of 1923–1924.

that Heisenberg was drawn into studying systematically the mathematical techniques that were useful to him in those problems of theoretical physics on which he had already worked for several years. The first opportunity of doing so occurred in the winter semester 1922–1923 in Born's Seminar.

The Seminar, which Heisenberg attended, was the famous one on '*Struktur der Materie*' ['The Structure of Matter'], which Hilbert and Born organized regularly in the 1920s. However, Hilbert himself—unlike the previous years when he ran the Seminar jointly with Peter Debye—rarely attended it at this time, partly because of his ill health and partly because he had turned his attention increasingly to the problem of the foundations of mathematics.[114] But the Seminar was attended by Lothar Nordheim, Hilbert's assistant for physics, and the mathematician Béla von Kerékjártó, as well as by Friedrich Hund, Pascual Jordan and Werner Heisenberg, all from Born's Institute. When Heisenberg first attended it, the Seminar was devoted to the study of the methods of the perturbation theory of celestial mechanics and their possible application to atomic systems. From the topics discussed in the very first seminar, Heisenberg gathered that Born approached atomic theory in a manner entirely different from Sommerfeld's. In some sense, the starting point of the investigations in Göttingen seemed to be way behind Munich. Born and his collaborators were not so well acquainted with the experimental facts as Sommerfeld, and they could not apply the latter's principal method: namely, to fit different facts intuitively into mathematical formulae. 'When I came to Göttingen,' Heisenberg recalled, 'Born appeared to me as an extremely good mathematician, who was interested in the mathematical methods of physics, but who did not know many [physical] details' (Heisenberg, Conversations, p. 152). Now the Seminar on perturbation theory that Heisenberg attended

> was meant by Born as an occasion to learn all the techniques available for the modern quantum theory, and possibly for inventing new methods. So he not only studied with us the technique of Hamilton and Jacobi, which was fairly well known to us at that time, but he studied the methods given in Charlier's *Mechanik des Himmels* [1902, 1907], especially the method of Bohlin and very refined schemes of perturbation theory, always with the idea of seeing what we could do to apply these schemes in quantum theory. However, one saw that in applying quantum theory to many-body problems one [encountered] horrible difficulties, because the many-body problem even in astronomy [i.e., in classical mechanics] is a hopeless affair. (Heisenberg, Conversations, pp. 152–153; also AHQP Interview)

[114]For the winter semester of 1922–1923, Born and Hilbert announced a Seminar on the Structure of Matter ('*Struktur der Materie*'), which was gratis and the participants were supposed to meet two hours per week (*Phys. Zs.* **23** (1922), p. 532). Heisenberg reported to Sommerfeld about a '*Privatseminar bei Born*,' which meant that the Seminar took place at Born's house (Heisenberg to Sommerfeld, 4 December 1922). However, this does not exclude the fact that the Seminar was identical with the one announced by Born and Hilbert, for Born frequently held lectures and seminars for smaller audiences at his home.

It is possible that Hilbert occasionally attended the Seminar at Born's house because Heisenberg mentioned in a letter that 'Born and Hilbert think that I can lecture well' (Heisenberg to Sommerfeld, 15 January 1923). However, this remark may also refer to a talk on the anomalous Zeeman effects that Heisenberg gave in the colloquium (see Heisenberg to Sommerfeld, 4 December 1922).

Heisenberg recalled that towards the end of 1922 certain people felt that many-body systems in atomic physics were not periodic systems, and that Born thought: 'We shall learn the methods of celestial mechanics from the book of Charlier, and it may be that there we shall hit on something which may be useful, say for the quantization of the helium atom, and so on' (Heisenberg, Conversations, p. 153).

The reasons why Born went into the more refined mathematical methods of celestial mechanics arose from his previous work on the perturbation methods in quantum theory. In particular, he had extended the methods used by Niels Bohr and Hendrik Kramers to a systematic approximation scheme for atomic systems, which applied both to cases in which the unperturbed system—i.e., the approximation to the system under consideration, the solution to which was already known—was nondegenerate as well to cases in which it was degenerate. However, a special role was played by those cases in which the periods of two or more motions were commensurable, that is, in which the unperturbed system was degenerate only when the constants of motion assumed certain values. In summer 1922 Born had noticed—in a note on the hydrogen molecule (Born, 1922)—that this situation might actually exist in the description of atoms and molecules having several electrons, and it was necessary to develop suitable methods to obtain their energy states. He was therefore interested in examining systematically all the refined methods of perturbation theory that the astronomers had employed to determine whether any of them could be used for treating atomic systems of the type of the hydrogen molecule. For this purpose he arranged the Seminar on perturbation theory in the winter semester of 1922–1923.

While Born's involvement in perturbation methods was perfectly understandable, one wonders why Heisenberg followed him in these predominantly mathematical studies. In Munich, thus far, he had not been particularly concerned with approximation methods. Sommerfeld, in fact, liked elegant mathematical solutions of theoretical problems; for instance, he had a certain preference for the method of the separation of variables, also in quantum problems, and he transmitted this love to his students. However, Heisenberg had meanwhile gone through a different experience. In June 1922 he had attended Bohr's lectures in Göttingen and had been impressed by the different physical and mathematical approach that Bohr employed. This approach made use of perturbation methods, especially for treating the helium atom. The helium atom, a three-body system, had not yet been solved exactly, and the only possible way to calculate its spectrum seemed to be to apply perturbation theory. Bohr had talked about it and given the preliminary results on the energy of the ground state of helium obtained by Kramers. Later, in fall 1922 the discussion of the question whether the theoretical result agreed with observation became more and more exciting, and Heisenberg was drawn into it; and to contribute to this problem he needed to know the approximation methods better. As a result he joined Born's programme for the Seminar enthusiastically. Early in December he wrote to Sommerfeld from Göttingen: 'Since I have been studying Poincaré's *Celestial Me-*

chanics intensively and in detail (stimulated by private seminars at Born's house), I find that altogether there exists really no difficulty in principle for [solving] the many-body problems of quantum theory; for instance, I believe that the complete solution of the sodium atom is by no means hopeless. Incredibly much is contained in Poincaré's book' (Heisenberg to Sommerfeld, 4 December 1922). He was convinced that by learning the mathematical methods, which the astronomers had applied to the many-body problem, he would make rapid progress in the problems of atomic physics he worked on at that time: namely, the calculation of the states of many-electron atoms, whose Zeeman effects he had discussed earlier. In addition, he noticed that also in the problem of turbulence, on which he was working for his doctoral dissertation he had to improve his knowledge of approximation methods. Hence he found Born's Seminar very useful. He felt that his eager participation in it would contribute to the solution of his scientific problems, apart from increasing his general knowledge about mechanics and atomic theory.

Not every theoretical physicist liked Born's approach. For instance, Enrico Fermi, who participated in a continuation of Born's Seminar on perturbation theory in the winter semester of 1923–1924, was quite unhappy with it. As Heisenberg recalled: 'He disliked the mathematical subtleties, proofs of convergence, etc. For example, we spoke about the theorem of [Ernst Heinrich] Bruns, that in an [infinitely close] neighbourhood of a periodic solution [of a many-body problem] there are always solutions which are non-periodic. Fermi hated this kind of mathematical subtlety. He felt, "That is not physics"' (Heisenberg, Conversations, pp. 153–154). Heisenberg himself enjoyed the discussions at Born's Seminar. 'I felt it was extremely interesting. I learned a lot about classical physics and mathematics, and also about what one could do and, especially, what one could probably not do in quantum theory. From the mathematical subtleties I obtained a stronger conviction that [Born's approach] was in the [right direction] of attacking the problems of atomic theory' (Heisenberg, Conversations, p. 154). Of course, one could argue that theorems, such as the one of Bruns, had nothing to do with the real situation of atoms; after all, it was known empirically that the atoms were stable and no such instabilities or nonperiodic behaviour, as predicted by Bruns' theorem, should arise. However, it was exactly the problem of stability in atoms that Born and Heisenberg wished to treat. Heisenberg still remembered very well Pauli's question about the stability of the atomic-core model in fall 1921, which had remained unanswered, and the most obvious way to deal with it seemed to be Born's. Born felt that one must try one's best with the available methods and even 'a negative proof that something did not work was just as good as a positive proof' (Heisenberg, Conversations, p. 154). This was not Heisenberg's own attitude, but he quickly recognized that it could be useful and he therefore plunged into the work connected with Born's Seminar.

The first task which Born assigned Heisenberg was to discuss Bohlin's method in the Seminar. The Swedish astronomer Karl Bohlin had once proposed an approximation procedure for treating special cases of the many-body problem in celestial mechanics (Bohlin, 1888); it was discussed in Carl Ludwig Charlier's *Mechanik des Himmels* (Volume 2, 1907) and, in great detail, in Henri Poincaré's *Mécanique Céleste* (Volume 2, 1893). Bohlin had considered the specific situation that systems consisting of, say, three bodies occasionally possessed degenerate motions; i.e., the periods of two planets moving around the sun were commensurable or even the same. In that case, one could not expand the action function in terms of λ, a perturbation parameter describing the influence of one planet on the other, but rather in terms of its square root. While in astronomy this situation was supposed to happen very rarely or 'accidentally,' it was supposed to establish in general the stability of the motion of electrons in many-electron atoms or molecules.[115]

To understand Bohlin's method, Heisenberg studied the books of Charlier and Poincaré. He found it difficult to grasp all the details of this method, but ultimately he succeeded in mastering it and presenting it in the Seminar. 'After I gave the talk,' he recalled, 'Born suggested that we should work on this problem in quantum theory, and so we agreed to write the paper on the phase relations' (Heisenberg, Conversations, p. 184). The paper that resulted from this collaboration was the first of a series to be published between 1923 and 1926; it was received by *Zeitschrift für Physik* on 16 January 1923 (Born and Heisenberg, 1923a).[116]

The physical problem on which Heisenberg and Born worked in late 1922 originated from the remarks Bohr had made about the models of helium in his fourth lecture at Göttingen on 19 June 1922 (Bohr, 1977, pp. 381–387). According to Bohr's considerations, the phase relations might play a role in explaining the doublets of ortho-helium. This remark not only stimulated Born to publish his note on the hydrogen molecule, in which he concluded that the quantum conditions required the existence of phase relations between the motions of the two electrons in the hydrogen molecule (Born, 1922); he also continued to think about a generalization of this result. If phase relations of this kind would actually hold in all many-electron atoms or molecules, this would be a very welcome result for two reasons. First, the phase relations would make sure that the many-electron systems in quantum theory would remain periodic, in spite of the fact that the normal solution of the classical many-body problem was not periodic; and from there would follow the fact that the system was stable. Second, they raised the hope of explaining the closed shells of Bohr's atomic theory. Now the important result of Born and Heisenberg's paper was that by applying Bohlin's perturbation method they could *derive* the existence of phase relations between the motions of several electrons in atoms or molecules. 'We

[115]The necessity of the astronomical perturbation method of Bohlin arose in the case of small planets, e.g., the asteroids Hidalgo and the Trojans, which possess nearly the same period as Jupiter.

[116]Heisenberg had already reported to Sommerfeld: 'I am now working with Born to improve and refine the Born–Pauli method [Born and Pauli, 1922]; with its help, for instance, one can prove that the quantum theory demands phase relations between the electrons of an atomic system. This method, too, essentially comes from Poincaré [i.e., his books on celestial mechanics]' (Heisenberg to Sommerfeld, 4 December 1922).

believe,' they noted, 'that, based on our calculations, we are now in a position to state the conjecture that in every atom the *entire* system of electronic orbits is exactly in phase (i.e., it is purely periodic) in the ground state' (Born and Heisenberg, 1923a, p. 44). And they showed that this conjecture was true in certain simple cases in which the unperturbed system possessed only one 'accidentally' ('*zufällig*') degenerate degree of freedom—i.e., for a certain set of action variables one frequency was zero (in contrast to the 'proper' ('*eigentliche*') degeneracy where one frequency is zero for all values of the action variables).

To prove their conjecture, Born and Heisenberg considered the Hamilton–Jacobi equation, i.e.,

$$H(J_k, w_k) = W, \qquad k = 1, \ldots, f, \tag{33}$$

for a system of f degrees of freedom, the last one being degenerate. They solved this partial differential equation for the action function S (the J_k being identical with the partial differential coefficients $\partial S/\partial w_k$, where J_k and w_k are the action and angle variables, respectively) by perturbation expansions in the following way. While the Hamiltonian H and the energy W could be expanded in powers of the perturbation parameter λ, i.e., $H = H_0(J_k^0) + \lambda H_2(w_k^0, J_k^0) + \ldots$ and $W = W_0(J_k) + \lambda W_2(J_k) + \ldots$, the action function S could be expanded as

$$S = S_0(J_k, w_k^0) + \sqrt{\lambda}\, S_1(J_k, w_k^0) + \lambda S_2(J_k, w_k^0) + \ldots \tag{34}$$

In Eq. (34), S_0, the action function of the unperturbed system, is the sum $\sum_{k=1}^{f} J_k^0 w_k^0$ (with J_k^0 and w_k^0 the unperturbed dynamical variables). S_1, S_2, etc., are functions of the unperturbed angles w_k^0. By substituting these expansions in Eq. (33), Born and Heisenberg found immediately that S_1 did not depend on the angles $w_1^0, \ldots, w_{f-1}^0$, the coordinates of the nondegenerate degrees of freedom. The dependence of S_1 on w_f^0, on the other hand, was obtained by considering the second approximation (i.e., by going over to the power λ). After averaging over full periods (from 0 to 1) of the angle variables $w_1^0, \ldots, w_{f-1}^0$, the S_2-term dropped out; there remained a differential equation for S_1, which could be integrated to yield the result

$$S_1 = \int \frac{\partial S_1}{\partial w_f^0}\, dw_f^0, \qquad \text{with} \qquad \frac{\partial S_1}{\partial w_f^0} = \left[\frac{w_2 - \overline{H}_2(w_f^0)}{(1/2!)(\partial^2 H_0/\partial J_f^2)}\right]^{1/2}, \tag{35}$$

where $\overline{H}_2(w_f^0) = \int_0^1 \ldots \int_0^1 H_2\, dw_1^0 \ldots dw_{f-1}^0$. According to the quantum condition, the product $\sqrt{\lambda} \oint \partial S_1/\partial w_f^0$ (over a full period of the angle w_f^0) must be an integral multiple of Planck's constant, provided the fth coordinate of the system is a libration coordinate (in which case the path of integration is a closed one). Since the action variable J_k must be small ($\sim\sqrt{\lambda}$)—because for no perturbation the system is degenerate in this degree of freedom—the only value which J_f can assume in *quantum theory* is zero. It means that the libration then has a vanishing amplitude and one can take the coordinate w_f^0 to be a constant. Hence $\overline{H}_2(w_f^0)$ becomes equal to W_2, which does not depend on w_f^0; moreover, its derivative with respect to w_f^0, which is identical with the frequency associated with the fth degree of freedom, will be zero. From this result Born and Heisenberg concluded that strict phase relations held between the motions of electrons in an atom or molecule. Such phase relations, they pointed out, were caused by the fact that the quantum conditions forbade J_f to assume arbitrarily small values; hence it had to be zero. In classical theory this restriction did not apply; as a result the libration in the fth coordinate was still possible, even with extremely small amplitude.

Born and Heisenberg also noted that one could assume the term S_1 to be equal to zero and that, in higher-order approximations with arguments similar to the above, all terms of the action function having an odd power of $\sqrt{\lambda}$ became zero. Hence, the perturbation scheme for accidentally degenerate quantum systems finally assumed the form given by the previous investigations of Born and Pauli (1922). Altogether this was the first example in which it had been proved that in quantum theory the situation in many-body systems could be described in a simpler and more unified manner than in classical dynamics. It also gave an indication for the first time that these quantum systems might be more stable than the corresponding classical ones.

In spite of the importance of their result, Born and Heisenberg did not immediately go into an application of their formalism to a specific system, such as the helium atom; they left this to a later paper. Heisenberg again noticed that Born's attitude was very different from Sommerfeld's. 'Born was very conservative in some ways; he would only state things which he could prove mathematically. He thought that our statement about how to obtain phase relations was a correct one, but whether it would apply actually to quantum theory he would leave [as an open question]' (Heisenberg, Conversations, p. 180). Born hoped, of course, that the results implied a favourable answer to the problem of stability of complex atoms and molecules, and even concerning Bohr's concept of the periodic system of elements. 'Born is very enthusiastic about these results,' Heisenberg informed Pauli, 'for one now has a simple mathematical method to determine the symmetry properties and the length of periods in the periodic system of [chemical] elements' (Heisenberg to Pauli, 12 December 1922). However, 'Pauli criticized this idea of phase rather early and he told me several times, "These phase relations of Bohr and Born, that's a swindle, that's not the real reason for the closed shells"' (Heisenberg, Conversations, p. 180).[117]

[117]Heisenberg did not remember exactly whether Pauli dismissed the phase relations already in 1922 or later. From Pauli's own development one may conclude that he did so only after he became more concerned with the difficulties of Bohr's atomic theory, especially after studying the problems of the anomalous Zeeman effects; and this happened in 1923. Pauli wrote to Sommerfeld at that time: 'Such commensurabilities of the periods [of electron orbits] are assumed to exist in general in a paper published by Born and Heisenberg in *Zeitschrift für Physik*. This, however, is only a very vague ("*graue*") theory. One has never been able to discover a hint that such a situation occurs in nature (insofar as it is not a matter of complete convergence of periods)' (Pauli to Sommerfeld, 6 June 1923).

Heisenberg, on the other hand, tended to share Born's optimism. In particular, he followed Born in exploring the refined techniques of perturbation theory; and, certainly in the beginning of his stay in Göttingen, he was very fond of this subject. 'I vote for dealing only with perturbation theory in Munich during the summer semester,' he wrote to Sommerfeld early in 1923 (Heisenberg to Sommerfeld, 15 January 1923). In December 1923, nearly a year after submitting the paper on phase relations, Born and Heisenberg sent a paper to *Annalen der Physik*, in which they again treated the general question of perturbation methods and examined the principal steps of applying them to molecules (Born and Heisenberg, 1924a). The approach and style of this work were typically Max Born's; there was emphasis on elaborate and rigorous mathematical methods, and not a single particular example was discussed.[118] It was entirely different from the kind of papers Heisenberg wrote in Munich. With some sarcasm he wrote to Pauli: 'Now the paper of Born and myself on molecules is also ready; it contains bracket symbols having up to eight indices, and will therefore probably be read by nobody' (Heisenberg to Pauli, 7 December 1923). Still, the training which he obtained in Göttingen was useful and necessary; through it he learned the entire picture of classical mechanics and the sophisticated methods used in it. At Born's Institute he was forced to perform the perturbation calculations with all the rigour that was necessary to solve complex dynamical problems. Although it turned out that in the helium problem, to which he applied these methods, the result was quite unsatisfactory, Heisenberg claimed that, 'Later on, through this technique I was able to formulate the correspondence principle in such a refined manner as we needed it' (Heisenberg, Conversations, p. 155). Thus, the hard labour on mathematical work in his early association with Max Born paved the way for his future formulation of quantum mechanics.

II.3 Concern with the Helium Problem

In Göttingen, Heisenberg became acquainted with Born's approach of treating theoretical problems with the help of the most rigorous mathematical tools available, and of carrying through the steps to the very end. Born used to consider a negative proof, that the method employed did not give an answer or gave the wrong answer, to be as valuable as a positive result. This attitude stood in marked contrast to Heisenberg's previous manner of attacking problems, which was determined by seeking to exhaust all possibilities of calculation to reach the final goal of obtaining a result in agreement with experiment, even at the expense of employing apparently 'wild' ideas such as half-integral quantum numbers in the anomalous Zeeman effects. Still, he adapted quickly to Born's approach and interests. The new experience, in which he invested considerable effort, enabled Heisenberg to reach a new level of theoretical understanding. This

[118] See Volume I (Section III.5) for a discussion of the paper of Born and Heisenberg (1924a) on molecules.

transition could be seen from his involvement in the helium problem in winter 1922–1923. He had begun to work on it while he was still in Munich, but he completed it by writing a joint paper with Born dealing with the electron motion and the energy levels of an excited helium atom, which was received by *Zeitschrift für Physik* on 11 May 1923.

Heisenberg's interest in the helium problem was stimulated by Niels Bohr, who in his fourth lecture at Göttingen on 19 June 1922, had given a report on the latest status of the calculation of helium terms (Bohr, 1977, pp. 379–387). Bohr first gave the facts that two spectra had been observed, one consisting of narrow doublets and the other of singlets, which were attributed to ortho-helium and parhelium, respectively. He then talked about his and Kramers' investigations on the structure of the helium atom, which had yielded the following models: either the orbits of the two electrons were coplanar ellipses, i.e., both electrons moved in the same plane (ortho-helium), or the two orbits made a finite angle with each other (parhelium). The state of lowest energy, in which both orbits were circles (corresponding to the lowest orbit in a hydrogen-like atom with nuclear charge $Z = 2$) would not exist in the ortho-helium model. The ground state of the helium atom should rather belong to the system of parhelium states; in particular, the two electrons would move on the lowest circular orbits, whose planes make an angle of 120° with each other. Bohr also presented the results of Kramers' calculation of the ionization potential of helium, i.e., of the energy of the ground state; he had concluded that the resulting value was too small, but he left open the question whether this was due to an inadequate approximation or not. Heisenberg, who had been acquainted with Bohr's models for some time, listened to the latest developments of the helium calculation with great interest.[119]

Several months later, in October 1922, Heisenberg was surprised to learn from Sommerfeld that in America John Hasbrouck Van Vleck had found that Bohr's model of the helium ground state gave 22 V for the ionization potential, which deviated by about 3 V from the experimentally observed value. On receiving Sommerfeld's letter with this result, Heisenberg wrote back:

> That probably Bohr's helium model is wrong, is extraordinarily fine in my view, for now perhaps certain physicists will have greater trust in the Zeeman effects [i.e., in Heisenberg's theory of the Zeeman effects, including half-integral quantum numbers]. Nevertheless, I am almost sorry for the sake of beauty in physics, because

[119]Bohr had mentioned, without going into details, certain conclusions concerning the models of helium—but not the calculation of the energy of the ground state—in a lecture to the Copenhagen Physical Society on 18 October 1921; this lecture had been translated into German and appeared in *Zeitschrift für Physik* (see Bohr, 1921e). Heisenberg saw this paper and he wrote immediately to Pauli concerning Bohr's model of the ground state of helium: 'By the way, what do you think about helium having the angular momentum $+1h/2\pi$ [in the ground state]? According to Bohr one can explain the fact that it is diamagnetic only by the assumption that the angular momentum of helium around the axis of the [magnetic] field is always zero. I don't like that particularly, but, of course, one cannot doubt that Bohr is right' (Heisenberg to Pauli, 6 March 1922).

> what will now become of all the other so convincing ideas of Bohr on the structure of elements? One must hope that most of them will somehow remain intact. I also now have strongly the hope that the model [i.e., the model with coplanar orbits in which the azimuthal quantum numbers of the two electrons are] $\frac{1}{2}$, $-\frac{1}{2}$ (which, by the way, also stems from Bohr and not from me) is correct. If you would like to compute it I would be happy, for then one might learn in the fastest way as to what is up with the systems containing several electrons. (Heisenberg to Sommerfeld, 17 October 1922)

In spite of his desire to pursue the helium problem, which was evident from these lines, Heisenberg added that he would not be able to perform the calculation on the helium model with the half-integral quantum numbers in the near future because of his preoccupation with the work on turbulence, the problem selected for his thesis. However, ten days later the situation had changed, and he reported again to Sommerfeld:

> Against what I wrote to you in my last letter, I could not resist the curiosity and did make the calculation of the helium [model] $\frac{1}{2}$, $-\frac{1}{2}$. According to present approximation I obtain an ionization potential of 24.6 Volt with a possible error of about $\pm\frac{2}{3}$ Volt, but a probable error of $\pm\frac{1}{3}$ Volt. This agrees perfectly with the value of 24.5 Volt measured spectroscopically by [Theodore] Lyman [1922]. (Heisenberg to Sommerfeld, 28 October 1922)

Heisenberg knew that Sommerfeld had become interested in the helium problem, as was his friend Wolfgang Pauli from whom he had obtained the information about the experimental value.[120] He also expected that Sommerfeld —to whom Bohr had mentioned in Göttingen, without approving of it, the possibility of assigning to the ground state of helium the electron orbits whose angular momentum assumes the values of $\frac{1}{2}h/2\pi$ and $-\frac{1}{2}h/2\pi$—would perhaps go ahead and evaluate its energy. Still, he was extremely curious to find out what could be achieved in the helium problem with the help of his favourite half-integral quantum numbers. Heisenberg, therefore, did not wait for Sommerfeld's reply from faraway America, but sat down to calculate it himself. He arrived at the result in less than two weeks, and in his letter to Sommerfeld, dated 28 October, he explained the steps towards the solution. For simplicity, he assumed that the orbits of the two electrons in the ground state of the helium atom were in the same plane; that is, if the interaction of the two electrons with each other were neglected, the two electrons should move on two different Kepler ellipses of the same shape in opposite directions, so that one electron passed through the perihelion of its orbit while the other was at its aphelion. The system, without

[120] 'This paper of Lyman should be in *Nature*; Pauli has told me about it' (Heisenberg to Sommerfeld, 28 October 1922). Heisenberg had seen Pauli at the *Congress of Natural Scientists and Physicians* at Leipzig in September; at that time Pauli had already performed some calculations on Bohr's helium model.

perturbation, was then a degenerate one, and its energy would be given by twice the value of the corresponding hydrogen-like system, i.e., $2Z^2W_H$ (where W_H is the energy of the Kepler orbit in the hydrogen atom, denoted by the principal quantum number $n = 1$). The perturbation in the two-electron system was, of course, provided by the mutual potential energy of the electrons, which depended not only on the shape of the electron orbits but also on the angle ϕ between the major axes of the two orbits. Heisenberg took ϕ as the variable which was canonically conjugate to $\pm p_\phi$, the angular momenta of the orbits, and applied the phase integral condition,

$$\oint p_\phi \, d\phi = n_2 h \qquad \text{with } n_2 = \tfrac{1}{2}\,. \tag{36}$$

That is, the perturbation in the two-electron system had the effect that the system obtained an additional degree of freedom described by the pair of canonically conjugate variables, (p_ϕ, ϕ), and this additional degree of freedom was quantized with half-integral quantum numbers.

To calculate the perturbation energy, W_1, which he identified with the perturbation term in $\overline{H}_1$, the Hamiltonian of the helium atom averaged over a full period of the unperturbed motion of the electrons, Heisenberg decomposed $\overline{H}_1$ into two parts: the first, $\overline{H}_1^{(1)}$, represented the contribution to the interaction of the electrons if they moved on the unperturbed orbits, while the second, $\overline{H}_1^{(2)}$, denoted the smaller, remaining part. 'We then have,' he argued, 'the case treated also by Born and Pauli, where we have decomposed the perturbation function into several parts, where each successive term is smaller than the previous one' (Heisenberg to Sommerfeld, 28 October 1922). However, before applying Born and Pauli's (1922) perturbation scheme, he introduced a new angle variable, w_2, besides the angle variable w_1 describing the unperturbed Kepler orbits of the electron, and expanded the perturbation terms as functions of w_1 and w_2. He explained:

> Completely in accordance with the Born–Pauli method I have now expanded H_1' [where H_1' denotes the difference between H_1 and its average, $\overline{H}_1$, over a period of the system] and $\overline{H}_1^{(2)}$ approximately in Fourier series by simply computing H_1' and $H_1^{(2)}$ for as many values of w_1 and w_2 as possible. (In the present approximation I have calculated about 25 points in the unit square $w_1 w_2$, and then I have evaluated the single terms by graphical integration.) Thus one arrives without great computational labour at the same number of terms in the expansion as Kramers had calculated at the time of [Bohr's Wolfskehl lectures in] Göttingen. The next term would also in my case lead to long numerical calculations, which I have not yet dared to do; in principle there is no difficulty in improving on the accuracy [of the approximation for the helium energy]' (Heisenberg to Sommerfeld, 28 October 1922).

By this simple and effective numerical evaluation of the perturbation function,

Heisenberg obtained two terms—i.e., $-1.115Z$ and 0.116—which had to be added to $2Z^2W_{\mathrm{H}}$, the energy of the unperturbed system, and the total result indeed yielded the fine agreement, mentioned above, with Lyman's experimental determination of the ionization potential of neutral helium.

Heisenberg was proud of the result which he had obtained in a straightforward manner by using the most primitive version of perturbation theory. He did not worry whether he had made any assumptions that contradicted the accepted principles of atomic theory; the only assumption he was aware of was that of half-integral quantum numbers, and he was convinced that these made sense ever since the success of his atomic core model for explaining the anomalous Zeeman effects. However, he did worry considerably about the accuracy of the approximation in the helium problem, and it was on this question that he hoped to do better in Göttingen. After all, he knew that Born had become an expert on quantum-theoretical perturbation theory; and he hoped that perhaps there existed a perturbation scheme that would give better results. About a month after he had reported his first result on the calculation of the energy of the ground state of helium, Heisenberg again wrote to Sommerfeld. In the meantime he had become familiar with the more refined perturbation methods of the astronomers and was ready to attack the numerical task again, though now with the help of an assistant, whose services Born had promised to provide. Heisenberg was very optimistic about future progress and wrote: 'The hope of obtaining the ionization voltage by an accurate calculation of the three perturbation terms appears to me to be very great, since Van Vleck found 20.7 Volt and the first terms of Kramers yielded 20.63 Volt [for Bohr's model of the helium ground state with integral quantum numbers], which agree completely' (Heisenberg to Sommerfeld, 4 December 1922). That is, since the different approximation models of Van Vleck and Kramers yielded identical results for the same model of helium, one could trust the mathematical tools. Heisenberg indeed got a helper, a young lady, for assistance in the mathematical calculation on his model for the ground state of helium. In January 1923 it appeared certain that the calculation would lead to a result of an accuracy of about 2 *per mille*, and that it would essentially agree with the one that Heisenberg had reported to Sommerfeld in October 1922.[121]

The convergence of the calculations towards the desired goal was not the only feature which satisfied Heisenberg. He received further encouragement from Sommerfeld. The latter had observed in fall 1922 that in searching for the appropriate quantum conditions, which act in the case of the two-electron system, one should make use of the equations

$$\int p_{r_1}\,dr_1 + \int p_{r_2}\,dr_2 = n'h \tag{37a}$$

[121] For example, in a letter dated 15 January 1923, Heisenberg wrote to Sommerfeld that Pauli, who had visited Göttingen on 13 and 14 January, had brought along a new perturbation method of Kramers. This method converged so fast that it gave the energy term, to within 1 percent, in the first approximation. Naturally, Heisenberg and his assistant also tried the new method.

and

$$\int p_{\phi_1} d\phi_1 + \int p_{\phi_2} d\phi_2 = kh, \tag{37b}$$

involving the phase integrals of the radial and azimuthal variables, $r_1, p_{r_1}, \phi_1, p_{\phi_1}$ and $r_2, p_{r_2}, \phi_2, p_{\phi_2}$, of the two electrons (denoted by the subscripts 1 and 2, respectively). The integrals in Eqs. (37) extend over a full period of the motion, and the radial quantum number n' and the azimuthal quantum number k are associated with the total two-electron system; in the ground state of the atom they are supposed to assume the value 1. Heisenberg wrote to Sommerfeld about it that, 'When I heard about your idea on the helium quantization through Herzfeld, I was very enthusiastic about it. I then immediately formulated it into the language of perturbation theory, and the surprising result followed that one does not need any assumption at all to arrive at half-integral quantum numbers' (Heisenberg to Sommerfeld, 4 December 1922). In order to prove his assertion, he argued as follows. Because of angular momentum conservation the sum of k_1 and k_2, the individual angular momenta of the two electrons, is a constant. This is mathematically expressed by the fact that the perturbation term in the Hamiltonian describing the helium atom does not depend on the angle ϕ $(= (\phi_1 + \phi_2)/2)$, which is canonically conjugate to the total angular momentum of the system, but only on the angle ϕ' $(= (\phi_1 - \phi_2)/2)$, i.e., on the difference of the angular coordinates of the two electrons. By identifying the corresponding phase integral, $\int k' d\phi'$ (k' being the momentum conjugate to the angle ϕ'), with integral multiples of Planck's constant (i.e., with h for the ground state), and assuming the total angular momentum of the atom in the ground state to be zero, Heisenberg could indeed conclude that the average angular momentum of the individual electrons assumes the values of $\frac{1}{2}$ and $-\frac{1}{2}$ in units of $h/2\pi$, respectively. He interpreted this result physically in a letter to Pauli: 'The difference of the perihelion angles,' he argued, 'is conjugate to the angular momentum of one electron, as always when the total angular momentum is given. However, the motion is periodic in the perihelion angle with the period 4π, *not* 2π; after [the angle] 2π the two electrons are exchanged, and again after 4π, and then the original situation is restored' (Heisenberg to Pauli, 12 December 1922). Thus, he claimed that the anomalous behaviour of the electrons in the helium atom with respect to quantum numbers could be explained by considering their motion in detail. This intuitive argument would later on, with a completely different justification based on electron statistics, play a role in the final solution of the helium problem.[122]

[122] Heisenberg also had an explanation why from this treatment one could not derive half-integral quantum numbers for the electron in the hydrogen atom. He argued that in the case of hydrogen, which may be obtained from that of helium by separating the nucleus into two parts and letting the electron–electron interaction go to zero, one was not allowed to make the contact transformation to the variables ϕ' and k'. 'For then,' he wrote to Sommerfeld, 'I have no secular perturbations anymore [i.e., under the conditions that each nucleus acts only on its own electron] and it is more reasonable to treat each electron by itself as a non-perturbed atom [i.e., each electron in the Coulomb field of a nucleus] with relativity, i.e., without degeneracy' (Heisenberg to Sommerfeld, 4 December 1922).

While Sommerfeld supported Heisenberg strongly and urged him to publish his results—actually Sommerfeld announced them in a note himself[123]—Wolfgang Pauli, whom Heisenberg informed about his helium calculations in December 1922, remained very skeptical. The main reason for Pauli's skepticism was that Niels Bohr had taken a completely negative attitude towards half-integral quantum numbers after they came up in the treatment of the anomalous Zeeman effect; and Pauli, who had been in Copenhagen since September 1922, followed Bohr in this respect. Therefore, when Heisenberg invoked half-integral quantum numbers again in his theory of the ground state of helium, Pauli assumed the task of presenting critical arguments against them. Heisenberg was faced with the obstinate opposition of the Copenhagen group against his ideas, and he complained to Sommerfeld: 'I am rather unhappy about the fact that in all these papers I am always in contradiction with [the views of] Pauli and Bohr. According to Pauli's letters [to me] Bohr is still firmly convinced of his helium model and therefore believes that mechanics is wrong when one wants to calculate the correct ionization potential [on its basis]' (Heisenberg to Sommerfeld, 4 January 1923). He argued, however, that Bohr had no good reasons for his belief, especially since the empirical evidence contradicted it, and he tried to convince Pauli about the superiority of his helium model (with half-integral quantum numbers) when the latter visited Göttingen in the middle of January 1923. Heisenberg, assisted by Max Born, then achieved a compromise; he persuaded Pauli that his treatment of helium was at least not inconsistent. With this satisfaction achieved, he hoped to publish the results of his calculations very soon.[124] And yet, he never submitted a manuscript for publication, because the continuation of his work on the helium atom, namely, the calculation of the excited states, led to a new situation, which shook his belief in his earlier results.

Heisenberg had indicated his plan to calculate the higher energy states of the helium atom in a letter to Sommerfeld at the beginning of December 1922, but until the middle of January 1923 no progress had been made.[125] But then, Born

[123] Sommerfeld wrote to Heisenberg, encouraging him to publish his helium calculation, in late November. At the same time he sent a note of his own on the helium problem, which was published in the *Journal of the Optical Society of America* (Sommerfeld, 1923c). In this note Sommerfeld proposed the quantum conditions, Eqs. (37), and came to the conclusion that, in many-body systems, not only integral, but also half-integral quantum numbers arise. Moreover, he argued that other than half-integral quantum numbers might also occur; for example if one takes into account the motion of the heavy nucleus in the hydrogen atom, most of the integral azimuthal quantum number will be distributed to the electron, but a small fraction will go to the nucleus. Thus, finally the integral or half-integral quantum numbers observed in one or two-electron atoms should be 'inexact' ('*ungenau*'). However, Heisenberg did not agree with this interpretation. He believed that the quantum numbers in atoms must be strictly either integral or half-integral.

[124] Heisenberg wrote to Sommerfeld on 15 January 1923: 'Hence I shall, in the near future, also "release" the He [-paper] (Born also thinks that is the right thing to do), after I have calculated [its energy] according to the new method of Kramers to about 1 *per mille*—unless you object to it.'

[125] Heisenberg wrote to Sommerfeld: 'Further I would also like to calculate the higher [i.e., excited] He-terms and then publish everything together [i.e., the results on the ground state and the

and Heisenberg, on the completion of their paper on the phase relations (Born and Heisenberg, 1923a), joined forces to attack the helium problem in the systematic manner characteristic of Born. They studied *all* the perturbation solutions of the equations of motion of a system consisting of two electrons and a doubly-charged nucleus in the approximation that one (the outer) electron moves mainly outside the other (the inner) electron. Working with great concentration, they arrived at conclusive results within a month.[126] Heisenberg immediately informed Pauli:

> Born and I have now calculated the most general model of helium in all details—even in still higher approximation than Bohr used for ortho-helium, taking into account besides the dipole term the term in r^{-3} [where r denotes the distance of the outer electron from the nucleus]—and we find that the energy certainly comes out to be wrong. [Going into greater detail, he reported:] It was rather easy to have an overview of all possible types of motion [of the two-electron system]; there are, as was to be expected, only three classes of motion possible in quantum theory (Bohr's ortho-helium and parhelium types, besides the ortho-helium in which the two electrons move in opposite directions) in addition to a couple of quantum solutions in which the inner electron goes through the nucleus (or does so at least for k [angular momentum] $\rightarrow \infty$). (Heisenberg to Pauli, 19 February 1923)

Heisenberg told Pauli that he and Born had been able to calculate the excited states in all three admissible models—those with pendulum orbits (*Pendelbahnen*) of the electrons were, of course, excluded—and that in no case did the calculated and the observed term systems agree with each other. The disagreement was not changed when corrections due to relativity were taken into account. 'The result,' he concluded, 'appears to me—provided no error is hidden somewhere—to be very bad for our present views; one will probably have to introduce totally new hypotheses—either new quantum conditions or assumptions to alter mechanics' (Heisenberg to Pauli, 19 February 1923).

The conclusions, at which Born and Heisenberg arrived in February 1923, confronted them with an entirely new situation. They had both been quite optimistic earlier that, with detailed perturbation calculations, they would finally be able to obtain an appropriate theoretical description of all atomic systems in agreement with the experimental observations. They had thought that perhaps they would have to refine available methods, or to go to higher approximations,

---

excited states]' (Heisenberg to Sommerfeld, 4 December 1922). However, writing again to Sommerfeld on 15 January 1923, he remarked: 'About the excited states, however, I don't know anything at all.'

[126]Concerning the status of the helium calculations with Born, Heisenberg informed Niels Bohr on 2 February 1923: 'The other work, about which I wanted to write to you, is a general investigation of all mechanically possible types of motion of the excited helium. If then finally the experimentally observed terms are not contained among them, one knows that mechanics is wrong.' This indicates that at the beginning of February the Göttingen collaborators were still engaged on the investigation, but had not yet found the results.

in order to find the desired solution of any given problem in atomic theory, but they had never expected a negative result—especially one that they could not improve upon.[127] They found the situation—which Heisenberg described in a letter to Pauli on 19 February with the words, 'it is a misery' ('*es ist ein Jammer*') —so desperate that they requested help from Copenhagen. 'I would be enormously interested to hear your and especially Bohr's opinion,' Heisenberg wrote. 'You must know what one should do now' (Heisenberg to Pauli, 19 February 1923). Since Pauli did not answer immediately, Born addressed Bohr personally. In a letter, dated 4 March 1923, he summarized the results of the calculation of the excited states of helium and asked Bohr whether he would agree to the publication of a short note on the material. Bohr agreed; on 17 April, Born sent the manuscript of the joint paper with Heisenberg to Copenhagen. On 2 May, Bohr wrote that Pauli and Kramers had meanwhile checked the calculations and confirmed their conclusions; Born immediately thanked Bohr and, in particular, Pauli, who had written a letter suggesting a few changes to improve the manuscript. 'We shall take into account all his suggestions,' promised Born, 'and shall change the text accordingly. I have sent his [i.e., Pauli's] letter with my own remarks to Heisenberg in Munich, who would make the text ready for publication' (Born to Bohr, 5 May 1923). Heisenberg did so, and on 11 May 1923 the paper of Born and Heisenberg (1923b) on the state of the excited helium atom was received by *Zeitschrift für Physik*.

Similar to the previous paper on the phase relations, Born and Heisenberg's paper on helium bore Born's hallmark. However, the difference was that now the same authors were not concerned with developing a general scheme for atomic problems, but treated a definite system: an atom with two electrons. Heisenberg had already performed a lot of specific calculations on it and had arrived at a well-defined conclusion concerning the ground state of helium. Yet, nothing of this was contained in the published paper, in which the procedure was dominated by Born's style and interests. For example, the authors began by writing down the three-body problem in full generality and searched not only for the specific solutions (which had been considered before as the most probable candidates for describing the physical situation in ortho-helium and parhelium), but for all solutions. Such a complete treatment of a similar problem had rarely been attempted previously, with the exception perhaps of Pauli's treatment of the hydrogen molecule-ion in his thesis. Compared with this earlier investigation, Born and Heisenberg very quickly obtained a definite answer by systematically using the perturbation scheme, the reason being that their mathematical methods were clear and clean and—other than employing classical mechanics and the

[127]Of course, Born and Heisenberg did try to avoid the negative conclusions of their helium calculations, and Heisenberg wrote to Pauli: 'Born and I still wish to discuss—I do not know whether you believe it to be reasonable—the following problem: to calculate, under the assumption that the formula for δ [i.e., the Rydberg correction which determines the energy terms of the helium atom] is at least *qualitatively* correct, what the orbits must look like in order to yield the energy approximately correctly' (Heisenberg to Pauli, 19 February 1923). Apparently, the result of this exercise was again negative, for neither Heisenberg nor Born returned to this suggestion again.

known quantum conditions—they made no arbitrary assumptions. Indeed, their reliance on well-tested physical and mathematical principles, as well as their conservative attitude, lent credibility to their conclusions. In carrying out their calculations, they proceeded straightforwardly, without invoking prior knowledge of their result—as Heisenberg had done so often before. Heisenberg was very impressed with this pursuit of mathematical rigour; he knew that this calculation was truly in the spirit of Göttingen and he would have been unable to carry it out in Munich. He found this venture exciting and followed Born's leadership without reservation; and, once the final result had been obtained, he did not play with 'improving' upon it or avoiding the negative conclusions.

Born and Heisenberg's procedure for analyzing the possible motions of the excited helium atom and calculating the associated energies was the following. They first wrote down the Hamiltonian H, describing a system of two electrons in the Coulomb field of a nucleus with charge Z, in polar coordinates. Since the total angular momentum of the system must be conserved, H does not depend on the angle which is canonically conjugate to it. Hence they introduced a new set of dynamical variables: the total angular momentum; ψ_1 and ψ_2, the distances of the perihelions of the two electrons from the nucleus—in case the two orbits are not coplanar, ψ_1 and ψ_2 are measured from the point where the orbits intersect; their conjugate momenta p_{ψ_1} and p_{ψ_2}; and, finally, the radius and the radial momentum, r and p_r, of the outer electron. The corresponding radial variables of the second, inner electron did not enter because the authors considered only those motions in which the excited electron mainly moved outside the orbit of the inner electron; hence, in the Hamiltonian, to a very good approximation, one could average over the radial variables of the second electron. Born and Heisenberg then expressed H as the sum

$$H = \text{const.} + \frac{1}{2m_e}\left(p_r + \frac{1}{r^2}\, p_{\psi_1}^2\right) - \frac{e^2(Z-1)}{r} + \frac{\Delta_1}{r^2} + \frac{\Delta_2}{r^3}\,. \tag{38}$$

The constant denoted the product of the energy of the ground state of hydrogen with Z^2, the square of the charge of the nucleus under consideration (in units of the absolute electron charge e), and represented the energy of the inner electron in the absence of the outer one. The following two terms (with m_e, the mass of the electron) gave the contribution of the outer electron when moving in the field of a nucleus, whose charge is reduced by one unit due to the screening effect of the inner electron. Finally, the last terms described the perturbation of the motion of the excited electron. The quantities Δ_1 and Δ_2 were functions of all variables except the radial variables of the inner electron.

The authors now observed that the right-hand side of Eq. (38) could be considered as an expansion in powers of r^{-1}, the inverse distance of the excited electron from the nucleus. Hence they proposed to apply perturbation theory; they took as the unperturbed system the one described by a Hamiltonian consisting of the first three terms of the sum in Eq. (38) and identified the term

Δ_1/r^2 with the first-order perturbation and Δ_2/r^3 with the second-order perturbation. The solution of the unperturbed system was, of course, well known, its energy states being identical with H_0,

$$H_0 = \text{const.} + \frac{W_H(Z-1)^2}{n_1^2}, \tag{39}$$

where the second term expressed the energy states of a hydrogen-like atom with nuclear charge $Z-1$. Clearly, W_H was the energy of the ground state of hydrogen, and n_1 the principal quantum number (composed of the azimuthal and the radial quantum numbers) of the excited electron. In the calculation of the states of helium, Born and Heisenberg had to go to the second order of the perturbation scheme because the unperturbed system was degenerate. For the resulting energy they found an equation which was formally identical to Eq. (39), except that the denominator in the second term, n_1^2, was replaced by the quantity $(n_1^2 + \delta)$, where δ was a nonzero correction. By comparing the theoretical expression for the energy terms with the empirical description of spectroscopic data, they could interpret δ as the well-known Rydberg correction.

In the most general model of the helium atom, δ depended on the variables of the outer electron, and on the perihelion distance ψ_2 and the momentum p_{ψ_2}, i.e., the azimuthal quantum number k_2, of the inner electron. In order to get a systematic idea about the values of δ, the authors classified the different types of motion of the two-electron atom, as admitted in the second-order perturbation approximation. To that end, they investigated the relation between the variables k_2 and ψ_2; i.e., they analyzed the shape of the real solutions of the function $k_2 = f(\psi_2)$. In particular, Born and Heisenberg were interested in the *conditionally periodic* solutions of the helium problem, which were the only ones that could possibly describe a stable motion of the three-body system. These solutions were characterized by the fact that the curve $f(\psi_2)$ in the (k_2, ψ_2)-plane was either closed or periodic in the variable ψ_2; in the first case the motion in the coordinate ψ_2 was a libration, in the second it corresponded to a rotation. With this restriction, the authors managed to organize their solutions into three types: in the first, the perihelion of the inner electron—whose orbit may or may not be in the same plane as the orbit of the excited electron—described full rotations in the course of time; in the second type of motion, ψ_2 librated between the points $\pi/2$ and $3\pi/2$; and, in the third type, ψ_2 librated around the points $\pi/2$ or $3\pi/2$. However, they immediately excluded the last type of solutions from physical consideration because the inner electron would, in that case, be able to approach the nucleus infinitely closely, and that was forbidden according to the accepted principles of quantum theory. To avoid pendulum orbits (*Pendelbahnen*) in general, Born and Heisenberg required further that the quantum number k_2 of the inner electron should assume the value unity. Then, for the Rydberg correction δ, they found the formula

$$\delta = \frac{9}{8k_1^2 Z^2}\left(-p + \frac{1-p^2}{2k_1}\right) + \frac{(Z-1)}{8k_1^3 Z^2}(3p^2 - 1). \tag{40}$$

In Eq. (40), the quantity p denotes $j - k_1$, the difference of the total angular momentum of the helium atom, j, and the angular momentum of the excited electron, k_1 (both in units of $h/2\pi$); it can take on the values $+1$, 0, and -1, respectively, and these values also describe different classes of helium models. For $p = 1$, the orbits of the two electrons were coplanar and the electrons moved in the same direction; for $p = 0$, they were perpendicular to each other; for $p = -1$, the electrons again moved in the same plane, but in opposite directions. By evaluating the right-hand side of Eq. (40), with $Z = 2$ and k_1 assuming the values 2, 3, 4, etc., and comparing the theoretical values of δ thus obtained (for each of the above p-values) with the experimental values of δ for ortho-helium and parhelium, the authors came to the conclusion that 'the result of our investigation is completely negative' (Born and Heisenberg, 1923b, p. 242).[128]

The virtue of Born and Heisenberg's analysis of the helium problem lay in an important fact: they were able, for the first time, to classify the perturbation-theoretical solutions of the excited two-electron atoms, which described multiply periodic motions. Although the periodic solutions were the only ones which had a chance to represent stable atoms, two questions that were related remained open. One was whether the calculation up to the second order really yielded an approximate solution, i.e., whether the perturbation series for the energy and other quantities converged or not. From the corresponding situation in astronomy the authors were aware of the fact that, according to the theorem of Bruns, close to any periodic solution of the three-body problem, for which the perturbation converged, there existed many solutions for which this was not so. 'We must therefore expect,' they concluded, 'that all the types of motion found by us do really exist; but in the close vicinity of each solution [found by us] there are infinitely many other, *not* conditionally periodic, solutions of the equations of motion, which are not given by our formulae' (Born and Heisenberg, 1923b, pp. 239–240). However, they did not answer the question whether there also existed solutions, which could not be obtained by evaluating the perturbation series, but might represent the empirical data better.[129] The reason was that Born and Heisenberg believed strongly that the motion of electrons in stable atoms was multiply periodic, and all multiply periodic motions could be accounted for by perturbation methods.

The second question was whether the periodic solutions obtained for the excited two-electron atom were dynamically stable against small variations of the

[128] The relativity correction to the motion of the outer electron contributes an additional term to Δ_2 in Eq. (38); hence it produces only an additional constant in δ, which does not change the results. The relativity correction to the inner electron's motion cannot play any decisive role with reasonable physical assumptions. According to an estimate, there arise possible corrections only for quantum numbers n_1 greater than 100.

[129] Adolf Smekal, in a review of Born and Heisenberg's paper on the excited states of helium, criticized that, 'The missing convergence of the representation obtained is emphasized, but not followed further despite its importance for the application of the correspondence principle to the helium problem.' He further added the comment: 'It would, for instance, be conceivable that the correspondence principle excludes all types of motion following from the perturbation theory, which are not the only possible ones; this the authors have not taken into account' (Smekal, 1925c, p. 1258).

orbit. Born and Heisenberg succeeded in showing that a stability condition was satisfied in the following sense: if one varied slightly the initial values of the variables ψ_2 and k_2 and studied the periodic orbit connected with this change, it was found to be close to the previous orbit determined by the initial values. One could, of course, argue against such a stability condition, for it was not the one used in classical mechanics. In the latter, one considered continuous variation of the orbits, which also included the simultaneous discussion of *all* nonperiodic solutions. Pauli (1922) had analyzed the stability of the hydrogen molecule-ion more in the spirit of the classical theory and, therefore, raised certain objections against the procedure of Born and Heisenberg. Heisenberg replied:

> Your theorem, "If one cannot treat the problem of stability [in atomic theory] classically, then the same is true of the energy," you will have to prove. But let us not quarrel unnecessarily. Basically, we are both convinced that all present helium models are wrong, as is all of atomic physics. Let us hope that the present splendid spring . . . will "change everything, everything". (Heisenberg to Pauli, 26 March 1923)

II.4 Half-Integral Quantum Numbers and the Difficulties of Atomic Theory

The year 1923 brought forth a new situation in atomic physics. The theoretical physicists in Copenhagen and Göttingen, in particular, changed their attitude towards the Bohr–Sommerfeld semiclassical quantum theory and declared it to be insufficient to account for the outstanding problems of atomic structure. Serious difficulties arose when one tried to translate the intuitive views, which Niels Bohr had developed since 1920, into a mathematical theory, that is, when one tried to *prove* that the electrons in atoms behaved exactly in the way that followed from the building-up principle (*Aufbauprinzip*) and the correspondence principle; moreover, one encountered serious difficulties in deriving the empirically observed properties of atoms—such as the ones characterized by their spectral lines—from the theoretical equations. Just the year before, in 1922, the prospects of atomic theory had appeared to be more favourable. When Max Born, after Bohr's lectures in Göttingen, joined the other theoreticians in working on the problems of atomic structure, he expected to achieve his proper share of the ultimate success that seemed to be close at hand. And the initial results, especially the treatment (with Heisenberg) of the phase relations in many-electron atoms appeared to support Bohr's model of atoms and their organization within the periodic system of elements. Although he had been aware of the fact that there were some difficulties in the application of the known mechanical and quantum-theoretical principles to certain problems, Born had expected that the helium calculations—in which he had joined his student Werner Heisenberg in January 1923—would be crowned with success. He had thought that all that was needed was to refine the mathematical methods and to define more clearly the underlying physical assumptions. But, contrary to all expectations, the results of the careful calculations spelled a disaster for the existing theory.

Some of the difficulties in obtaining the correct empirical properties from calculations of atomic models had already arisen long ago and had become more pronounced since 1921. The first indication could be found in Wolfgang Pauli's treatment of the hydrogen molecule-ion, a system consisting of two nuclei and an electron (Pauli, 1922). Apart from the fact that Pauli had been able to find only a metastable ion, he had also had to introduce another hypothesis to carry out his calculations. This hypothesis, which he called the 'mechanical correspondence principle,' seemed to Bohr to contradict the principles he had used in quantum theory.[130] On the other hand, in the description of the anomalous Zeeman effect with the help of the atomic-core (*Atomrumpf*) model and the half-quantum numbers, there had entered hypotheses which also went beyond the established theory; Niels Bohr had, therefore, criticized these attempts as well. 'We must conclude from the occurrence of the anomalous Zeeman effect,' he had explained in his Göttingen lectures, 'that the classical theory is inadequate' (Bohr, 1977, p. 391). In spite of Bohr's negative reaction, it was not clear to what extent the assumptions made by Pauli in the work on the hydrogen molecule-ion, or by Landé and Heisenberg in describing the anomalous Zeeman effects, were really incompatible with the accepted and tested foundations of atomic theory. After the *Naturforscher-Versammlung* (Congress of Scientists and Physicians) at Leipzig in September 1922, Heisenberg reported to Landé about his discussions with Wolfgang Pauli and Richard Becker there, and said: 'Gradually it is becoming the general conviction that one must really give up much of the present mechanics and physics if one wants to arrive in a different manner [i.e., not by using half-integral quantum numbers and the atomic-core model] at [a description of] the anomalous Zeeman effects' (Heisenberg to Landé, undated, around beginning of October 1922).

While Bohr's conviction about the breakdown of mechanics in problems of atomic theory became stronger because of Kramers' failure, at the end of 1922, to reproduce the correct ionization potential of helium, other people were still more optimistic. However, the discussions of Pauli, who was now a Copenhagen spokesman, with Born and Heisenberg in the middle of January 1923 showed that the views held in Göttingen had also moved closer to Bohr's. The first indication of such a change was expressed in a letter of Heisenberg to Sommerfeld, who was then in America. He wrote:

> But I hold Bohr's view to be correct, that mechanics is wrong, though in a special sense. We agree with Pauli (and Bohr) in stating the following theorem (which you have already proposed yourself in your letter): *Quantum theory deviates from mechanics in the fact that the phase integrals are invariant not only with respect to adiabatic, but also with respect to rapid change*. In this theorem the Zeeman effect

[130]See N. Bohr, 1923a, English translation, p. 12, footnote.

> and the Stern–Gerlach effect are included. (Heisenberg to Sommerfeld, 15 January 1923)

The main point of Heisenberg's remark was that in this way he could counter Bohr's criticism, expressed in the Göttingen lectures, against one of the basic assumptions of the atomic-core model of the anomalous Zeeman effects—namely, that the orbital planes of the electrons in complex atoms were arranged either parallel or perpendicular to the axis of the core—for that assumption contradicted energy considerations. In rapid, nonadiabatic changes, energy may be dissipated; hence Bohr's argument would cease to hold. Born and Heisenberg employed the same idea of rapid changes in atomic systems in the last part of their paper on phase integrals, which they submitted for publication in the middle of January (Born and Heisenberg, 1923a). There they also discussed the difficulties, which Albert Einstein and Paul Ehrenfest had noticed when they attempted to explain the orientation of atoms parallel and antiparallel to the direction of an inhomogeneous magnetic field (Stern–Gerlach effect) by dynamical considerations (Einstein and Ehrenfest, 1922). Born and Heisenberg now argued that these discrete orientations came about by a nonadiabatic, fast transition of the originally statistically oriented atoms. In this transition, they argued, the mechanical equations were violated and energy and momentum conservation might not hold in individual processes, though they still held on the statistical average. The last argument strongly resembled the one which Heisenberg had used about a year earlier to justify the violation of Rubinowicz's selection rules in the anomalous Zeeman effects. At that time, however, he had put the blame primarily on a breakdown of classical electrodynamics; but in January he and Born suggested that classical mechanics was also not necessarily satisfied in atomic processes. A month later their doubts about the validity of classical mechanics became overwhelming. From the failure of their calculation of the excited states of helium Born and Heisenberg concluded that: 'Evidently there are only two ways out of this difficulty: either the quantum conditions are wrong . . . or the motion of the electrons no longer satisfies the equations of mechanics even in the stationary states. Both points of view lead to fundamental difficulties for the hitherto substantiated quantum-theoretical results' (Born and Heisenberg, 1923b, p. 243).[131]

It should be noted that with the above remarks Born and Heisenberg did not wish simply to appease the great Niels Bohr. The letters exchanged between Göttingen and Copenhagen during the first five months of 1923 clearly confirm that Born and Heisenberg's conversion to Bohr's view, i.e., the assumption that basic parts of the existing atomic theory were wrong, was genuine. From then on

[131] The Göttingen theoreticians maintained their belief in the nonvalidity of the existing quantum theory for the helium problem. When, in 1924, F. von Wiśniewski claimed that he had derived from this theory the correct values of the parhelium spectrum (Wiśniewski, 1924), Heisenberg submitted a note to *Zeitschrift für Physik*, in which he isolated the nonmechanical assumption in his approach (Heisenberg, 1924b, p. 175).

they believed that *both* the violation of classical mechanical equations and the existence of half-integral quantum numbers pointed to the necessity of creating a new atomic theory. However, there still remained differences between the respective attitudes in Copenhagen and Göttingen. The principal difference was that Bohr and his associates tried to banish half-integral quantum numbers from the theoretical description, while the people in Göttingen tended to follow Heisenberg in assuming their existence. There were several reasons in favour of half-integral quantum numbers. One was the success, which Alfred Landé had achieved in fall 1922 by explaining the structure of most of the complex line multiplets and their anomalous Zeeman effects in terms of half-integral quantum numbers (Landé, 1922a). Heisenberg, after a talk he gave on this subject during his first semester at Göttingen, reported to Sommerfeld: 'Since my lecture on the Zeeman effect, all physicists here believe in the theory [of half-integral quantum numbers]' (Heisenberg to Sommerfeld, 4 December 1922). The second reason was that the explanation of the energy of the ground state of the helium atom also seemed to require the use of half-integral quantum numbers. Though Heisenberg had not published his results, for he believed in spring 1923 that something was basically wrong with atomic models, the helium model with half-quantum numbers still remained the most plausible model. Even Wolfgang Pauli, who had been very critical towards this work, wrote in summer 1923: 'With respect to Heisenberg's model of the helium ground state, I think that one should carry through an accurate calculation of the energy as long as one does not possess a better model' (Pauli to Sommerfeld, 6 June 1923).[132]

While Heisenberg did not publish anything on the ground state of helium because of the general difficulties occurring with mechanical calculations in atomic models, he still hoped for an experimental confirmation of the half-integral quantum numbers. 'A strict proof for or against half-integral quantum numbers may be derived from the Stark effect of helium,' he had written to Sommerfeld already in December 1922, adding that he had requested experimental data from Wilhelm Steubing of Aachen (Heisenberg to Sommerfeld, 4 December 1922). Since Steubing did not send any information, he turned to Sommerfeld a month later. 'I would therefore like to request you,' he asked, 'to look for an American or Japanese (Takamine?), who will be able to measure the first-order Stark effect of He and Li^+ as accurately and as fast as possible' (Heisenberg to Sommerfeld, 4 January 1923). Since the term-splitting by an electric field 'turned out,' in atomic theory, to be proportional to the angular momentum of the individual electrons, Heisenberg hoped to decide the question quickly: namely, that with half-integral quantum numbers the Stark pattern would not show a middle component, but only lines situated symmetrically about the centre (i.e., the position of the line in the absence of an electric field).

[132]However, in the same letter Pauli added right away that the dynamics of Heisenberg's model for the ground state of helium appeared to be very involved and to lead to similar difficulties with the principles of classical mechanics and quantum theory, as already had shown up in the simpler case of the excited states of helium.

Although the experimental result did not become available until several years later, Heisenberg was convinced that it would confirm his assertion.[133]

However, before the question of the quantum numbers of helium could be decided, the Göttingen theoreticians discovered another strong indication of the existence of half-integral quantum numbers. It came up in a general survey of the properties of atomic systems, especially in certain investigations connected with Max Born's lectures on atomic mechanics in winter 1923–1924. In the course of these lectures Born came across many questions which, to his taste, had not yet been answered satisfactorily; and he requested his collaborators, especially the assistants Friedrich Hund and Werner Heisenberg, to clean them up. One such question was the extension of perturbation-theoretic methods to apply also to arbitrary molecules—a continuation of the work of Born and Erich Hückel (1923)—which Born and Heisenberg completed and submitted to *Annalen der Physik* (Born and Heisenberg, 1924a). At the end of this paper, which was highly theoretical and lacked specific examples, the authors announced further investigations, in particular of molecules whose nuclei were arranged on a straight line (Born and Heisenberg, 1924a, p. 31). Heisenberg would turn to this problem in a paper on the stability and reaction energies of molecules and ions consisting of three atoms and submit his results in June 1924 to *Zeitschrift für Physik* (Heisenberg, 1924d).[134] Meanwhile, however, Hund analyzed the data on term series of atoms and, in January 1924, arrived at an interesting conclusion, about which Heisenberg informed Landé immediately. He reported:

> From the [energy] term values, Mr. Hund once did calculate systematically for the entire periodic system of elements as to which orbits dive in the sense of Schrödinger and which do not. At the same time he made very general calculations of the size of the ions by making use of the entire empirical data. Now one must be able to derive theoretically from the size of ions also as to which orbits dive and which do not. When he used integral [quantum] numbers, there arose the grossest contradictions . . . [However,] if one takes half-integral quantum numbers, everything turns out to be reasonable. Mr. Hund has established really *sine ira et studio* [impartially] all the tables for integral and half-integral k [k being the azimuthal quantum number of the series electron], and at the end they [the tables] decide convincingly in favour of half-integral k. [He concluded triumphantly:] Bohr also will have to believe in it. (Heisenberg to Landé, 19 January 1924)

Heisenberg was very happy about Hund's result, and he wanted to explore its consequences a little further. By talking to Hund, who was much more familiar

[133]The Stark effect of helium was examined in detail by J. Stuart Foster (1927a) and Jane M. Dewey (1926, 1927). The results revealed a complicatedly shifted pattern, which would have been difficult to analyze with the help of Heisenberg's 1922 formula. However, the observed shifts were in perfect agreement with the quantum-mechanical calculations of a helium atom in an electric field (Foster, 1927b, c).

[134]In the paper on triatomic molecules Heisenberg discussed, in particular, the origin of the electric moment in the case that the three atoms are arranged in a straight line; he attributed it to the deformation of one of the atoms.

with the problems of molecular theory and crystal physics, he hit upon an interesting idea. 'From my discussions with Hund,' he recalled later, 'I came to understand that when an atom is deformed by an electric field in a crystal, the deformability actually arises because of the auto-electric field, and this is the quadratic Stark effect' (Heisenberg, Conversations, p. 188). This conclusion followed from the assumption that the deformation of the ions was caused by the action of an electric field, which induces in each ion an electric momentum proportional to the field strength; the proportionality constant, α, is identical to the polarizability of the ion and can be regarded as a measure of its deformability. Now the quantity α, say for an alkali ion, can be derived from the spectroscopic data in the following manner. One assumes that the derivations from the Coulomb field of the atomic core are created by an electric dipole moment induced in the core by the series electron. This deviation, of course, gives rise to the terms of the Rydberg correction; and these terms, according to quantum theory, depend on the azimuthal quantum number k of the series electron. Hence it is possible to obtain the polarizability α directly from the Rydberg corrections. Heisenberg was intrigued by the opportunity of combining the spectroscopic problem with the problem of crystal physics, and he mentioned the point to Born who was an expert in the latter field. They decided to collaborate on the problem of the deformability of ions and its influence upon their general optical and chemical properties. In the course of this investigation they compared the polarizability of alkali ions, as determined from the Rydberg corrections, with the observed polarizability of the neighbouring noble gases to the left of the alkali elements in the periodic table. Due to the smaller nuclear charge of the noble gas atoms their polarizability should be larger than that of the corresponding alkali ions; but this conclusion could be drawn only if, in the above-mentioned calculation of the polarizability α, one used the half-integral quantum numbers k. Born and Heisenberg therefore declared: '*From the deformation constants of noble gases and of noble-gas-like ions there follows the half-integral valuedness of the quantum number k*' (Born and Heisenberg, 1924b, p. 397).

Hund and Heisenberg presented their new confirmation of half-integral quantum numbers at the meeting of the Lower Saxony Section (*Gauverein Niedersachsen*) of the German Physical Society, held at Braunschweig on 9 and 10 February 1924. Pauli, who also attended this meeting, gave the following report to Bohr:

> In Göttingen they have now computed that, *apart from all questions of complex structure*, the penetration of the orbits into the atomic core and the term values in the series spectra agree better with half-integral than with integral k-values. I shall first make a general comment on this: The atomic physicists in Germany now fall into two classes. The members of one calculate a given problem first with half-integral values of the quantum numbers, and, if the result does not agree with experience, they then calculate nevertheless with integral quantum numbers. The members of the other class calculate first with integral numbers, and, if the result

> does not agree, then they calculate with half-integral numbers. Both classes of atomic physicists share the characteristic that one cannot derive *a priori* any argument from their theories that says for which quantum numbers and which atoms one must take half-integral values and for which integral values. This they can only decide *a posteriori* by comparing with experience. (Pauli to Bohr, 21 February 1924)

In spite of his general criticism, which was definitely directed towards the results obtained in Göttingen, Pauli added in his letter to Bohr that he would not include Heisenberg among the two classes of theoreticians he had mentioned. In fact, Max Born had also been excluded, for he was about to search for a justification of the half-integral quantum numbers on the basis of his new idea of the discretization of mechanical equations.

Thus, at the end of 1923 and in the beginning of 1924, the existence of half-integral quantum numbers was regarded in Göttingen, similar to what Bohr had thought a year earlier, as a strong indication of the breakdown of the theory. However, the Göttingen theoreticians went beyond this pessimistic conclusion; they claimed that the half-integral quantum numbers, in fact, would show how to construct a more satisfactory atomic theory. Born was beginning to look for the mathematical scheme to formulate the new atomic mechanics, and Heisenberg was concerned about testing the basic ideas by deriving from them a theory of the anomalous Zeeman effects. However, whatever the prospects of this enterprise, the Göttingen theoreticians insisted on the fact that they had demonstrated beyond any doubt that several phenomena could be described more correctly by formally introducing half-integral quantum numbers rather than integral quantum numbers, which followed from the principles of the quantum theory of conditionally periodic systems. Whether their claim was justified in the case of the deformation of ions (Born and Heisenberg, 1924b) was doubted seriously a year later by Erwin Schrödinger. He criticized the calculation of the polarizability by Born and Heisenberg, and wrote about it to Sommerfeld:

> I have now studied the matter and find that it fits rather inaccurately, if one can call this fitting at all. I cannot see any reason how on p. 395, line 6 from bottom [of the Born–Heisenberg paper, 1924b, where the authors compared their theoretical Rydberg corrections with the experimental values], one can speak about "the inaccuracy of data", if one is dealing with [the data from] spectroscopy! The apparently small disagreements of the table [with the experiments] are punches in the face of experimental spectrometry. (Schrödinger to Sommerfeld, 7 March 1925)

Schrödinger then presented in his letter to Sommerfeld an example in which experiment and the theory of Born and Heisenberg disagreed by a factor of two. He concluded that a new analysis of the data was necessary and suggested that a theory with integral quantum numbers might still reproduce the data if one took into account a resonance effect between the atomic core and the series electron. Schrödinger worked out the details of his analysis in a paper which was published in *Annalen der Physik* in June 1925 (Schrödinger, 1925b). In this paper he noted:

> All calculations have been carried out with *integral* [angular] momentum quantum number, but the results mentioned do not depend materially on this decision; *in favour of it* one may then count the improved constance of the polarizability in several series, especially in the Bergmann series of Cs, which has been measured by K. W. Meissner with great precision; *against it*, one may count the fact that the polarizabilities thus calculated turn out to be too large in comparison to the optical refractive indices. (Schrödinger, 1925b, pp. 69–70)

However, remarked Schrödinger, the argument against integral quantum numbers might be explained because of the inhomogeneity of the electric field of the series electron, which causes the polarizability of the atom.

The difficulties and ambiguities connected with the analysis of the atomic data, to which Schrödinger addressed himself in the specific example of the polarizability of ions, constituted a general feature of atomic theory between 1922 and 1925. During this time the theoretical physicists approached many problems by various methods based on the quantum theory of multiply or conditionally periodic systems; that is, they employed in particular the equations of classical dynamics. The resulting formulae depended on certain quantum numbers; and if one could not describe the data by taking integral quantum numbers, one would try to improve the agreement with the help of half-integral quantum numbers. However, in doing so, one was not able to decide about the origin of half-integral quantum numbers because—as Pauli remarked to Bohr—the available theory did not provide any information about it. Later on, after quantum mechanics would replace the quantum theory of multiply periodic systems, the situation would be completely different; then it would turn out that certain half-integral quantum numbers arose from replacing classical mechanical equations by the corresponding quantum-mechanical ones. Other half-integral quantum numbers would reflect directly the fact that electrons possess an intrinsic angular momentum (or spin) of magnitude $\frac{1}{2}h/2\pi$. In many cases, in which half-integral quantum numbers had made their appearance before the advent of quantum mechanics, the changes brought in by replacing classical equations by quantum equations, and those by including electron spin, could not be easily disentangled. Hence it is often impossible to connect uniquely the old half-integral quantum numbers and the new ones such as electron spin.[135] In any case, in 1924 Born and Heisenberg were primarily interested in isolating clearly

[135] In the case of ionic polarizability, the first quantum-mechanical calculation was carried out by Ivar Waller in 1926. He used the wave mechanical approach to obtain a formula relating the polarizability of the ion to the Rydberg corrections of the atomic terms (Waller, 1926, p. 645, Eq. (40)). Although this formula was not identical with the corresponding formula of Born and Heisenberg (1924b, p. 394, Eq. (7)), half-integral quantum numbers appeared in it. However, the evaluations from both formulae yielded values for α that agreed rather well.

the deviations of empirical data from the theoretical description in order to derive the modifications by which a better argument could be achieved. In the beginning, certainly, these authors were not too much concerned with a detailed analysis of the physical origin of these deviations; they had, in fact, assumed that all failures of the old theory had a common origin and could be cured by *one* principal modification of atomic theory, i.e., the reformulation of mechanics. An example of how this assumption worked in practice was provided by Heisenberg's renewed treatment of the anomalous Zeeman effects from fall 1923 to summer 1924.

II.5 Return to the Anomalous Zeeman Effects

On returning to Göttingen in October 1923—after a two-month vacation, including a trip to Finland—Heisenberg found that Max Born was engaged in pursuing new ideas. In Heisenberg's absence, Born had thought about the reasons why the available theory of atomic structure was not able to account for such systems as the helium atom, and he came to the conclusion that this failure was not surprising. He decided that the principles governing the existing approach were far from consistent. For example, although one described the interaction of atoms with radiation with the help of Bohr's frequency condition, which was a *difference* equation, in the interaction of electrons (and nuclei) one still employed the *differential* equations of classical dynamics. This inconsistency had to be removed, and Born wanted his co-workers to collaborate in that investigation. As Heisenberg informed Pauli: 'Born has formulated our working programme for the immediate future in the phrase "discretization of atomic physics"' (Heisenberg to Pauli, 9 October 1923). At the same time Heisenberg also informed Pauli that he was about to make use of Born's ideas to construct a 'new Göttingen theory of the anomalous Zeeman effects' and mentioned the outline of this scheme. With this, he launched a new attack on his favourite problem of atomic theory. Before discussing the new approach, let us give a summary of the progress that had been made on this subject since Heisenberg last worked on it.

Between summer 1922 and fall 1923 the status of the theory of anomalous Zeeman effects had changed considerably. The change had been initiated by Miguel A. Catalán's investigations of the spectra of manganese and chromium, revealing the existence of line groups or multiplets, which consisted of four, six, and eight components (Catalán, 1922). These results, and the independent ones of Hilde Gieseler in Tübingen (Gieseler, 1922), which became known after mid-1922, demanded a modification or extension of the previous theoretical description of line multiplets and their anomalous Zeeman effects. By August 1922, Arnold Sommerfeld submitted a paper to *Annalen der Physik*, in which he described the complex line multiplets formally with the help of his method of the inner quantum numbers j; he showed that the new data obeyed the selection rule $\Delta j = \pm 1, 0$, and that the frequency 'displacement law' ('*Verschiebungssatz*') was obeyed in the spectra of neighbouring elements in the periodic system (Sommerfeld, 1923a). In the same paper, he also interpreted the different multiplicities of lines of a given atom by assuming that its energy terms had increasing multiplicity, starting with 1 for the *s*-term, 3 for the *p*-term, etc., until a characteristic 'permanent' level number or 'permanence' number was reached, which all the higher terms assumed. Alfred Landé, however, went beyond these steps and introduced a new systematics of the multiplet spectra and their Zeeman effects in winter 1922–1923 (Landé, 1923a). In this scheme, an extension of his previous vector model, he described the angular momentum structure of the atoms by three vectors: **R**, the angular momentum of the atomic core; **K**, the angular momentum of the series electron; and **J**, the total angular momentum of the atom. The latter was obtained, of course, by taking the (vector) sum of the two partial momenta, **K** and **R**, both of which were associated with half-integral quantum numbers (i.e., $K, R = \pm\frac{1}{2}, \pm\frac{3}{2}$, etc., in units of $h/2\pi$); however, in contrast to the semiclassical vector model, the largest value of the quantum number J (of the total angular momentum **J**) was smaller by half a unit than the sum of the quantum numbers K and R, thus excluding parallel positions of the vectors **K** and **R**. Landé's 1923 vector model sufficed to explain the multiplicities of the complex spectra of many-electron atoms. Moreover, when a weak magnetic field was applied to a given atom, the energy terms were displaced, the shifts ΔW being given by the product, $\omega_L hmg$, where ω_L is the Larmor frequency, h Planck's constant, and m the magnetic quantum number (which was supposed to assume the values $J - \frac{1}{2}, J - \frac{3}{2}, \ldots, -J + \frac{1}{2}$). The factor g, which Landé called the 'splitting factor' ('*Aufspaltungsfaktor*'), and which was later on called 'Landé's g-factor,' was obtained from the equation

$$g = 1 + \frac{J^2 - \frac{1}{4} + R^2 - K^2}{2\left(J^2 - \frac{1}{4}\right)}\,. \tag{41}$$

(In Eq. (41), the quantum numbers are denoted by the same symbols as the corresponding angular momenta.) This formula fitted the experimental data very well, as Ernst Back, Landé's experimental colleague in Tübingen, proved immediately in the case of manganese spectra (Back, 1923b).

It should be noted that Landé was perfectly aware of the fact that his results violated the accepted principles of atomic theory at many places, and he summarized these violations in his lecture at the Bonn meeting of the German Physical Society on 19 September 1923 (Landé, 1923e). Besides the introduction of half-integral quantum numbers he listed the following violations: the usual composition rule of angular momenta had to be abandoned in order to make the multiplicities agree with the data; also the cases did not occur in which the direction of the total angular momentum of an atom was parallel or antiparallel to the direction of the magnetic field; Larmor's theorem was violated because the splitting factor g, Eq. (41), was not equal to unity—e.g., it equalled 2 for *s*-states;

the building-up principle (*Aufbauprinzip*) broke down because the angular momentum of an ion did not coincide with the angular momentum of the atomic core of the corresponding atom, being smaller by $\frac{1}{2}$ (in units of $h/2\pi$)[136]; in the formula for the Landé g-factor, Eq. (41), the cosine law was only approximately satisfied, i.e., the splitting energy in weak magnetic fields was not proportional to the cosine of the angle between the directions of the angular momenta **R** and **J**, and the additional term $-\frac{1}{4}$ in the denominator destroyed the usual mechanical interpretation. Because of the presence of these violations, Landé made a plea for a still unknown 'substitute mechanics' ('*Ersatzmechanik*') in order to describe the dynamics of complex atoms.

Wolfgang Pauli approached the anomalous Zeeman effects, to the extent they were known in 1922 and 1923, in a completely different manner. Following the suggestions of Niels Bohr, he sought to adhere in his calculations to those principles which Bohr considered indispensable in atomic theory. In particular, Pauli stuck to the use of integral quantum numbers, allowing the half-integral quantum numbers to occur only in the formal, intermediate steps of the derivations. He began by considering complex atoms in strong magnetic fields (Paschen–Back effect) because this situation offered better chances for carrying out the (classical) analysis. However, he had to deviate right away from the usual principles by introducing an anomalous magnetic moment of the atomic core; that is, he interpreted the Paschen–Back pattern as arising from the fact that the atomic core contributed a magnetic moment that was twice the amount obtained from classical electrodynamics. He thus arrived at a formula for the energy terms of complex atoms in strong magnetic fields, which contained only integral quantum numbers, even if the core's angular momentum was assumed to be half-integral. Since no data were available to test this equation, Pauli turned to the case of weaker magnetic fields by means of the following procedure. He proposed to derive the anomalous Zeeman splittings from the Paschen–Back (i.e., strong magnetic field) terms with the help of a sum-rule. The latter stated that the sum of all energy states, for given magnetic and azimuthal quantum numbers of the series electron, but different total angular momentum quantum numbers of the atom, was a linear function of the field strength during the transition from weak to strong magnetic fields. This could also be expressed by assuming the 'permanence of the g-sums'; that is, the sum of all g-factors belonging to the states under consideration was a constant. With the help of the permanence rule, Pauli was indeed able to reproduce the formula for the g-factor, Eq. (41) (Pauli, 1923a).

Shortly afterwards, Landé used Pauli's procedure to determine the frequencies of the multiplet components of the free atom by starting from the atom in a strong magnetic field (Landé, 1923f). He found that the frequency shift from the centre could be expressed as the product of a common factor ω_i (a frequency

[136]The breakdown of the building-up principle was first stated by Landé in a letter to *Naturwissenschaften* dated 15 July 1923 (Landé, 1923c).

which was a measure of the strength of the internal field acting on the series electron) and an interval factor, γ, which depended on the quantum numbers. In a mechanical model, the product $\omega_i \gamma h$ represented the energy of interaction between the series electron and the atomic core; here γ was identical with the cosine of the angle between the directions of the angular momenta **R** and **K**. In strong fields **H**, this cosine could be replaced by the product of two cosines, $\cos(\mathbf{R}, \mathbf{H})$ and $\cos(\mathbf{K}, \mathbf{H})$, for the angular momenta of the atomic core and the series electron then performed precessions around the axis of **H**. For weak fields Landé obtained the formula

$$\gamma = \frac{J^2 + \frac{1}{4} - R^2 - K^2}{2RK}, \tag{42}$$

which deviated (from the equation derived in the mechanical model) by the additive term $\frac{1}{4}$ in the numerator. (The letters J, K and R denote the quantum numbers of the corresponding quantities.) Pauli's procedure did not allow one to determine uniquely the relation between Paschen–Back terms and the multiplet components in weak fields; yet Landé succeeded in establishing this relation by employing an additional assumption, stating that the expression, $\sum_j |(mg_s - mg_w) + (\gamma_s - \gamma_w)|^2$, where the subscripts denote the presence of strong (s) and weak (w) magnetic fields, respectively, should be a minimum (Landé, 1923f, p. 120, Eq. (13)). However, Pauli proved in a subsequent paper that this assumption was unnecessary; he showed that one could determine from the quantum theory of multiply periodic systems as to which anomalous (i.e., weak magnetic field) Zeeman component would pass into which Paschen–Back component in the case of strong fields (Pauli, 1924a).

The unsatisfactory feature of Pauli's treatment of the anomalous Zeeman effects was that, although he had tried to use only Bohr's principles of atomic theory, he had to abandon the interpretation of his results in terms of any mechanical model if he wanted to fit the observations. Hence these results could be made to agree either with the empirical data or with the models of atomic structure. For example, Pauli arrived at the following conclusion: if he used a suitable mechanical model for a many-electron atom, he did not arrive at Landé's g-formula, Eq. (41), but at a similar formula, in which the term $J^2 - \frac{1}{4}$ appearing in the numerator and denominator of the second term on the right-hand side of Eq. (41) was replaced by $(J + \frac{1}{2})^2$. Since Pauli preferred to introduce the quantity $j = J + \frac{1}{2}$ instead of Landé's total angular momentum quantum number J, this replacement meant that in the formula derived from the mechanical model the term that emerged was j^2 rather than $j(j-1)$. As Pauli wrote to Sommerfeld:

> The structure of the expressions [i.e., of the expressions in the formula for the g-factor with j^2 and $j(j-1)$, respectively] is rather similar but one *cannot* make the difference disappear by changing the normalization of the indefinite additive constants involved in the R, K, and j. For in one case the Runge denominators

> consist of two equal factors $(j \cdot j)$, while in the other case, of two different factors $j(j-1)$. One may also say that the two expressions are related to one another like the differential quotient $(d/dj)(1/j)$ and the difference quotient $1/j - 1/(j-1)$), which seems to hint at something that is non-mechanical (Pauli to Sommerfeld, 19 July 1923).[137]

Pauli concluded his letter with the remark that the theory of the anomalous Zeeman effects was as much in a 'greatly lamentable state' ('*großer Jammer*') as the theory of many-electron atoms.

Although from summer 1922 to fall 1923 Heisenberg was mainly concerned with problems other than the multiplet structure of complex atoms and their anomalous Zeeman effects, and although he did not publish any paper on these questions from September 1922 to July 1924, he had not put aside these problems altogether. In fact he constantly thought about them and reported his ideas and conclusions in letters to Landé, Pauli and Sommerfeld. Heisenberg closely followed the journals, which reported on more and more complex types of multiplets and their Zeeman patterns. He noticed that the new multiplets did not fall into the doublet and triplet types he had considered earlier, and he began to think about an extension of his previous theory. Thus he wrote to Landé in September 1922: 'Have you read the Tübingen dissertation of Miss Gieseler [Gieseler, 1922]? It contains the Zeeman effects of chromium, and I do have the faint suspicion that from it there will once follow a unification of Bohr's and my own points of view. Some points of the chromium model are already certain in my mind. Others are completely in the dark' (Heisenberg to Landé, 15 September 1922). At the same time Landé kept him informed about his investigations of the anomalous Zeeman effects by sending him the manuscript of a paper, which he was just about to submit to *Zeitschrift für Physik* and in which he dealt with the connection between the spectroscopic effects and the so-called magneto-mechanical effects (Landé, 1922a). Heisenberg discussed the content of this paper with Pauli and Richard Becker at the Congress of Natural Scientists and Physicians (*Versammlung Naturforscher und Ärzte*) at Leipzig in September 1922. During the following months, while Heisenberg was deeply engaged in the work on helium, Landé reported great progress on the problem of the anomalous Zeeman effects. 'Landé has found a complete theory of *all* multiplets and their Zeeman effects, in analogy to the theory of doublets and triplets!!! Always using half-integral quantum numbers,' Heisenberg wrote to Sommerfeld on 4 December 1922. At the same time Pauli sent to him in Göttingen a manuscript on the theory of anomalous Zeeman effects from Copenhagen. While Heisenberg greeted Landé's results with enthusiam—he found much to criticize in Pauli's treatment, as he disclosed in some detail in writing to Sommerfeld. He com-

[137] In this quotation we have replaced Pauli's notation, k and i for the angular momenta, by K and R, respectively. The last sentence of Pauli's remark means that the inverse of j^2 can be obtained by differentiating $-1/j^2$, while the inverse of the product $j(j-1)$ is found by taking the difference of $1/(j-1)$ and $1/j$, which corresponds to a difference quotient.

plained, for example, about Pauli's 'arbitrary' theoretical assumptions and his failure to explain empirical facts, such as the Runge denominator, which seemed to Heisenberg to be essential in order to arrive at a physical understanding:

> On the whole [he concluded] I must write as the final criticism of Pauli's work that I cannot understand at all how Bohr and Pauli insist so forcibly on such a fruitless consequence of general quantum principles. And, as much as I am convinced about the incorrectness of Pauli's arguments, equally uncomfortable do I feel with the inflexible consistency with which Bohr believes everything to be right that comes out wrong [from the theory in comparison to experiment], and, what comes out right to be wrong. This state of physics is really uncongenial to me. (Heisenberg to Sommerfeld, 4 January 1923).

A little more than a month later Heisenberg assumed a more positive attitude towards the Copenhagen approach to the anomalous Zeeman effects. As a result of the failure of his work with Born to explain the observed spectra of the excited helium atom, he was even prepared to listen to Bohr and Pauli's arguments, which he had previously declared to be simply wrong. And yet, he remained convinced that in working on problems of atomic theory one had to pay special attention to the agreement of the theoretical results with experimental observation—a point which Pauli had not properly taken into account in his work. He wrote to Pauli: 'I shall soon publish the following axiom as motto for the *Zeitschrift für Physik*: A theory may still be wrong if it yields something right, but it can never be right if it yields something wrong' (Heisenberg to Pauli, 21 February 1923). Of course, he noticed that in Landé's new description of complex multiplets the original '*Rumpf*' (atomic-core) model had to be replaced by the vector model; however, he was still confident that his old doublet theory would not be invalidated in spite of Landé's and Pauli's differing opinions. But the debacle of the helium calculation prevented him from going to work on these questions immediately. 'At the moment I believe the helium problem to be the most important one,' he declared in February 1923. 'For the Zeeman effects it is still too early' (Heisenberg to Pauli, 21 February 1923). During the following spring and summer Sommerfeld struggled in Munich with the helium problem without reaching any satisfactory conclusion.[138] Heisenberg witnessed these efforts and decided that the problem was too hard to be solved at the moment; in fall 1923 he, therefore, turned again to the question of the anomalous Zeeman effects.

Heisenberg's response to the respective attempts of Landé and Pauli was very different. In Landé's work he admired—and envied—the enormous skill of the author in describing the empirical data. He was particularly happy that in this description the half-integral quantum numbers, which he regarded as his favourite brainchild, played a central role. In general, concentrating as he always

[138] Sommerfeld gave an account of his work on the helium problem in the fourth edition of his *Atombau und Spektrallinien* (1924, pp. 200–206).

did on the positive aspects of other people's work, he was inclined to emphasize the agreement of Landé's formulae with his own theoretical ideas and to take the contradictions less seriously. Still he noticed that Landé's successful vector model lacked a deeper theoretical foundation. While he remained confident that an extension of his atomic-core model would provide this foundation, he became increasingly aware of the fact that the future theory of the anomalous Zeeman effects would deviate from the principles of classical dynamics. Therefore, in spring 1923, he welcomed Pauli's thorough analysis, which at least showed clearly that besides the classical principles certain nonclassical assumptions, such as the double magnetic efficiency of the core (i.e., the fact that the magnetic moment of the core assumes twice the value obtained from the mechanical angular momentum consideration) would have to be included. Moreover, he admitted that Pauli's particular point of view, expressed in the 'magnetic' interpretation of the anomalous Zeeman effects—which stated that in strong fields the magnetic quantum number denotes the components of the total angular momentum (of the atom) in the direction of the field, rather than those of the series electron's angular momentum, as Heisenberg had assumed—made some sense. He wrote to Pauli:

> The main difficulty of my doublet theory is without doubt the restriction $m < j$. I have not yet found a practical way out, and I now believe that it is probable that the "magnetic" point of view, as you call it, still contains a formal truth. In my opinion, for the calculation of the energy only the "mechanical" point of view [i.e., his own] is appropriate, while for the explanation of the selection rules only the "magnetic" one. I have pondered for long about it and I suspect that both are simultaneously correct in this sense. (Heisenberg to Pauli, 26 March 1923)

He added that the last conclusion was only a 'crazy notion' ('*Bieridee*'), but he thought that it might work in the future. In any case, he was prepared to consider certain aspects of Pauli's approach seriously and to give them a place in his own attempts at deriving a more consistent theory—to which, however, he returned only later in 1923.

The starting point of Heisenberg's work on the anomalous Zeeman effects in Göttingen could be traced to Pauli's remarks, in his letter of 19 July 1923 to Sommerfeld, concerning his failure to obtain an interpretation of the data in terms of a classical model.[139] Since Born had meanwhile independently arrived at a similar conclusion, namely, that in quantum theory difference equations had to replace the classical differential equations, Heisenberg expected—supported by Born's new motto about the 'discretization of physics'—that he could now make quick progress towards a consistent theory of the anomalous Zeeman effects. The goal he wanted to achieve was different from the one which Landé had reached previously, and also different from what Pauli had attempted to do; Heisenberg

[139] Heisenberg had probably seen Pauli's letter of 19 July to Sommerfeld, but since it arrived at the time of his doctoral examination not much of it may have stuck in his mind.

desired to accomplish the goals of both Landé and Pauli with one coup. That is, he wished—because of his insistence on agreement with experiment—certainly to arrive at formulae that were as successful as Landé's g- and γ-formulae, Eqs. (41) and (42). On the other hand, he recognized the virtue of sticking to basic principles similar to those of Bohr's atomic theory. To realize his purpose he was prepared to consider—in contrast to Pauli—drastic changes in the theoretical description. While he worked on his new approach, Heisenberg stayed in close contact with the two persons who had contributed most to the understanding of complex structure and anomalous Zeeman effects; he regularly informed Wolfgang Pauli and Alfred Landé (who had become *Extraordinarius* in Tübingen) about his ideas and results. While Pauli served as a severe critic, who would constantly remind him about what he had not achieved and how inconsistent were his concepts, Landé supplied Heisenberg with arguments and counterarguments that proceeded from the empirical data available in Tübingen, then perhaps *the* centre of experiments on the anomalous Zeeman effects.

In the beginning of his renewed concern with the subject, Heisenberg reported to Pauli about the outlines of what he called the 'new Göttingen theory' of the anomalous Zeeman effects. Its motto was: 'The model representations have, in principle, only a symbolic sense; they denote the classical analogue of the "discrete" quantum theory' (Heisenberg to Pauli, 9 October 1923). In the Bohr–Sommerfeld theory, the energy of a multiply periodic system of f degrees of freedom was obtained by taking $H_{\mathrm{cl}}(J_1, \ldots, J_f)$, the Hamiltonian of the classical model, and substituting the quantized values for the action variables $J_1, \ldots, J_f$. In the 'new' theory, however, H_{cl} had to be replaced by a different function, whose structure might depend on the problem being treated. In the case of the anomalous Zeeman effects, Heisenberg proposed to take the integral invariant F, that is,

$$F(k, r, j, m) = \int\int H_{\mathrm{cl}}(k, r, J, m)\, dJ\, dw, \tag{43}$$

where the integration over J extends from 0 to the value of the total momentum j (in units of $h/2\pi$), while the integration over w extends over a full period of the normalized angle w (i.e., from 0 to 1), the variable canonically conjugate to j; m is the projection of j in the direction of the magnetic field. The quantum-theoretical Hamiltonian, H_{qu}, was then obtained from the difference equation,

$$H_{\mathrm{qu}} = \Delta F = F(j) - F(j-1), \tag{44}$$

which had the same mathematical structure as the frequency condition, $\nu_{\mathrm{qu}} = \Delta H$.[140] By assuming for the classical model of the anomalous Zeeman effect

[140]In the published paper, submitted in June 1924, Heisenberg took the difference of $F(j+\frac{1}{2})$ and $F(j-\frac{1}{2})$, instead of the right-hand side of Eq. (44), which, of course, does not change the conclusions. (See Heisenberg, 1924e, p. 292, Eq. (3). Heisenberg also used capital letters, J, K, R, for the momenta, which we have replaced by small j, k, r, in order to distinguish his notation from Landé's.)

the one used previously by Pauli and Landé—i.e., an atom consisting of a core (which exhibits the anomalous, double magnetic moment) and a series electron—he arrived in the case of weak external magnetic fields (i.e., the inner precession frequency ω_i is much larger than the Larmor frequency ω_L) at the function F,

$$F = \omega_i h \frac{\frac{1}{3}j^3 - r^2 j - k^2 j}{2kr} + \omega_L hm\left(j + \frac{j^2 - r^2 + k^2}{2j}\right), \tag{45}$$

where j, r and k denote, respectively, the total angular momentum of the atom, the angular momentum of the core and that of the series electron; m is the projection of j in the direction of the magnetic field. From F the quantum-theoretical energy states could be obtained with the help of Eq. (44), after inserting the proper quantum numbers—i.e., the core momentum values 1 (always in units of $h/2\pi$) for doublets, $\frac{3}{2}$ for triplets, etc.; the series electron momenta being $\frac{1}{2}$ for the s-state, $\frac{3}{2}$ for the p-state, etc. Heisenberg assumed the values of j to run from $k + r$, $k + r - 1, \ldots$ to $|k - r|$, while the value of the magnetic quantum number m extended from j to $-j$ in unit steps. Both assumptions differed from those of Landé's vector model (Landé, 1923a); in particular, Heisenberg took into account those cases in which the direction of the total angular momentum of the atom was either parallel or antiparallel to the direction of the external magnetic field. Still, the total number of different states, because of the difference condition, Eq. (44), was equal to $2k$ or $2r$, in agreement with experiment. With these assumptions, Heisenberg was able to reproduce the known empirical results on the anomalous Zeeman effects and complex structure; for example, in triplet p-states he again obtained the ratio 2 : 1 for the frequency differences of the components in the absence of an external magnetic field, which had been one of his main earlier successes in Munich (Heisenberg, 1922a, p. 295).

The theory satisfied all the principal requirements. First, Bohr's building-up principle (*Aufbauprinzip*) seemed to hold, because the momentum of the atomic core of the element next in the periodic table was obtained from the core of the previous element by adding the angular momentum $\frac{1}{2}$ (in units of $h/2\pi$); as a consequence, the rule of alternating spectra (spectroscipic *Wechselsatz*) was obtained. Second, the g-sum-rule (i.e., the permanence of g-sums) could be proved. Third, there existed no arbitrary assumption, such as the exclusion of the orientations of angular momentum parallel to the direction of the magnetic field. Still, Heisenberg was aware of a weak point of his theory. 'The important negative aspect of the theory is,' he wrote to Pauli, 'that one now ceases to understand the quantum theory at all. This, however, I find rather congenial. Now the real goal must be to arrive from the symbols *in a unique way* at the discrete states; whether the formulae thus obtained make comprehensible sense, is doubtful to me' (Heisenberg to Pauli, 9 October 1923).

Heisenberg quickly worked out all the necessary details of his theory. He also soon presented it publicly: on 5 November 1923 he spoke at a meeting of the

Lower Saxony Section (*Gauverein Niedersachsen*) of the German Physical Society in Göttingen on his new quantization procedure and its application to the anomalous Zeeman effects. A month later he completed the manuscript of a paper on it; on 7 December he sent two versions of the paper to Alfred Landé and Wolfgang Pauli, respectively.[141] To Landé he wrote: 'Allow me to send you the original manuscript of my paper; unfortunately the new formulation looks somewhat different, but you can also easily see from the old version what I have really done. I hope thereby to clear myself of the suspicion that I want to keep my work a secret' (Heisenberg to Landé, 7 December 1923). He mentioned further that the publication of this work would take some time, for he planned to send the manuscript first to Niels Bohr. 'I believe,' he wrote to Landé, 'that Bohr will agree with it, as Pauli also did. Hopefully you agree, too; you can, together with [Ernst] Back judge so much better whether the formulae represent the real behaviour of the Zeeman effects of elements like Pb, Ne, etc.' Finally he added: 'However, in the new version of the paper, my opinion about the possible Zeeman effects has changed, insofar as I now believe that still other Zeeman effects might arise . . . if at the periphery [of the atom] *two* electrons move on equivalent orbits' (Heisenberg to Landé, 7 December 1923). At the moment, however, Heisenberg had not yet gone very far in this respect, except that he had been able to prove the following theorem: the Zeeman effects of all noble gases and of all ionized alkali atoms should be similar to those of helium. He concluded his letter to Landé by expressing the hope that he had not interfered with Landé's own plans for publishing on this subject.

While Landé agreed in general with the ideas on which Heisenberg had built the new theory and the results derived from it, Pauli, on reading the manuscript, arrived at a different opinion.[142] 'I do not at all share your point of view regarding Heisenberg's new theory,' he wrote to Landé, and went on to explain: 'I even consider it to be an ugly theory. For, in spite of radical assumptions, it does not provide an explanation of the half-integral quantum numbers and of the failure of Larmor's theorem (especially of the factor two in the magnetic anomaly). I don't think much of the whole thing' (Pauli to Landé, 14 December 1923). Pauli's own analysis had shown that one had to concentrate exactly on the explanation of these facts. 'Since I regard such an explanation to be the most important point, and since the thing [i.e., Heisenberg's new theory] is purely formal and does not contain any new physical idea,' Pauli wrote a few days later

[141] The draft manuscript, which Heisenberg sent to Pauli, was probably identical to the manuscript entitled '*Über ein neues Quantenprinzip und dessen Anwendung auf die Theorie des anomalen Zeemaneffekts*' ('On a New Quantum Principle and Its Application to the Theory of the Anomalous Zeeman Effect') preserved in the *Bohr Archives* at Copenhagen (AHQP, 45, 8). Because of the Gothic script, Heisenberg did not send a copy to Bohr in December 1923, but Pauli may have taken it to Copenhagen the following April.

[142] Pauli's agreement, which Heisenberg mentioned in the letter to Landé dated 7 December 1923, probably referred to the outline of Heisenberg's ideas contained in the letter of 9 October 1923. However, after seeing the details of Heisenberg's theory, Pauli withdrew his support.

to Hendrik Kramers in Copenhagen, 'his is not the theory I am waiting for' (Pauli to Kramers, 19 December 1923). Meanwhile he had learned from Heisenberg that he would give a talk on his new theory in Sommerfeld's Seminar; hence Pauli added in his letter to Kramers: 'Heisenberg, who is in Munich during the Christmas vacation, will give a talk there tomorrow afternoon at 5:30 P.M., and I shall just arrive there in time to make many nasty comments (*um dort tüchtig zu schimpfen*). It will be a stormy session' (Pauli to Kramers, 19 December 1923).[143]

Through Pauli's vehement protestations Heisenberg was reminded strongly about what he had not achieved in his new theory. 'Nevertheless I believe,' he wrote to Landé, who had joined Pauli in arguments against his theory, 'that the principle is formally all right, since now even the infamous cosine-law and the change of the doublet separations in a vertical column [of the periodic system], which is approximately $(Z-\alpha)^3$ [where Z is the nuclear charge and α a screening constant] (e.g., for alkalis and alkaline earths, Li, Na $\rightarrow$ Cs, etc.) can be derived' (Heisenberg to Landé, 22 December 1923). The cosine law, mentioned here by Heisenberg, stated that in order to describe the natural splitting of complex multiplets, one had to assume an interaction between the series electrons and the core, which was approximately proportional to the cosine of the angle between their respective angular momenta. The nonclassical deviations, such as the ones occurring in Landé's g- and γ-formulae, Eqs. (41) and (42), followed automatically from Heisenberg's treatment. Heisenberg noticed, however, that the cosine law was obeyed only when the core was in the s-state; in other cases one had to expect a violation of the law and, as an immediate consequence, a violation of Landé's formula, Eq. (42). He wrote to Landé:

> The derivation of the general g-formula [i.e., of the formula describing the natural and magnetic splitting of the multiplet components for arbitrary external field strengths] of the second type still causes me some logical difficulties up to now, but it might perhaps work. How nice it would be if one had just to look up one of your papers and take out the final and definite g-formula, but that will evidently come soon. (Heisenberg to Landé, 22 December 1923)[144]

Indeed, the prospect of deriving from his theory other anomalous Zeeman effects, which were not covered by Landé's results, softened in Heisenberg's mind the impact of Pauli's harsh criticism. On the same day as he wrote to Landé, he summarized the new theory in a long letter to Niels Bohr and asked for his opinion about the matter. He hoped to receive some encouragement or helpful criticism from the master in Copenhagen. However, before the answer could arrive, he went ahead, with the help of Landé, to explore some special cases of

[143]The exact date of Heisenberg's seminar in Munich was either 20 December or 21 December 1923. From Pauli's letter to Kramers it appears to have been 20 December; however, Heisenberg wrote in a letter to Landé, dated 22 December, that he had spoken the previous day and Pauli and Lenz were present.

[144]As early as March 1923 Pauli had claimed that Landé's g-formula, Eq. (41), applied only to those multiplets whose core was in an s-state, i.e., the last core-electron was bound in an s-orbit. (See Pauli to Landé, 10 March 1923.)

complex atoms in greater detail, in order to obtain a hint for the generalization of formulae (41) and (42).

The possibility of a closer collaboration with Landé had already begun to interest Heisenberg in connection with a different problem. In a letter to Landé, he referred to his (i.e., Landé's) previous work on X-ray spectra and expressed his general agreement with the results.[145] He wrote:

> In the next few days I shall now study carefully the X-ray papers of yours and others (Bohr–Coster, etc.). I have indeed the suspicion that especially in your notation of K and J [i.e., the quantum numbers associated with the angular momenta k and j, respectively] the energies H_{qu} might be exchanged in such a way that they agree with the present [i.e., the relativistic] formulae . . . If I find some positive result, then I would most prefer to write a paper with you on the X-ray spectra. (Heisenberg to Landé, 13 December 1923)

In this connection he inquired whether Landé would be passing through Munich during the Christmas vacation, so that he could meet him. Landé would not, but he invited Heisenberg to visit Tübingen on his way back from Munich to Göttingen; Heisenberg accepted and spent the afternoon of 9 January 1924 in discussions with Landé in Tübingen.[146] During this meeting they decided to collaborate on the multiplet structure and the anomalous Zeeman effects of neon. On the other hand, Landé continued to work on his analysis of X-ray spectra alone and submitted a paper on the nature of the relativistic doublets in March 1924 (Landé, 1924a).[147]

[145] In his paper on the theory of X-ray spectra, submitted in June 1923 and published in July, Landé had derived an analogy between the X-ray and optical doublets by proposing to associate the X-ray terms with a specific set of quantum numbers (Landé, 1923b). As a result, it seemed that both doublets had a common origin; i.e., the X-ray doublet arose from an interaction between the core and the series electron, and not from Sommerfeld's relativistic theory. This point interested Heisenberg in December 1923.

[146] Landé invited Heisenberg to visit Tübingen in his reply to Heisenberg's letter of 13 December 1923. Heisenberg thanked Landé for the invitation on 22 December. In his letter to Landé, dated 5 January 1924, Heisenberg announced that he would arrive in Tübingen on Wednesday, 9 January, at 2.07 P.M., and would leave at 6.06 P.M. to continue his journey to Würzburg and Göttingen.

[147] In the paper on relativistic doublets, Landé assembled further evidence that the optical and the X-ray doublets had a common origin, namely, the interaction between the core and the series electron. One difficulty with this explanation was that the relativistic doublet splitting (by which one accounted previously for the X-ray spectra) increased with the fourth power of the core's effective charge, while the magnetic splitting (by which Heisenberg and Landé had explained the optical doublets) increased only with the third power. However, in case the core's effective charge was close to unity, the difference did not matter very much. Another difference between the two descriptions was that the product $k(k-1)$ in the relativistic formula was replaced by $(k-\frac{1}{2})^2$ in the 'magnetic' formula (k being the quantum number of the series electron); hence, in the latter formula, there occurred again an additive term of $\frac{1}{4}$ (which had already played a role in Eq. (42) describing the complex structure). Of course, Heisenberg was happy that half-integral quantum numbers again played a role in Landé's considerations of the X-ray spectra. After the appearance of the work of R. A. Millikan and I. S. Bowen (1924a), which showed that there was a continuous transition from the optical to the X-ray doublets, Heisenberg took it as a confirmation of Landé's theory and congratulated him on his success (Heisenberg to Landé, 13 February 1924).

Landé had analyzed the optical spectra of neon in summer 1923 and had arrived at the conclusion that it could not be accounted for either by the usual core model or his new vector model (Landé, 1923d).[148] In particular, he had found that the spectra consisted of one singlet system, two triplet systems and a quintet system; because of their relation to the terms of ionized neon, he had grouped together the singlet system and one triplet system, on the one hand, and the other triplet system and the quintet system, on the other. Now, in winter 1923–1924, Landé hoped that Heisenberg's new theoretical ideas might help to understand the complicated structure of the neon spectra better. Heisenberg's interest in neon spectra and their Zeeman effects also arose from the fact that they did not obey Landé's formulae (41) and (42). But he suggested a reason for the discrepancy. 'It is very understandable that in the case of neon, where the separation of the atomic core and series electron really makes no sense, complicated g-values result,' he argued. 'The deeper reason is probably this: For complicated atoms (with several "outer" electrons) the "jump" $j \rightarrow j - 1$, which defines the energy [of the line separation], can occur not only such that the directions of k and r change, but also, e.g., such that $r \rightarrow r - 1$; in general, inner orientations [i.e., those in the core] may change. Then, of course, quite different [from the ones considered formerly] Zeeman effects will result' (Heisenberg to Landé, 13 December 1923). Therefore, he hoped to profit from Landé's experience in organizing the observed spectra and Zeeman effects for the purpose of developing a more consistent theory of many-electron atoms.

In the beginning the collaboration with Landé on the neon spectra did not move very fast, especially because Heisenberg became involved in the further discovery of half-integral quantum numbers in connection with Friedrich Hund's work on diving orbits (*Tauchbahnen*) and his own on ions. But in February 1924 Heisenberg began to work on the material provided to him from Tübingen. On 23 February he summarized his results in a long letter to Landé. 'First, what concerns your Paschen–Back terms,' he wrote, 'I have found that they follow as a consequence of the theory if—corresponding to the *two* outer electrons—one applies the recipe $\int H\,dj$ to both electrons.' From this assumption he immediately derived the g-value for the s-terms of neon. With respect to the p- and d-terms Heisenberg found that their g-values would depend crucially on the mutual couplings of the outer electrons. 'Only in the limiting case, in which one of these couplings strongly dominates the other,' he wrote, 'the g-values are simple rational fractions like [the ones considered] up to now.' However, a rough estimate of the coupling ratios in neon from the magnitude of the multiplet splittings yielded the result that all the couplings in question had the same order of magnitude. 'One must therefore expect,' he concluded, 'that the g-values are no rational fractions and that, in particular, Preston's rule is not valid anymore.' The consequence of this argumentation was that in the case of very high principal quantum numbers the g-values would approach rational fractions, but

[148] Landé's interest in the spectra of neon arose because its Zeeman effects did not seem to be described by Eq. (41) for the g-factor. (See Pauli to Landé, 10 March 1923.)

for small principal quantum numbers they would deviate. 'Now this is such a crazy result that I did not dare to continue my calculations,' Heisenberg remarked. 'On the other hand, it gives me the hope to understand the difference between neon and lead; besides, the result is independent of the special theory [model] used as the basis' (Heisenberg to Landé, 23 February 1924). Finally he asked Landé whether the empirical data confirmed these conclusions and to inform him about the results, for he wanted to discuss these matters with Bohr during his forthcoming visit to Copenhagen. Five days later, on 28 February, Heisenberg sent a postcard to Landé containing his 'guesses' for the g-values of neon in the limit of infinitely high principal quantum numbers. Since there were not enough data available to test the predictions, Landé decided not to include the Zeeman effects in their joint paper.[149] The authors confined themselves to discussing only the structure of the neon spectra with no external magnetic field present. Because of Heisenberg's visit to Copenhagen and the following spring vacation, the submission of the paper for publication took some time. Finally, on 12 May Heisenberg returned the manuscript with his corrections to Landé; the latter sent it to *Zeitschrift für Physik*, where it was received on 18 May 1924 (Landé and Heisenberg, 1924).

The main content of Landé and Heisenberg's paper on the term structure of multiplets was that they introduced a new organization into the field of complex spectra. In particular, they distinguished between multiplets of 'first order' ('*erster Stufe*') and those of 'higher order' ('*höherer Stufe*'). The multiplets of the first order exhibited the usual anomalous Zeeman effects; for weak fields they were described by the formulae (41) and (42). According to Landé and Heisenberg, these spectra should be emitted by atoms in whose core all electrons were bound in an s-state, and the core had a g-value of 2.[150] If, however, one of the core electrons (which contributes to the core momentum) was bound in a p- or d-state, multiplets of 'second order' would arise; if two electrons were bound in higher orbits, multiplets of 'third order' would arise. The authors then analyzed the spectra emitted from the neon atom as an example of second-order multiplets. They claimed that in order to describe the terms of higher-order multiplets, one had to associate several angular momentum quantum numbers with the core. In practice, they obtained these quantum numbers by employing what they called the 'branching-rule' ('*Verzweigungsregel*'). This rule arose from Bohr's procedure of building-up the atoms of the periodic system of elements (*Aufbauprinzip*) and stated the following: If the core of an atom is constructed from an ion (having the total angular momentum quantum number j) by adding an electron and binding it in an s-state, then it possesses two angular momentum values, namely, $j + \frac{1}{2}$ and $j - \frac{1}{2}$, respectively. This branching rule fitted the observations beautifully; it satisfied the requirement of the building-up principle because the ion with angular momentum j could be considered as the core of an

[149]Only later, in fall 1924, did Ernst Back supply the necessary data (Back, 1925a). Landé then analyzed them according to the principles developed in early 1924 (Landé, 1925a).

[150]Pauli had pointed out this fact already in March 1923. (See footnote 144.)

element preceding the element under consideration in the periodic system. Of course from the point of view of the existing atomic theory Landé and Heisenberg's rule could not be understood. The authors noted that 'just this momentum-branching process, so simple to describe formally, has become one of the main objections against the applicability of the quantum rules for conditionally periodic systems to the coupled systems' (Landé and Heisenberg, 1924, p. 286).[151] On the other hand, the rule seemed to be consistent with the formal modification of atomic theory, which Heisenberg had begun in fall 1923.

Meanwhile, between the end of 1923 and May 1924, the Göttingen theory of the anomalous Zeeman effects had made some progress. Niels Bohr, to whom Heisenberg wrote about his new ideas on 22 December 1923, had invited the young man to spend some weeks with him in Copenhagen to discuss the problems of atomic theory. Heisenberg accepted the invitation with pleasure and went to Copenhagen in March 1924. In spite of the persistent difficulties of understanding the anomalous Zeeman effects, Bohr encouraged Heisenberg to publish his results. On 15 May, Heisenberg sent him the manuscript of his paper on the 'formal rules' governing the magnetic splitting of terms, and wrote: 'I have composed it as well as I could, but I am not yet satisfied with many points. But even if many details are different from what I have, I believe nevertheless that the basic idea is correct; I have worked hard to write the paper in such a way that I do not claim too much in detail' (Heisenberg to Bohr, 15 May 1924). Three weeks later Bohr visited Göttingen to talk to Max Born about the implication of the latter's new quantum-mechanical ideas, which had led to a consistent derivation of Hendrik Kramers' recent dispersion theory. On this visit Bohr also gave his approval to Heisenberg's work on the anomalous Zeeman effects, which seemed to be a natural application of Born's ideas. 'The latter,' Heisenberg wrote triumphantly to Pauli, 'I shall publish now (without physical interpretation, because it does not exist) with papal blessing (*päpstlichen Segen* [i.e., Bohr's approval])' (Heisenberg to Pauli, 8 June 1924). A week later he submitted the paper, entitled 'On an Alteration of the Formal Rules of Quantum Theory in the Problem of Anomalous Zeeman Effects' ('*Über eine Abänderung der formalen Regeln der Quantentheorie beim Problem der anomalen Zeemaneffekte*'), to *Zeitschrift für Physik*; it was received on 13 June 1924 and was published in August (Heisenberg, 1924e).

In this paper Heisenberg gave an account of the consequences of the basic ideas, as conceived in fall 1923, for solving the problem of the anomalous Zeeman effects of complex atoms. After outlining the general idea, the so-called 'new quantum principle' (in section 1), he evaluated the quantum Hamiltonian of Pauli's classical model ('*Ersatzmodell*') of the anomalous Zeeman effects, in which the atoms consisted of a core (with double magnetic moment) and a series electron. Heisenberg then showed in some detail how the known results of Landé

[151] A special case of the branching-rule, namely, if one of the resulting core momenta was zero—and had therefore to be excluded—was already expressed in Landé's rule of 1923 quoted earlier (Landé, 1923c, e).

and Pauli concerning the g- and γ-factors (sections 2 and 3) and the permanence principle (section 4) followed. The advantage of his treatment was that it yielded the magnetic splittings of first-order multiplets for arbitrary magnetic field strengths, not only in the limiting cases of very strong and very weak fields. Moreover, he argued that the branching-rule, given earlier (Landé and Heisenberg, 1924), was equivalent to applying the averaging rule, Eq. (43), to the quantum states of atoms and the corresponding ions. For example, in the case of neon, the ion emitted lines having a doublet structure, hence its core had a quantum number equal to 1; if to this core one added an electron in the p-state, there resulted a neon ion whose angular momentum assumed the values $\frac{5}{2}$, $\frac{3}{2}$ and $\frac{1}{2}$. According to the new quantum principle, each pair of neighbouring values, i.e., $\frac{5}{2}$ and $\frac{3}{2}$, and $\frac{3}{2}$ and $\frac{1}{2}$, belonged together, representing the states p_1 and p_2, respectively, of the neon ion; these were also the states of the core of the neon atom. Finally, the angular momentum values of the neon atom were obtained by adding, according to the branching-rule, the angular momentum of the outer (series) electron to the angular momenta of the p_1- and p_2-states of the core (i.e., the ion considered above). Now the states of the neon atom, whose core had the angular momentum quantum number $r = \frac{5}{2}$, had quintet multiplicity, those whose core had $r = \frac{3}{2}$ had triplet multiplicity, and those whose core had $r = \frac{1}{2}$ had singlet multiplicity. Hence the branching-rule was satisfied *in agreement with the building-up principle* (*Aufbauprinzip*). Now, however, Heisenberg became even more ambitious; he wanted to derive from his theory the building-up principle itself, including the change of statistical weight in atoms. For this purpose he discussed (in section 6 of the paper) the case of atoms possessing several outer electrons.[152]

For atoms with n outer electrons one had to introduce the momenta, $k_1, \ldots, k_n$, of all these electrons in the equivalent (*Ersatz*) mechanical model; and, if a magnetic field acted on the atom, one had also to introduce their projections, $p_{k_1}, \ldots, p_{k_n}$, on the direction of the magnetic field vector. Heisenberg now claimed that one had to calculate the quantum-theoretical Hamiltonian corresponding to this system again according to equations of the type (43) and (44). Especially, when a very strong external field was applied to the atom, one had always to average the classical Hamiltonian 'simultaneously' over the momentum variables $p_{k_1}, \ldots, p_{k_n}$. As a particular case, Heisenberg discussed the Zeeman effect of the neon spectra, arriving basically at the results he had reported to Landé in February. A special case in the derivation was the validity of the cosine law, namely, the fact that even in the presence of several electrons the interaction between the core and an outer electron might be proportional to the cosine of the angle between the corresponding angular momenta (Heisenberg, 1924e, section 7). This fact could hardly be explained from classical theory, but Heisenberg found it to hold under special conditions. For example, if of two

[152] Heisenberg had started to think about this question early in 1924 in connection with the problem of the anomalous Zeeman effects in neon and lead (Heisenberg to Bohr, 3 February 1924).

series electrons the inner one could be considered as bound to the core in an *s*-orbit, it would act on the other (after performing the averaging procedure) with a spherically symmetrical potential; hence it would give rise to different magnitudes of splitting of the different higher-degree term multiplets, say, of the quintet and triplet systems, but still the *g*-factor of each multiplet system would obey Landé's formula, Eq. (41). However, if the inner electron was not bound in an *s*-orbit, the cosine law would not hold; moreover, the structure of the terms would then be very complicated and the multiplets completely messed up. In such cases one had to expect, as Heisenberg had written to Landé, irrational *g*-factors and a breakdown of Preston's rule (which claimed that the *g*-values did not depend on the principal quantum number of the atom). Heisenberg also noted the possibility that even for first-degree multiplets the *g*-values might not be rational, as when the couplings between the electrons caused a considerable perturbation of the orbits; and this would be true, *a fortiori*, for multiplets of higher order, like the ones in the neon spectra.

With his paper of summer 1924 on the anomalous Zeeman effects Heisenberg hoped to have come closer to the two goals which he had had in mind since fall 1923: to describe the available empirical data by a set of specific theoretical rules *and* to relate these rules to the general changes that were necessary in the existing atomic theory (i.e., in the quantum theory of multiply periodic systems). As for the full description of the data, he was aware of the fact that he had only proposed certain steps in the direction of a solution. 'Of course, the real Zeeman model will be *more complicated* than the present one,' he wrote to Landé, 'but it must also contain the essential features of the present [model]' (Heisenberg to Landé, 15 June 1924). More than the question of the final model to explain the anomalous Zeeman effects, however, Heisenberg was bothered by a more fundamental problem: whether the reformulation of quantum theory attempted meanwhile by Max Born and his co-workers on the basis of the programme of discretization would serve as an adequate tool for removing the erstwhile difficulties. 'I see a hope,' Heisenberg had written to Bohr already in February 1924, 'of obtaining the half-integral quantum numbers from the formalism $\int H\, dj$' (Heisenberg to Bohr, 3 February 1924). In June, Born had completed the systematic reformulation of perturbation theory in terms of difference expressions, and the problem of the anomalous Zeeman effects could be attacked consistently by calculating the second-order perturbation energy; this was necessary because the unperturbed systems were degenerate (Born, 1924b, p. 394–395). In this scheme Heisenberg's model Hamiltonian, H_{cl}, entering Eq. (43), would be identical to W_2, the second-order perturbation energy, derived from averaging the corresponding term H_2 (in the classical Hamiltonian of the atomic system in the external magnetic field) over the angle variables of the nondegenerate degrees of freedom. By substituting in H_2 the quantum values for the corresponding action variables, and by performing the averaging procedure towards W_2, one could indeed arrive at Heisenberg's Hamiltonian *with half-integral quantum numbers*. While Heisenberg interpreted these observations as so much 'grist to my Zeeman mill' ('*Wasser auf meine Zeemanmühle*,' Heisenberg to Pauli, 8 June 1924), he was fully aware that some of the basic assumptions of his theory still remained to be justified. As he stated in his paper:

> The provisional character of the preceding calculations shows up especially clearly from the fact that we had to use an *Ersatz* model deviating from ordinary mechanics. In a final theory of the Zeeman effects the non-mechanical properties of our *Ersatz* model must follow, as the g-formula in §3, as a consequence of the quantum mechanics which is demanded by retaining stability even in the case of coupling of the electrons. No step has yet been taken to achieve this goal. But one may suspect that the fundamental difficulties confronting the interpretation of the half-integral quantum numbers, the relativistic doublet formula and the double magnetism, originate from the same root as the quantum mechanical modification attempted by the g-formula in §1. (Heisenberg, 1924e, p. 300)

In spite of the hope expressed in these lines, it was clear that Heisenberg had not come closer at all to the final goal, the achievement of which Pauli had demanded as being essential for a consistent theory of the anomalous Zeeman effects. Consequently Pauli remained completely dissatisfied with the results, for as Heisenberg reported to Bohr: 'I received a letter from Pauli. He has obtained something about the intensities; otherwise he reviles everybody, but especially my atomic physics' (Heisenberg to Bohr, 25 August 1924).

In spite of Pauli's continuing opposition the Göttingen theoreticians, encouraged by Niels Bohr, remained quite hopeful.[153] Heisenberg immediately turned to the next big problem, which had remained unsolved thus far: the helium problem. 'I have calculated quite a lot about excited helium,' Heisenberg informed Bohr. 'On a purely formal level it seems that one may get the correct terms for ortho-helium, if one averages over the k of the outer electron (not over the k of the inner electron, because it is only $k = 1$) and over j and applies in addition Born's formalism to the principal quantum number.' It followed, said Heisenberg, that the ortho-helium doublet would be 'quasi-relativistic,' in agreement with the magnitude of its splitting. In order to justify his averaging procedure, he proposed to make use of the following assumption: one has to average, in the Hamiltonian of a given system, over all quantum numbers belonging to those degrees of freedom that are degenerate if higher-order perturbation terms are neglected. 'The following physical argument seems to speak in favour of this procedure,' he argued:

> If two atoms of *different* frequencies are coupled, then the dispersion formula is valid and the atoms *very rarely* induce jumps in each other. If, however, the frequencies are *the same*, then in the presence of strong interaction a steady energy

[153]On receiving the proof-sheets of Heisenberg's paper on the anomalous Zeeman effects (Heisenberg, 1924e), Bohr wrote: 'I am convinced that it provides a very important contribution to the discussion of the question treated [here], and I believe that also the formulation of your views has assumed a suitable form' (Bohr to Heisenberg, 5 July 1924).

> exchange through jumps occurs. The latter [i.e., the energy exchange] might—if only jumps of 1 unit occur, as in the case of *k* and *j*—give rise to such an averaging process. (Heisenberg to Bohr, 5 July 1924)

However, he concluded that he did not believe that these ideas made any sense yet; they were only an attempt to reformulate the problem in order to make some progress. Some days later he reported to Bohr again; he had meanwhile seen Hermann Schüler's measurements of the spectra of ionized lithium in a recent issue of *Naturwissenschaften* (Schüler, 1924), and thought that they 'formally' supported his new helium calculations. 'But in order really to find out something about helium,' he wrote, 'I am, I think, too stupid this semester; it is high time that we had vacations' (Heisenberg to Bohr, 14 July 1924). However, before the summer vacation began, Heisenberg took a step forward in his scientific career: he obtained his *Habilitation* at the University of Göttingen.

It had been Max Born's plan, ever since he got to know Heisenberg well during the winter semester 1922–1923, to make him a *Privatdozent* in Göttingen as soon as possible because he needed someone to assist him in teaching the courses. A year later, as a result of the invitation to Born to visit the United States in the winter semester 1924–1925, his desire to make Heisenberg a *Privatdozent* became even stronger. So he tried to push Heisenberg through the *Habilitation* procedure in summer 1924; when his work on the anomalous Zeeman effect obtained Bohr's blessings, he immediately accepted the paper 'On the Alteration of the Formal Rules' as the *Habilitation* (or inaugural) dissertation. There were still minor problems to be settled. For example, as Heisenberg recalled later, in the beginning it took the *Pedell* (the University porter) a long time to carry the dissertation around to the members of the faculty (who had to see and approve it), until he discovered the possibility of expediting the procedure by presenting the man a box of cigars. Ultimately the formal requirements were fulfilled in time, and Heisenberg gave his *Probevorlesung* (trial lecture) on 28 July 1924, a few days before the summer semester ended. Thus, only a year after his near failure in the doctoral examination at Munich, Heisenberg became a *Privatdozent* at Göttingen. Concerning this fast promotion Friedrich Hund recalled the following conversation between Arnold Sommerfeld and Max Born, which occurred in summer 1924: 'Sommerfeld: "A challenge to the Munich faculty!" Born: "Just wait, he [Heisenberg] will certainly justify it!" Sommerfeld (laughing): "I know, I know"' (Hund, 1961, p. 3). Heisenberg had every reason to be happy with his professor, Max Born, and the University of Göttingen for the opportunity he had received. After a long vacation, including a hiking tour through Württemberg with his comrades of the Youth Movement, he would finally arrive in Copenhagen in September 1924, there to work for seven months under the direction of Niels Bohr.

Chapter III
The Penultimate Sharpening of the Correspondence Principle

In spring 1924 Werner Heisenberg visited Copenhagen for the first time, and this visit brought him under the influence of Niels Bohr. Under Bohr's guidance Heisenberg not only acquired a new approach to atomic problems, but he also completed his education in atomic theory by learning more about physical argumentation than he had from Sommerfeld in Munich and from Born in Göttingen. He had gone to Copenhagen from Göttingen, where the theoreticians had been working on the programme of changing the mathematical structure of classical mechanics in such a way that it would account better for the observed properties of atomic systems. Although Born and his co-workers did follow—more than the Munich physicists—the Copenhagen approach to atomic theory, including ample use of the correspondence principle, their work was concentrated on the mathematical formulation of problems. In Copenhagen, Heisenberg experienced a new atmosphere. He found that Bohr's attitude differed from that of his two previous teachers. In particular, he noted that 'Bohr would always say, "First we have to understand how physics works; only when we have completely understood what it is about can we hope to represent it by mathematical schemes"' (Heisenberg, Conversations, p. 230). Bohr was very much concerned with the empirical data, at least as much as Sommerfeld, but he took a different view than the Munich physicist. This had a strong impact on Heisenberg who later summarized his impressions in the following words:

> Bohr never looked on problems from the mathematical point of view [as Sommerfeld did]; he looked at them from the physical point of view. I learned more from Bohr than from anybody else that the new type of theoretical physics was almost more experimental than theoretical. That is, one had to cover the experimental situation by means of concepts which fitted. Later on one had to put the concepts into mathematical forms, but that was a more or less trivial process which had to be solved. But the primary thing here was that one had to find the words and concepts to describe a strange situation in physics that was very difficult to understand. (Heisenberg, Conversations; also AHQP Interview, First Session, p. 4)

In early 1924, after his return from America, Bohr was filled with an optimistic spirit. In addition, the arrival of John Slater in Copenhagen, who brought along an idea about how to reconcile the conception of the light-quantum with the classical description of radiation, fostered this optimism. Slater's idea was transformed by Bohr and Kramers into a new radiation theory (Bohr, Kramers and Slater, 1924), which for the first time offered the hope of understanding the interaction of radiation with atomic systems completely. Many physicists at different places welcomed the Bohr–Kramers–Slater theory as a major success. Born, to whom Heisenberg reported about it on returning from Copenhagen, wrote enthusiastically to Bohr: 'Although I only have Heisenberg's brief report, I am fully convinced that your new theory has hit upon the truth and that, in a certain sense, it might also be the last word that can be said about these questions' (Born to Bohr, 16 April 1924). Heisenberg shared Born's positive opinion and, even after listening to the serious criticism of Albert Einstein, who passed through Göttingen on 7 June 1924 (Heisenberg to Pauli, 8 June 1924), he remained optimistic that the new Copenhagen radiation theory could be suitably adapted. He was very interested in making use of it in his further work on atomic problems, especially on his return to Copenhagen in fall 1924 for a longer period of time. In the papers, which he submitted from Copenhagen, he endeavoured to derive new theoretical consequences from the correspondence principle in conjunction with the new radiation and dispersion theory.

From 1924 onwards the theoretical work in Copenhagen and Göttingen marched side by side for several years and, at least in the early period, it is possible to isolate the common trend in the publications, such as Kramers' notes on dispersion theory (Kramers, 1924a, b), Born's paper on quantum mechanics (Born, 1924b), Heisenberg's paper on the anomalous Zeeman effect (Heisenberg, 1924e), and Kramers and Heisenberg's paper on the extension of dispersion theory (Kramers and Heisenberg, 1925). Heisenberg later recalled the opinion with which people in Copenhagen and Göttingen viewed these papers: 'People said, "One must try in this direction, and all these papers probably contain some tools, but it is clear they do not give the final solution. It's just something moving in the right direction, but the final solution must be deeper than that"' (Heisenberg, Conversations, p. 232). People shared the impression that each new paper went a little farther than the previous one. As Heisenberg said:

> It reminded me of the game which children play, in which you hide something in the room and then the child walks around to find it, and you say, "There you are quite cold, now you are getting warmer, and so on." And as soon as a paper got warmer, so to say, one had the impression that it was more satisfying. One would say, "All right, that helps." . . . Born's paper [1924b] was pretty good; it gave a proof of Kramers' dispersion formula. Everybody had the impression that one must try in this direction. Some people would perhaps criticize, "After all, it [Born's paper 'On Quantum Mechanics,' 1924b] is not so much more than Kramers' paper [1924a]. Kramers did all the right things that did mean anything. But still it definitely goes in the right direction." So there was a clear progress, step by step, from the Ladenburg paper [1921a], Kramers' paper, Born's paper, then the paper of Kramers and myself; every paper got a little bit farther in the right direction. (Heisenberg, Conversations, pp. 233–234; also AHQP Interview)

The line of research, which Heisenberg joined in spring 1924 under the influence of Bohr and Kramers, was different from the conventional one. He recalled: 'It [indicates] a rather unusual state in physics that people would tell you to write papers of that kind. It was that strange situation that everybody agreed by that time that physics would contain contradictions . . . In such a situation one is willing [to accept] papers which only go somewhere in the right direction' (Heisenberg, Conversations, pp. 232–233). Still, not everyone agreed with this trend. For example, Pauli criticized the work being done in Copenhagen and Göttingen. Also Sommerfeld, who was in favour of clear and well-defined mathematics, did not appreciate the dark intermediate steps. Recalled Heisenberg about his former professor: 'He certainly did not like Born's paper, nor did he much like my paper with Kramers. But he had to agree that he could not do any better' (Heisenberg, Conversations, pp. 234–235).

In Copenhagen, Heisenberg moved away from the influence of both Sommerfeld and Born and came under the decisive influence of Niels Bohr. He believed that he could learn more by following the physical ideas of Bohr than from anybody else. Max Born, quite unselfishly, let his favourite student pursue his chosen path of development. In April 1924, when he learned about the invitation to Heisenberg to spend the winter semester 1924–1925 in Copenhagen, he wrote to Bohr: 'I shall, of course, miss him (he is a charming, worthy, very bright man, who has become very dear to my heart), but his interest precedes mine, and your wish is decisive for me' (Born to Bohr, 16 April 1924). During the seven months that Heisenberg spent in Copenhagen he completed his education in atomic theory. He was introduced to a new range of problems connected with radiation and dispersion theory—in which, as well as in his old acquaintance, the anomalous Zeeman effects of complex atoms, he learned to make refined applications of the correspondence principle. The experience and insights, which he accumulated in Copenhagen, helped him to achieve a new, higher level of his scientific development. From that level, then, he was able to take the decisive step towards the discovery of quantum mechanics within a few months. In this sense he could truly claim that he had *learned physics from Niels Bohr*.

III.1 Getting to Know Niels Bohr

The first time that Heisenberg met Bohr was on the occasion of the lectures on atomic structure, which the latter delivered from 12 to 22 June 1922 in Göttingen. Heisenberg recalled later:

> Sommerfeld, my teacher in Munich, had taken me to Göttingen, although I was at that time only a 20-year-old student in my fourth semester. Sommerfeld was warmly interested in his students, and he had noticed how strongly Bohr and his atomic theory interested me. The first impression of Bohr still remains quite clearly in my memory. Full of youthful excitement, but a little self-conscious and shy, his head a little to one side, the Danish physicist stood on the platform of the auditorium, the strong Göttingen summer light streaming in through the open windows. He spoke softly and with some hesitation, but behind every carefully chosen word one could discern a long chain of thought, which eventually faded somewhere in the

> background into a philosophical viewpoint which fascinated me. (Heisenberg in Rozental, 1967, p. 94)[154]

Heisenberg enjoyed Bohr's lectures. Although he already had learned the content of what Bohr had to say in Sommerfeld's courses, he noticed that, 'it all sounded quite different from Bohr's own lips. We could clearly sense that he had reached his results not so much by calculation and demonstration as by intuition and inspiration, and that he found it difficult to justify his findings before Göttingen's famous school of mathematics' (Heisenberg, 1971, p. 38). There were discussions after each lecture, in which—especially after the third lecture—Heisenberg participated. Bohr had talked about the calculations of his collaborator Hendrik Kramers on the Stark effect of the hydrogen atom, in particular when the strength of the electric field was very weak and the electric splitting of the components was of the same order as the fine-structure splitting (Kramers, 1920b). Bohr had concluded by saying:

> No experiments have yet been performed on the transition of the fine-structure to the usual Stark effect by the gradual increase of an electric field. The quantum theory yields very many details of the phenomenon to be expected. Even if we really should not be unprepared to find that the quantum theory is false, it would surprise us very much if such a detailed picture obtained from the quantum theory should not be valid; for our belief in the formal reality of the quantum conditions is so strong that we should wonder very much if experiments were to give a different answer than what is demanded by the theory. (Bohr, 1977, p. 371)

Now Heisenberg, who knew Kramers' 1920 paper quite well because he had reviewed it earlier in Sommerfeld's Seminar, dared to dissent from this opinion and this gave rise to his first discussion with Niels Bohr.

The question of the Stark effect, especially the quadratic Stark effect which Kramers had treated in his calculations, had also interested Sommerfeld at about the same time as Kramers, and in January 1921 he had submitted a note on it (Sommerfeld, 1921b) without having noticed Kramers' paper (1920b), which had appeared several weeks earlier. The methods used in the respective approaches of Kramers and Sommerfeld were different: Kramers had based his calculation on correspondence arguments, while Sommerfeld had referred to earlier calculations of his student Paul Epstein, who had used just Sommerfeld's phase integrals (Epstein, 1916c). Sommerfeld had, therefore, asked Heisenberg to present the content of Kramers' paper in the Seminar. Heisenberg was thus acquainted with the objection that could be raised against Kramers' correspondence treatment of the problem. The main point was that the quadratic Stark effect could be thought of as the limiting case of the scattering of light with very long wavelengths.[155] Then, however, one had to conclude that Kramers' calculation must

[154] Heisenberg obtained the travel expenses for Göttingen from Sommerfeld.

[155] In the case of the dispersion of light, radiation is absorbed by an electron, bound in the atom, and reemitted; while in the quadratic Stark effect two external (static) electric fields act on the electron bound in the atom.

yield the orbital frequency of the electron in the hydrogen atom, a result which was certainly wrong. The physicists in Sommerfeld's group had therefore come to the following conclusions: it was accepted that the spectral line was due to the emission of a light-quantum in the transition of an electron between two energy levels; on the other hand, one could see from the correspondence principle (or from Rubinowicz's principle) that the emitted light was a spherical wave, except that it did not have the orbital frequency of the electron; hence there was difficulty in the application of the correspondence principle. The same difficulty did not, however, arise with the first-order Stark effect; there one could argue that the hydrogen atom in the absence of the electric field was degenerate, and that a slow precession of the electron started when the field was switched on. Hence the states of the electron, especially the transition frequencies, had nothing to do with the frequencies occurring in the (classical) dispersion formula in that case, and Heisenberg's objection could not be raised in connection with the linear Stark effect. However, in the quadratic Stark effect the correspondence principle applied; i.e., one had definitely to determine why Kramers' calculation gave frequencies different from those of electrons in their orbits. Therefore Heisenberg, remembering fully well the earlier discussions in Sommerfeld's Seminar on this question, felt strongly about raising this objection.

Bohr was not prepared to deal with Heisenberg's objection. 'Bohr answered that one should take here into account the reaction of the radiation on the atom, but he was obviously worried by this objection. When the discussion was over, Bohr came to me and suggested that we should go for a walk together on the Hainberg outside Göttingen. Of course, I was very willing' (Heisenberg in Rozental, 1967, pp. 94–95). Heisenberg was very happy that Bohr took the time to talk to him, and he was very impressed by the manner in which these private discussions went on. As he said later on: 'I remember that in our discussions on the Hainberg, Bohr first tried to use the [above] excuse, but then he talked himself away from it a bit. He felt that this was only a lame excuse, not a really good thing' (Heisenberg, Conversations, p. 104). Bohr then explained to Heisenberg that the difficulties arising in the description of the quadratic Stark effect represented a very common situation in atomic theory, and that in order to understand anything—or just to believe that one had understood something—one had to find excuses. If one did not use such excuses, one would face serious contradictions that would block progress for a number of years. During the walk, which took Bohr and Heisenberg on one of the trails—passing the *Café* (*Kaffeewirtschaft*) '*Zum Rohns*'—to the top of the Hainberg, from where one had an excellent view of Göttingen and its surroundings, Heisenberg learned more about Bohr's ideas than from his previous study of the papers. He recalled:

> That discussion which took us back and forth over Hainberg's wooded heights, was the first thorough discussion I can remember on the fundamental physical and philosophical problems of modern atomic theory, and it has certainly had a decisive influence on my later career. For the first time I understood that Bohr's view of his atomic theory was much more skeptical than that of many other physicists—e.g. Sommerfeld—at that time, and that his insight into the structure of the theory was

> not a result of a mathematical analysis of the basic assumptions, but rather of an intense occupation with the actual phenomena, such that it was possible for him to sense the relationship intuitively rather than derive them formally. (Heisenberg in Rozental, 1967, p. 95)[156]

The discussions about atomic physics between Bohr and Heisenberg on the Hainberg in Göttingen ended by Bohr inquiring about the young man's background and plans. 'Bohr invited me to come to Copenhagen for a few weeks the following spring, and perhaps later, possibly on a scholarship, to work there for a longer period,' Heisenberg recalled many years later (Heisenberg, 1964, see Rozental, 1967, p. 95).[157] He was extremely flattered by Niels Bohr's personal interest in his future; to be invited by Bohr meant a great honour. He knew that his friend Wolfgang Pauli was about to go to Copenhagen in fall 1922; he would be able to follow him very soon. However, it took some time before Heisenberg was able to go to Denmark. In the meantime he remained in contact with Bohr's Institute, mainly through Pauli, discussing in letters the progress of his work on the anomalous Zeeman effects and the helium atom.[158] When he was about to complete his doctorate in Munich, Sommerfeld wrote to Pauli asking whether it would be possible for Heisenberg to spend some time in Copenhagen during summer 1924. Pauli replied: 'Independent of whether I shall be able to return later [i.e., after fall 1923] again to Copenhagen, your question is whether Heisenberg might come here in summer 1924. Bohr is considering this seriously (the money for that is in any case available). He has told me and others repeatedly what a favourable impression he obtained of Heisenberg when he discussed with him in Göttingen. However, he cannot make a final decision at the moment' (Pauli to Sommerfeld, 6 June 1923). After the doctoral examination Heisenberg proceeded to Göttingen, where he again became involved in the problem of the anomalous Zeeman effects. He wrote a long letter, dated 22 December 1923, to Niels Bohr, reporting the results of his work and asking for the latter's opinion. Instead of giving the desired opinion, Bohr in his reply made a proposal. Concerned about resuming his work on atomic theory after his visit

[156] In his memoirs, *Physics and Beyond*, Heisenberg tried to reconstruct the 1922 discussion with Bohr on the Hainberg in some detail (Heisenberg, 1971, Chapter 3). According to this account Bohr immediately began to express doubts about the usual conception of an atom being a microscopic planetary system; he emphasized that he rather considered the property of the stability of atoms as the primary fact. Bohr pointed out that clearly the available mechanical description of atoms was very unsatisfactory; hence the calculations, such as Kramers' on the quadratic Stark effect, could at best serve to indicate vaguely certain empirical relationships, and the available pictures of atoms would not give more than a vague representation of the actual situation. In a future theory these pictures would be replaced by new concepts.

[157] It is not clear whether Bohr actually fixed the date for Heisenberg's visit to Copenhagen. In the account which Heisenberg gave in the Bohr memorial volume (Rozental, 1967), he claimed that his first visit to Denmark took place in spring 1923; this is certainly wrong, for the visit took place in spring 1924, as is proved by the letters exchanged between him and Bohr.

[158] Between fall 1922 and summer 1923, when Pauli was in Copenhagen, Heisenberg and Pauli exchanged many letters. On the other hand, Heisenberg addressed only one letter to Bohr, dated 2 March 1923.

to England and the United States, and wondering about which problems to investigate, Bohr thought of a paper which he had left unpublished the previous summer. Thus he wrote to Heisenberg:

> In the next few days I intend to occupy myself again with questions related to the anomalous Zeeman effect in connection with the completion for publication of the second part of my paper on quantum theory and atomic structure, the first part of which appeared a year ago in the *Zeitschrift für Physik* [Bohr, 1923a]. Recently, before he left here, Pauli actively helped me in my efforts to make the second part ready. Although a manuscript finally became available, I had, however, to postpone the publication, and until now I was unfortunately hindered from occupying myself with it. In this second part I have attempted to investigate the problems of atoms having several electrons more exactly than before and, especially, to discuss the associated fundamental questions of quantum theory. In connection with the completion of this work, as well as the discussion of new points of view proposed by you, it would give me special pleasure to have the opportunity of discussing these questions personally with you; therefore I wish to ask you whether it would suit you to come to Copenhagen for a few weeks. (Bohr to Heisenberg, 31 January 1924)

Bohr added that he would be able to pay the expenses for Heisenberg's travel and stay, and concluded: 'I often remember with great joy our meetings in Göttingen, and I very much hope that we shall once be able to collaborate here in Copenhagen for a longer period. In this connection I shall be grateful if you would write to me about your future plans. Now, however, I hope first that you will be able to accept my invitation for a shorter visit in the near future' (Bohr to Heisenberg, 31 January 1924).

This kind letter from Bohr made Heisenberg very happy. While he expected severe criticism of his new theory of the anomalous Zeeman effects, similar to the one he had received from Pauli in December 1923, Bohr had offered him possibilities of which he had hardly dreamt before. Not only had he been invited to visit Copenhagen, the mecca of atomic theory, Bohr also wanted him to assist him in writing the second part of his essay, 'On the Application of the Quantum Theory to Atomic Structure,' containing the theory of spectra of complex atoms.[159] This subject interested him deeply, but he was even more attracted by the prospect of personal discussions with Niels Bohr about the various aspects of

[159]The first part of the essay, with the subtitle 'The Fundamental Postulates' ('*Über die Anwendung der Quantentheorie auf den Atombau, I. Die Grundpostulate der Quantentheorie*'), was received by *Zeitschrift für Physik* on 15 November 1922 and appeared early in 1923 (Bohr, 1923a). An English translation, made by L. F. Curtiss, appeared as a supplement to the *Proceedings of the Cambridge Philosophical Society* in 1924. The second part, subtitled 'Theory of Series Spectra,' was supposed to contain the application of the fundamental principles to explain the series spectra in detail, especially the complex spectra of many-electron atoms and the influence of electric and magnetic fields upon them. Bohr had worked on it since fall 1922; when he left Copenhagen for England (to attend the British Association meeting in Liverpool) and the United States in September 1923, he had completed only a preliminary manuscript. The German version of this manuscript was prepared by Hendrik Kramers later in that year, but it was not published (see Bohr, *Collected Works*, Volume *3*, 1976, pp. 501–558).

his own work. Finally, Bohr had indicated that he would succeed Pauli in the role of assisting him. Heisenberg grasped the opportunity offered by Bohr and replied immediately. 'For your kind letter and so friendly invitation I wish to express my most grateful thanks. You can hardly imagine how delighted I am with this invitation, for I have desired nothing more than to discuss once with you the physical questions, which occupy me so much, and to hear from you a judgment about my papers. Thus I shall come to Copenhagen with pleasure, if it suits you at the end of this month, or in the beginning of March' (Heisenberg to Bohr, 3 February 1924). Heisenberg selected the earliest possible date on which he could leave Göttingen, as the winter semester closed at the end of February and he had still to complete a paper with Born on ions (Born and Heisenberg, 1924b). After emphasizing again how much he looked forward to the forthcoming meeting with Bohr, he remarked: 'As for my plans for the future, only this much is certain that I want to get my *Habilitation* here with Professor Born before long. If, in between, I could study a semester with you, I would be very happy; Professor Born also agrees with it' (Heisenberg to Bohr, 3 February 1924). Thus, Heisenberg immediately expressed his willingness to work under Bohr for a longer period.

After this, several letters were exchanged between Copenhagen and Göttingen. Bohr and Heisenberg finally decided that the latter's visit should take place during the weeks following 15 March 1924; before that date Bohr was going to be very occupied, but afterwards he expected to be able to take some time off for his guest. Bohr sent 500 Danish crowns to cover the expenses, which Heisenberg gratefully acknowledged, and also arranged for the visa to Denmark. Armed with his boundless optimism and the good wishes of the physicists in Göttingen, Heisenberg was finally ready to go to Copenhagen.[160]

The person, however, who most wanted Heisenberg to profit by a stay with Niels Bohr in Copenhagen was neither Born nor Sommerfeld, but Wolfgang Pauli. Ever since he had left Munich, and in particular since he had himself gone to Bohr, Pauli had carefully watched his friend's scientific development. Pauli had been the most severe critic of Heisenberg's attempts to solve the problem of the anomalous Zeeman effects. He had argued against the use of half-integral quantum numbers, both in the description of complex spectra and the calculation of the ground state of the helium atom; moreover, he had constantly pointed out the inconsistencies in Heisenberg's treatment of atomic problems, such as the neglect of stability considerations. In promoting these negative arguments, Pauli had strictly followed the principles of Bohr's atomic theory, rather than violating them whenever the empirical data made it necessary. By assuming this extremely conservative point of view, Pauli had hoped to isolate the fundamental difficulties of atomic theory, which Heisenberg and others hid behind arbitrary assumptions. Pauli was convinced that drastic changes had to be made in the existing

[160] For example, James Franck wrote to Bohr: 'Here we are very happy that now Heisenberg will go to you, and we hope that you also will have joy with the wonderful youngster' (Franck to Bohr, 15 March 1924).

quantum theory, and he was against easy solutions which only formally covered the difficulties. As one such formal and unsatisfactory solution he considered Heisenberg's theory of the anomalous Zeeman effects, which the latter had developed in fall 1923. Pauli felt certain that Heisenberg was on the wrong path, and he did everything to persuade him about the futility of these endeavours. But he was even more concerned about Heisenberg's future scientific development. Since Pauli was convinced that theoretical atomic physics in Göttingen was too formal and too mathematically oriented, he believed that Copenhagen would be the right place to improve Heisenberg's physical thinking. Thus, when in the beginning of 1924 the possibility arose that Heisenberg might go to Copenhagen, Pauli wrote a letter to Bohr, in which he candidly described the strong and weak qualities of Born's collaborator. He said:

> Recently I saw Heisenberg at a physicists' meeting in Braunschweig.[161] I always feel strange about him. When I think about his ideas, they appear terrible to me and I grumble about them to myself. The reason is that he is very *unphilosophical* [our emphasis], and he does not care about the clear formulation of the basic assumptions and their connection with the erstwhile theories. However, when I talk to him, he pleases me very much, and I see that he has quite a few new arguments—at least in his mind. I then regard him—apart from the fact that he is personally also a very nice human being—to be very important, even gifted with genius, and I believe that he will one day advance science greatly. (Pauli to Bohr, 11 February 1924)

Pauli's accusation that Heisenberg's approach to physics was unphilosophical may sound surprising at first. After all, fresh out of secondary school Heisenberg had gone to Sommerfeld because he was particularly interested in the philosophical questions underlying physics. He had derived this interest from reading Plato's *Timaeus* and Hermann Weyl's *Space*, *Time*, *Matter*. It is true that Sommerfeld had then persuaded the young man first to do a 'wagoner's work,' i.e., to start his study of physics by carefully solving little problems in physics before getting ready to think about the underlying philosophy of physical theories. Heisenberg grew in Sommerfeld's school by following his advice, and he continued to do the 'wagoner's work' under the supervision of Born. But he also tried to solve problems of great importance in theoretical physics, like the turbulence problem in hydrodynamics or the problem of anomalous Zeeman effects in atomic theory. Hence one may ask as to what Pauli meant by an unphilosophical attitude in physics.

Pauli, who like Heisenberg had been a student of Sommerfeld's, was a different kind of personality. He had had some connection with philosophy since birth: Ernst Mach had been a friend of his family and at the same time his

[161] This was the meeting of the *Gauverein Niedersachsen* (Lower Saxony Section) of the German Physical Society, held on 9–10 February 1924 at Braunschweig. Heisenberg talked there on the deformability of ions (Born and Heisenberg, 1924b). But he certainly also discussed privately with Pauli the continuation of his work on the anomalous Zeeman effects. He also mentioned to Pauli Bohr's invitation to visit Copenhagen.

godfather. Pauli had read voraciously in philosophy and literature, just as in physics. He had entered physics by writing a few papers on problems of general relativity; he was so widely read in the literature of this subject that Sommerfeld had asked him first to assist him in writing the review article on relativity for the *Encyklopädie der mathematischen Wissenschaften* and, finally, to write it alone (Pauli, 1921b). Pauli, very early in his scientific development, came into intense scientific contact with the leading philosophical thinkers in physics, especially Albert Einstein, Hermann Weyl, and Niels Bohr. They exerted a decisive influence on him, balancing the more pragmatic influence of his teacher Arnold Sommerfeld.

It was indeed much easier to recognize the presence of philosophical ideas in relativity theory than in the quantum theory around 1920. After all, special relativity theory had arisen from applying a few guiding principles and assumptions, which could be described as being philosophical themselves. This trend was even stronger in general relativity theory and its creator, Albert Einstein, was perhaps the greatest philosophical thinker of physics in the early twentieth century. As the mathematician Felix Klein characterized him in a letter to Pauli: 'Now, Einstein's development [of general relativity], as one would expect from a genius, is an irrational one: a fusion of philosophical needs with a strong physical instinct and a gradual penetration into the foundations laid by the mathematicians' (Klein to Pauli, 8 May 1921). Pauli admired these qualities of Einstein, and he wanted to become a physicist in this sense himself. After quickly leaving behind the mathematical environment of Göttingen, he tried to attack the problems of physics, especially atomic theory, by following the example of Einstein: that is, first setting up the underlying philosophical and physical principles and only then developing the mathematical methods. This procedure also coincided perfectly with that of Niels Bohr, and it was therefore not surprising that Pauli selected Bohr as his guide in atomic physics; he immediately became 'Bohrian' ('*verbohrt*') after his arrival at Bohr's Institute in fall 1922.

What was now the special philosophical attitude towards atomic theory that Pauli hoped Heisenberg might learn from Niels Bohr? Atomic theory in the early 1920s was a field in which people frequently made arbitrary and ad hoc assumptions in order to fit the empirical data. For example, Pauli had himself introduced such an assumption—which he called the 'mechanical correspondence principle' in his work on the hydrogen molecule-ion; another one consisted in the half-integral quantum numbers used by Landé and Heisenberg in the description of the anomalous Zeeman effects. While Sommerfeld tolerated such assumptions to some extent, Bohr—as Pauli noticed soon enough in Copenhagen —was very suspicious of them. The Copenhagen attitude, at least in 1922 and 1923, was to set up the principles of atomic theory and to discuss the physical phenomena, including the anomalous Zeeman effects, according to these principles; if this treatment led to a disagreement with the empirical data, it had to be stated openly. Pauli had followed this procedure closely in his work on the complex spectra; he had avoided introducing any assumption which could not be justified on the basis of Bohr's principles of atomic theory, and he had not succeeded in finding a consistent and satisfactory description of the anomalous Zeeman effects.

On the other hand, Heisenberg, who had treated the same problem of atomic theory, had deviated strongly from this procedure. First, he had employed new assumptions—without worrying whether they agreed with the accepted principles of atomic theory—just to arrive at a successful description of the data. Second, he had applied certain mathematical methods without a proper understanding of their physical meaning, a procedure which contradicted the attitude of Niels Bohr. Third, Heisenberg had mixed together assumptions and hypotheses, whose consistency was not proven at all. For example, in his new (1923) theory of the anomalous Zeeman effects he had combined, without anlayzing their compatibility, Landé's successful phenomenological description with Pauli's results (which had been obtained on the basis of Bohr's atomic theory). Hence Pauli felt that Heisenberg had either forgotten about the *real* difficulties of atomic theory or had buried them in a formal approach. This procedure was clearly in contrast to the one on which the theory of relativity had been developed. In the latter, Einstein had started out with a clear notion of the principle of relativity and all other consequences had followed. Hence, inasmuch as Einstein was philosophical, Heisenberg was not; only Niels Bohr, the greatest philosophical thinker in atomic physics in Pauli's view, could help him to improve.

Pauli deeply regretted the fact that Heisenberg was not philosophical. On the other hand, in spite of the fact that he disagreed with most of what Heisenberg had done until then in atomic theory, especially on the anomalous Zeeman effects, he had great faith in his friend's abilities. He intuitively believed that more was contained in Heisenberg's work than he was able to criticize. Thus he wrote to Bohr: 'Also in his latest thing, about which, as I learned, he wrote to you, there will be much truth, although it is not clearly thought out' (Pauli to Bohr, 11 February 1924). However, what was true and correct in Heisenberg's theory had still to be brought out with the help of someone who was more philosophical. 'I am therefore very pleased,' he wrote to Bohr, 'that you have invited him to Copenhagen.' And he expressed his great expectation connected with Heisenberg's visit to Copenhagen as follows: 'Hopefully you will then take atomic theory forward in good measure and solve several of the problems with which I have tormented myself in vain and which are too difficult for me. I hope also that Heisenberg will then bring back home a philosophical attitude in his thinking' (Pauli to Bohr, 11 February 1924). Niels Bohr was thus prepared to receive Heisenberg; he knew that the most important point on which he could influence Heisenberg was to acquaint him with a more philosophical attitude towards atomic physics.

III.2 Spring Visit to Copenhagen

In March 1923, after the final failure of calculations on the helium problem was proven, Heisenberg wrote to Pauli in Copenhagen expressing the hope that the beautiful spring might alter the dark prospects of quantum theory. This hope did not come true, and Pauli was completely disillusioned about the most important problems of atomic theory when he left Bohr's Institute in fall 1923. The following spring Heisenberg finally went to Copenhagen, filled with optimism about the progress of quantum physics, which might come about by discussing his work with Niels Bohr. He recalled later on about his arrival:

> During the Easter vacation of 1924 I finally boarded the Warnemünde ferry for Denmark. Throughout the trip I feasted my eyes on a host of colorful boats, including four-masters in full rig. At the end of the First World War, a large part of the world's merchant fleet had ended up at the bottom of the sea, with the result that the old sailing boats had to be brought out again, and the seascape looked all the brighter for it—much as it had done a hundred years before. When I eventually disembarked, I had some trouble with customs—I knew no Danish and could not account for myself properly. However, as soon as it became clear that I was about to work in Professor Bohr's Institute, all difficulties were swept out of the way and all doors were opened to me. And so from the very outset I felt safe under the protection of one of the greatest personalities in this small but friendly country. (Heisenberg, 1971, p. 45)

The burden of scientific questions which Heisenberg wanted to ask Bohr was certainly greater than of the luggage with which he arrived in Copenhagen. Indeed, he had many irritations, which had accumulated over the past two years and which now had to be straightened out. There was, for example, the old question of using half-integral quantum numbers in connection with the anomalous Zeeman effects. Bohr, already in his 1922 Göttingen lectures, had dismissed this possibility as not being in agreement with the principles of atomic theory; these principles required that only integral quantum numbers should occur in the quantum theory of multiply periodic systems. Even in the case of many- or several-electron atoms, such as the helium atom, Bohr had not made an exception; he had not accepted Sommerfeld's suggestion about distributing an integral quantum number among a pair of degenerate degrees of freedom. In fact, Bohr had assembled evidence from experiments to rule out half-integral quantum numbers. For example, in the paper with Dirk Coster on X-ray spectra, he had studied the empirical data for the screening constant associated with K-, L-, M- and N-levels. Bohr and Coster noted that the values thus obtained increased by one unit in going from one period in the system of elements to the next. They had concluded:

> This agreement is of essential significance in that, quite apart from the still open question of the interpretation of the relativistic doublets, it provides a direct support for the correctness of the quantum numbers used in the classification of the electron orbits in the atoms. If, for example, as proposed in several recent papers, the subordinate quantum number k should be assigned half-integral rather than integral values, matters would appear quite differently. Because of the much larger value of the relativity term for the K-levels corresponding, according to the formula,

> to the assumption $k = \frac{1}{2}$, this assumption would mean that the screening constant, instead of the increase of one unit from noble gas to noble gas as claimed by the theory, would show an increase of about 5 units from krypton to xenon, and even an increase of more than 20 units from xenon to niton. (Bohr and Coster, 1923, p. 367; Bohr, 1977, p. 542)

In a later paper Bohr had made this argument against half-integral values for the quantum numbers even more cogent. He had stated:

> In fact, it is possible to show by a simple calculation that every electron in an n_k-orbit in a nuclear atom would fall into the nucleus if $N/k = hc/2\pi e^2 = 137$ [with N the charge of the nucleus in absolute units of the electron's charge e, k the subordinate quantum number, h Planck's constant, c the velocity of light in vacuum]. Thus, it is seen that, for the heavier elements, the penetration of the series electron into the interior of the atom would lead to the result that orbits with $k = \frac{1}{2}$ cannot exist. (Bohr, 1923c, p. 266, footnote; *Collected Works*, *4*, 1977, p. 639, footnote)

In connection with this denial of half-integral quantum numbers, Bohr had even encouraged Pauli, who spent the academic year 1922–1923 with him, to write a paper on the anomalous Zeeman effects from the Copenhagen point of view, in particular, using only integral values for the quantum numbers.

Heisenberg had been unhappy, at times even angry, with the Copenhagen insistence on the 'fruitless' quantum principles. After Pauli's failure to produce a theory of the anomalous Zeeman effects, which could explain the empirical data, he had sought again to obtain a description of the data with the help of half-integral quantum numbers. With respect to Bohr and Coster's arguments against the use of half-integral quantum numbers in X-ray spectra, a counterargument was available in early 1924: Landé's new considerations, which seemed to be supported by empirical evidence, claimed that the relativistic and nonrelativistic doublets had the same origin; hence the entire theory of atomic structure, not just the validity of half-integral quantum numbers, was at stake. Moreover, the theoretical argument concerning the orbits with quantum number $k = \frac{1}{2}$ falling into the nucleus appeared to be shaky as long as the nature of the relativistic effect on spectral lines was not clarified. Finally, new evidence for the occurrence of half-integral quantum numbers had been found in Göttingen in early 1924 on the basis of the analysis of the spectroscopic properties of ions by Friedrich Hund and Born and Heisenberg. In spring 1924 Heisenberg was therefore convinced that Bohr would ultimately have to accept the half-integral quantum numbers. He felt very strongly about this matter, just as he had felt earlier with respect to the question of the quadratic Stark effect—which Bohr had not been able to answer either.

However, the main thing, which Heisenberg wanted to discuss with Bohr, was his new theory of the anomalous Zeeman effects. Would Bohr assume a less negative attitude towards this attempt than Pauli? The friendly letters from Bohr had raised Heisenberg's expectation that this was so. Hence, following his arrival in Copenhagen on Saturday, 15 March 1924, he eagerly awaited the opportunity of talking to Bohr. But this opportunity did not come right away, for Bohr was very occupied with administrative matters.[162] In the beginning, when he did not know anyone at Bohr's Institute, Heisenberg felt more lonely than he had ever felt before. Moreover, he was for the first time at a non-German research institute, and this fact raised its own particular problems. As he recalled later: '[In] the first few days I was deeply depressed by the superiority of the young physicists from all over the world who surrounded Bohr. Most of them could speak several foreign languages, while I could not express myself reasonably in even one; they knew of the world outside, of many various people's culture and literature, they played various musical instruments extremely well, and above all they understood much more of modern atomic physics than I. That I should be able to find a place in such a circle seemed quite hopeless' (Heisenberg in Rozental, 1967, pp. 95–96). Heisenberg felt quite lost, for this was a situation he had not encountered before. He had frequently travelled to new places in his life; he had changed from the University of Munich in Bavaria to the University of Göttingen in Northern Germany; he had attended conferences in Leipzig and Innsbruck; and he had hiked with friends and comrades of the Youth Movement in various parts of Germany, Austria, and even Finland. But in all these movements he had had company, people with whom he could talk and communicate. For the first time, this was not so in Copenhagen. Moreover, at German universities Heisenberg had become known as being extraordinarily talented in theoretical physics, while in Copenhagen he quickly recognized that several other young scientists were at least as capable if not better. Even in playing music, where he was proud of performing well on the piano, some of Bohr's visitors surpassed him in versatility. In short, Heisenberg felt that he had every reason to be depressed. Instead of obtaining a privileged position immediately upon arrival, he noticed that he had to struggle to find a place in Bohr's circle. Still, he tried to do as well as he could, and within a few days he became acquainted with some of the people at Bohr's Institute. Besides Hendrik Kramers, he met the Americans John Slater, Harold Urey and Frank C. Hoyt, the Norwegian Svein Rosseland, and the Danes Christian Møller and Sven Werner. Heisenberg soon became friends with several of them. 'I remember with particular pleasure,' he said later, 'my first discussions with Kramers from Holland, Urey from U.S.A., and Rosseland from Norway. They all seemed to know Bohr well and to respect him highly, and they were full of optimism with regard to the development of Bohr's theory' (Heisenberg in Rozental, 1967, p. 96).

[162] In his previous letters Bohr had already mentioned the possibility that he might be very occupied in the beginning of March; hence he had proposed to shift the date of Heisenberg's arrival in Copenhagen from 12 March, as had been planned earlier, to 15 March. Heisenberg then confirmed that he would arrive on 15 March and added: 'If you then still have too much other work, I can also very well do some sightseeing in Copenhagen on Saturday and Sunday' (Heisenberg to Bohr, 11 March 1924). Apparently Bohr's preoccupation lasted longer than the weekend of 16 March 1924.

After a few days, during which Heisenberg had been wondering what to do in order to force a discussion of all kinds of issues with Bohr, the latter entered his office and proposed that the two of them take a walking tour through the island of Zealand, as there was little opportunity in the Institute of having long conversations and of getting to know each other better. This proposal perfectly suited Heisenberg, who was an enthusiastic outdoorsman. Now he could have Niels Bohr all to himself, separated from the crowd of people who had somehow intimidated him. The walking tour lasted three days. On the first day they reached Helsingör; on the second day they hiked to Gilleleje, the northern tip of the island; and on the third day Bohr and Heisenberg passed through Tisvildeleje and Fredriksborg and returned to Copenhagen.[163] This tour served exactly the purpose which Bohr had had in mind: to relax from his strenuous duties at the Institute and to get into a deeper contact with his young visitor by discussing many aspects of life, history, culture, and civilization in Denmark and Germany, respectively, before turning to the problems of atomic physics. Heisenberg recalled later:

> Bohr was obviously glad that this way he could show me some of the places in Denmark which meant particularly much to him, Hamlet's castle Kronborg at the northern end of the Sound between Denmark and Sweden, the elaborate Renaissance palace Frederiksborg in the lake near Hillerød, the great forest which stretches northwards to Esrum Lake and the small fishing villages on the Kattegat from Gilleleje to Tisvildeleje. On the course of this tour, Bohr told me much about the history of the country and its palaces, and of the events from the earliest times with the connections to the Icelandic Sagas which he knew so well. (Heisenberg in Rozental, 1967, p. 96)

Indeed, Bohr attempted to give Heisenberg a detailed impression about his native land and tried to find out from him, at the same time, all he could about Germany. Bohr asked questions with genuine sympathy about Heisenberg's boyhood, adolescence and upbringing. Many years later Heisenberg tried to reconstruct these discussions (Heisenberg, 1971, Chapter 4, pp. 46–57). He recalled, in particular, that Bohr began by inquiring about how young people in Germany had experienced World War I. Heisenberg reported about the excitement and enthusiasm connected with the outbreak of the war, whereupon Bohr mentioned the quite different feelings which the same event had aroused in neutral Denmark. Bohr listened patiently to Heisenberg's appraisal of the virtues, which the war had stimulated in people and which were still considered valuable later on, such as the readiness to pursue a common goal together, the fearlessness to risk one's life, and the purposefully controlled discipline. After that, Bohr carefully analyzed other, negative aspects of these virtues; he spoke about the moral blindness that went along with certain kinds of discipline, and about the

[163]The walking tour took place several days after 15 March, when Heisenberg arrived in Copenhagen, but before 24 March when Bohr answered Pauli's letter of 21 March. Most probably, Bohr and Heisenberg were away from Copenhagen between 20 and 23 March 1924.

dangers arising from neglecting the individuality of man. He then related the principal virtues of his beloved Vikings, such as their independence of thought and action, and their strong belief in personal freedom; he saw these virtues partly represented in modern times in the British way of life, including parliamentary democracy and the quality of mutual fairness of winners and losers in relation to each other. Bohr missed this element of fairness in the discussions about relativity theory in Germany, as he remarked in connection with Heisenberg's report on some events during the Leipzig meeting of natural scientists and physicians (*Versammlung der Gesellschaft Deutscher Naturforscher und Ärzte*) in fall 1922.[164] He tried to convey to Heisenberg his conviction that scoring a victory at all costs was not the most important thing in life, and one had to be able to accept defeats without bearing grudges. Bohr discussed all these matters seriously, candidly and compassionately, and he intended with his remarks and stories to open Heisenberg's eyes to a vision that lay beyond the narrow limits of national frontiers. Heisenberg was very appreciative of what Bohr said and was greatly impressed by Bohr's devotion to the pursuit of that politics which avoided international conflicts and allowed all people in the world to lead peaceful and independent lives.

Heisenberg learned about Bohr's interest in the story of Prince Hamlet, about his love for the Danish countryside—which did not contain high mountains but offered many views of the sea—and about many other aspects of his rich personality. He discovered that Bohr was not only a powerful rational thinker, but a warm-hearted human being. Bohr enjoyed listening to Heisenberg's reports of his tours with the boy scouts through Germany, and Heisenberg was glad to notice that Bohr also enjoyed all sorts of youthful pastimes. As he recalled: 'On the beach we often tried to see who could throw a stone furthest out, or whether we could hit a floating log. Bohr told [me] that he and Kramers had once found a mine left over from the war, and they had tried to see who could hit the detonator. After several vain attempts, they realized that they would never be able to enjoy the victory if they had hit it, for the explosion of the mine would have killed them both. After that they found another target' (Heisenberg in Rozental, 1967, p. 96). Heisenberg got to know Bohr more quickly and thoroughly than one could normally manage within a few days. For example, he discovered that Bohr's tendency towards philosophical generalization was stimulated by very simple games, and he later remembered the following story from the walking tour: 'Once, when on a lonely road I threw a stone at a distant telegraph post, and contrary to all expectations the stone hit, he said, "To aim at such a distant object and to hit it, is of course impossible. But if one has the impudence to throw in that direction, and in addition to imagine something so absurd as that one might hit it, yes, then perhaps it can happen. The idea that something perhaps could happen can be stronger than practice and will"' (Heisenberg in Rozental, 1967, p. 97). Naturally, besides all these things, Heisenberg also learned about Bohr's views on the problems of atomic theory, for

[164]See footnote 94 for the events at which the opponents of Einstein's theory sought to promote their case.

their discussion took a major place in their conversations. Yet, when they finally returned to Copenhagen, it was mainly the impression of Bohr's personality rather than of physics, which caused Heisenberg to write to Pauli: 'I am, of course, absolutely enchanted with the days I am spending here' (Heisenberg to Pauli, 26 March 1924).

Niels Bohr took an unusual personal interest in his young visitor, whose talents he had already gauged two years earlier in Göttingen and who now had been recommended to him so warmly by the hypercritical Wolfgang Pauli. Their meeting in Denmark confirmed Bohr's earlier view, for he reported to Rutherford: 'At present we are having a visit of Dr. Heisenberg from Göttingen, who is a very ingenious and sympathetic man indeed' (Bohr to Rutherford, 24 March, 1924). Heisenberg, on the other hand, had never met a person like Bohr with whom he could discuss just about everything. Arnold Sommerfeld, in spite of his warm human interest in his young disciples, was still '*der Herr Geheimrat*,' and, as for Max Born, 'it would not have occurred to anyone, certainly not his collaborators, to talk about such matters [as Heisenberg discussed with Bohr] with him' (Heisenberg, Conversations). Bohr had taken Pauli's request seriously and allowed Heisenberg to become familiar with his character and widespread interests before they even started to talk about physics.

The discussions on physics, which Heisenberg had with Bohr in spring 1924, were concerned mainly with general principles rather than with the details of atomic models, in accordance with Pauli's suggestion to Bohr in his letter of 11 February 1924. As Niels Bohr recalled later:

> Our discussion touched on many problems of physics and philosophy, and special emphasis was placed on requiring a unique definition of the concepts in question. The discussions on problems of atomic physics were devoted, above all, to the strangeness of the quantum of action for the formulation of concepts that were applied to describe all experimental results. In this context, we also talked about the possibility that, as in relativity, also here [in quantum theory] mathematical abstractions might perhaps turn out to be useful. At that time there did not yet exist such perspectives, but the development of physical ideas had already reached a new stage. (Bohr, 1961b, p. ix)

Indeed, Bohr did everything to open Heisenberg's eyes to a philosophy underlying atomic theory. Heisenberg, who had gone to Copenhagen to discuss certain specific questions arising from his recent work and other such matters of detail, discovered that Bohr preferred to resume the general discussions which they had initiated in Göttingen two years previously. He learned that in the meantime he (Heisenberg) had not paid much heed to philosophical arguments. He became aware of the fact that the pursuit of physics was more than the sum of a number of successful calculations using clever tricks. And he now realized that he needed a philosophy to guide him in his further work. At the same time he found the man who could help him acquire it. Gone from his mind were the problems that had seemed so important before coming to Copenhagen—the half-integral quantum numbers and the difficulties of understanding Kramers' treatment of the quadratic Stark effect. They now appeared to be minor disagreements not worth quibbling about; and Heisenberg, who had come determined to argue with Bohr to defend his point of view, became an ardent partisan of Bohr's views by the time he left Copenhagen. As Pauli had foreseen, the profound human concern and intellectual and spiritual probing to which Bohr submitted Heisenberg changed the latter's outlook completely. Many years later, Heisenberg described his meeting with Bohr in spring 1924 as a 'gift from heaven' (Heisenberg, Conversations), but the hidden mediator of this gift had been Wolfgang Pauli, Heisenberg's sharpest, if friendly, critic.

In addition to providing Heisenberg with a philosophical point of view, Bohr also reported some progress on a fundamental problem which had bothered him (Bohr) for many years. The problem was how to incorporate the light-quantum features—which showed up in the frequency condition for line spectra, and even more explicitly in the recently discovered Compton effect—into a consistent approach to quantum theory based on the correspondence principle. According to the latter, the quantum and the classical descriptions of atomic phenomena should coincide in the limit of high quantum numbers. But how could light-quanta ever turn into electromagnetic waves? Towards the end of 1923 an American visitor, John Slater, had arrived in Copenhagen and had brought a new idea to resolve the difficulties, at least the one concerning the proper incorporation of Bohr's frequency condition into quantum theory. In particular, Slater had postulated that every atom was surrounded by a 'virtual field of radiation having the frequencies of possible quantum transitions' (Slater, 1924a, p. 307); that is, with each given stationary state of an atom was associated a set of virtual oscillators, whose frequencies were identical with the frequencies of transitions from that state to all possible states which could be reached by emitting or absorbing radiation. Slater had talked about 'virtual' fields because he had not considered a specific transition; he had rather assumed that the virtual field stemming from an atom yielded the probabilities for the spontaneous transitions of the atom, in a way similar to that of an external field (i.e., the field created by the superposition of the virtual fields of other atoms, or by an external alternating electromagnetic field) which yields the probabilities for induced radiation. While Slater had planned to use his idea for combining the elements of the theory of classical radiation and light-quanta, Bohr and Kramers had taken it up as a proper tool for avoiding altogether the light-quantum features of radiation in the theoretical description. Bohr, Kramers and Slater succeeded finally in connecting the apparently discontinuous processes occurring in the atom (when discrete lines are emitted or absorbed) with the continuous character of the known radiation field (of classical electrodynamics) by paying what they thought was the minimum price: the replacement of energy and momentum conservation in individual atomic processes (such as emission, absorption, or scattering of radiation) by a statistical conservation, i.e., by conservation of these properties on the average for large ensembles of atoms. At the same time they had concluded a greater independence of transition processes in distant atoms. The existence of the virtual radiation field, the statistical conservation of energy and momentum, and the statistical independence of processes in distant atoms formed the main content of the so-called Bohr–Kramers–Slater theory of radiation, which was presented in a paper of the three authors submitted in January 1924 to *Philosophical Magazine* and, a little later, in German translation to *Zeitschrift für Physik* (Bohr, Kramers, and Slater, 1924). This theory filled Bohr and all of his collaborators with a new optimism, which Heisenberg sensed immediately during his spring visit. The reason was that it now seemed possible, for the first time in the development of quantum theory, to overcome the contradictory situation that had prevented progress in a fundamental question: namely, the difficulty posed by light-quantum-like phenomena, such as the Compton effect, to the application of the correspondence principle. With the Bohr–Kramers–Slater theory, however, one could quickly bridge the contradictions; during the time when Heisenberg visited Copenhagen, in March 1924, Kramers obtained the dispersion formula on the basis of the new theory (Kramers, 1924a).[165]

Thus, in spring 1924, Bohr appeared to Heisenberg as the wisest man in atomic physics, who, in spite of certain outstanding difficulties, possessed the most complete theoretical description of the observed phenomena; moreover, he had just been able, together with his collaborators, to obtain a major breakthrough in the fundamental question of radiation. The Copenhagen optimism infected Heisenberg; on his return to Göttingen, he passed on the good news to Max Born and others.[165a] As a result of the report which he received from Heisenberg, Born became enthusiastic about the Bohr–Kramers–Slater theory, as he wrote to Bohr in a letter dated 16 April 1924. In the same letter he indicated that he had started to connect the new theory with the quantum-theoretical perturbation theory and the Göttingen programme of discretization, with the goal of deriving Kramers' dispersion formula. Bohr was very happy about this development. In the beginning of June he visited Göttingen to inform himself about the progress, which Born and Heisenberg had achieved in the meantime, and he was very impressed by what he learned. An intense collaboration developed between Göttingen and Copenhagen, and Heisenberg reported the results in numerous letters to Bohr.[166] Around 21 June 1924 Kramers visited Göttingen, and soon afterwards he submitted another note to *Nature*, in which he derived his dispersion formula on the basis of the scheme that Born had

[165] It should be mentioned, however, that for the derivation of Kramers' dispersion formula it was not necessary to assume only the statistical conservation of energy and momentum, as was shown later by Max Born (1924b).

[165a] It should be mentioned that, before coming to Copenhagen, Heisenberg had not thought too highly of the Bohr–Kramers–Slater radiation theory. He had written to Pauli: 'Bohr's work on radiation is certainly very interesting, but I do not see in it a fundamental progress' (Heisenberg to Pauli, 4 March 1924). Yet, a few weeks later, while in Copenhagen, he referred jokingly to 'phase relations between our virtual oscillators' ('*Phasenbeziehungen zwischen unseren virtuellen Oszillatoren*') (Heisenberg to Pauli, 26 March 1924). It shows that he had taken the jargon of the Bohr–Kramers–Slater theory into his own vocabulary.

[166] Between April and the end of July, Heisenberg wrote five letters to Niels Bohr: 6 April, 15 May, 5 July, 14 July and 28 July 1924.

developed in his previous paper on 'quantum mechanics' (Kramers, 1924b).[167] The Copenhagen–Göttingen collaboration on atomic theory in spring and summer 1924 was so strong that serious objections against the Bohr–Kramers–Slater theory did not shake it. These objections were raised by Albert Einstein, who passed through Göttingen on 7 June 1924 and argued that statistical conservation of energy and momentum might lead to deviations from well-tested and accepted laws, such as Kirchhoff's law of emission and absorption.[168] As a consequence of Einstein's visit, Born detached himself from the assumption of statistical conservation of energy and momentum in the Bohr–Kramers–Slater theory (see Born, 1924b, p. 386); however, because of this possible defect of the theory, he and the other Göttingen physicists saw no reason not to follow Bohr's guiding principles.[169]

Bohr's greatest admirer in Göttingen was, of course, Heisenberg, ever since he had been won over by the Danish master in spring 1924. 'You can hardly imagine,' he had written to Bohr, 'how much these days in Copenhagen meant for me scientifically, but not only scientifically. Also please accept my special thanks for the hiking tour of Kattegatt' (Heisenberg to Bohr, 6 April 1924). But Heisenberg would obtain more opportunity to interact with Bohr; in June 1924, during Bohr's visit to Göttingen, it was decided that he should visit Copenhagen again in the fall, this time for several months. Nothing better could happen to him, and he expressed his feelings and hopes to Bohr in one sentence: 'I look forward with great pleasure to fall and winter in Copenhagen and I have great hope that one, i.e., even we, will be able to calculate something in quantum theory by then' (Heisenberg to Bohr, 5 July 1924).

III.3 In the Footsteps of Niels Bohr

It had been envisaged for a long time that Heisenberg would spend some time working in Copenhagen under the guidance of Niels Bohr. It had already been Sommerfeld's wish; then Pauli, who had preceded Heisenberg in Göttingen and Copenhagen, was convinced that he would benefit by it. When Heisenberg visited Bohr in spring 1924, the latter proposed that he should return soon for an extended period. 'Heisenberg told me,' wrote Born to Bohr, 'that you want him

[167] Born's paper on 'quantum mechanics' was received by *Zeitschrift für Physik* on 13 June 1924. He spoke on the contents of this paper at the Hamburg meeting of the Lower Saxony Section (*Gauverein Niedersachsen*) of the German Physical Society on 21 June 1924. Kramers was also present at the meeting, where he gave a talk entitled '*Über Streuungen und Absorption von Licht an Atomen*' (see *Verh. d. Deutsch. Phys. Ges.* (3) 5 (1924), p. 37). His second note on the quantum theory of dispersion was signed Copenhagen, 22 July 1924 (Kramers, 1924b).

[168] Heisenberg, in a letter to Pauli, dated 8 June 1924, reported about Einstein's visit and an argument of Einstein's against the Bohr–Kramers–Slater theory.

[169] In contrast to Born and Heisenberg, Pauli, whom Bohr had been able to persuade about the Bohr–Kramers–Slater theory during his visit to Copenhagen in April 1924, became very critical of it later in the year (see his letter to Bohr, 2 October 1924).

as assistant next winter' (Born to Bohr, 16 April 1924). Born agreed in principle with this plan—though it meant the temporary loss of his closest collaborator—primarily because he expected to be in the United States during the winter semester 1924–1925. During Bohr's visit to Göttingen two months later the matter was decided, and in his letter to Born, dated 18 June 1924, Bohr spoke about Heisenberg's visit to Copenhagen the following September. At the same time he started to make the financial arrangements, which came through very fast. 'A few days ago I was informed,' he reported to Heisenberg early in July, 'that your fellowship of the International Education Board is fully in order. You will receive a stipend of $1000 for one year; in addition, the American grant will reimburse you for your travel expenses to and from Copenhagen' (Bohr to Heisenberg, 5 July 1924). This was more than Heisenberg had hoped for, as he wrote back to Bohr: 'It is really great how the American grant takes care of my travel as well' (Heisenberg to Bohr, 28 July 1924). Knowing about the financial difficulties of scientists in Germany, Bohr had arranged all the details of Heisenberg's visit carefully. On 22 August he sent him a cheque for $250, the stipend for three months, and requested him to write about the date of his departure in order that arrangements could be made for the travel ticket. About a week later, Heisenberg, who was then in Munich with his parents, received the ticket with another letter from Niels Bohr. 'As you see,' he wrote, 'I have permitted myself to get for you a second-class ticket instead of a first-class one, of which we had talked originally; thus, instead of a ticket from Göttingen, I got a ticket for your entire travel from Munich via Göttingen and Berlin to Copenhagen' (Bohr to Heisenberg, 1 September 1924). Heisenberg was overwhelmed by Bohr's thoughtfulness and thanked him gratefully.[170]

After the financial arrangements had been successfully made, two questions still remained to be settled between Copenhagen and Göttingen: First, when should Heisenberg arrive at Bohr's Institute, and, second, how long could he stay? The answer to the first question depended only on Heisenberg, i.e., on his plans for the summer vacation. He wrote to Bohr: 'It is not yet quite certain when I will be able to come. In any case, I shall make sure that I shall be there when Kramers leaves Copenhagen, so that I can help you. Therefore I would very much like to hear from you *when* Kramers goes away from Copenhagen. (I think this will be around 12 September?) 1 September is a little too early for me, because I would like to spend a little more time in Munich and the local mountains' (Heisenberg to Bohr, 15 July 1924). He wanted to relax fully from his strenuous semester's work in Göttingen, to go on walking tours with the boy scouts, and to hike in the mountains before taking up physics again with Bohr.

[170]Never before in his life had Heisenberg had as much money as $250 in his hands. He confessed this in a letter to Bohr, in which he also thanked him for the arrangement of the travel ticket: 'That you have sent me the ticket directly from Munich [to Copenhagen] is absolutely wonderful, and I just hope that the exchange [of the first-class ticket from Göttingen to Copenhagen] has not caused you extra expenses. However, I am very happy about it; I had myself thought about the possibility of making this exchange, but I did not dare to write to you about it' (Heisenberg to Bohr, 4 September 1924).

At the end of August he proposed to arrive in Copenhagen on 16 September, to which Bohr replied: 'It would suit us very well in every respect if you would come here around 16 September' (Bohr to Heisenberg, 1 September 1924). Happy with this response, he set the time of his arrival by train in Copenhagen at 7 P.M. on 17 September.

With respect to the second question, namely, to arrange an academic year's leave of absence from the University of Göttingen, a difficulty had arisen in summer 1924. The reason was that Heisenberg had just received his *Habilitation* at that time, and, starting in the winter semester 1924–1925, he was supposed to give courses of lectures. It was possible, of course, to get a leave of absence, but not for several semesters. 'A little difficulty seems to arise because of the length of time which I shall spend in Copenhagen,' Heisenberg wrote to Bohr in July. 'Professor Born and I would prefer that I should be in Göttingen from 1 May to the end of July [1925] to assist Professor Born with the lectures. But, with your permission, I shall be able to make up for the time, which will be missing to complete a year [by spending the rest] with you during the summer vacation of 1925 (September or August)' (Heisenberg to Bohr, 14 July 1924). Bohr agreed with this proposal; as a result Heisenberg would spend several weeks in September and October 1925 in Copenhagen in order to fulfill the requirement of the International Education Board Fellowship.

Bohr and his collaborators also helped to solve the other problems connected with Heisenberg's stay in Copenhagen. The question of lodging was quickly settled, for as Bohr wrote already in July: 'Kramers has further arranged that you will have a room at Mrs. Maar's, who is looking forward with pleasure to your stay at her house' (Bohr to Heisenberg, 5 July 1924). Another minor problem concerned the visa. However, every detail was solved pleasantly and satisfactorily, and on the appointed date Heisenberg arrived in Copenhagen, now for the second time within a year and not a stranger anymore.

The arrangement with the accommodation was very convenient. Heisenberg stayed, together with the other visitors of Bohr's Institute, at the home of Mrs. Maar, the widow of a former university professor. She had a very nice house in Copenhagen and used to take in as boarders those young people from abroad who remained for a longer time in Copenhagen. She would also help the newcomers with language problems. It occurred to Heisenberg that, since he was going to live in Denmark for an extended period of time, he should learn Danish, and, since English was the common language at Bohr's Institute, he should learn it, too. First, he concentrated on Danish. For an hour or two every day after lunch, Mrs. Maar would talk to him in Danish and help him read the newspapers, and soon he acquired enough knowledge of it to be able to get along. Heisenberg took up the study of English as well. He had had some French at school—at the *Gymnasium* the emphasis had been on Greek and Latin—but no English. His mother had made an effort to teach him English, but without much success. He was able to read the scientific papers written in English with some difficulty, but that was about all. Now in Copenhagen he definitely wanted to improve on it. After a couple of months Bohr asked him to give a talk in the

Colloquium. Heisenberg expected that his talk would be in Danish. 'I was quite proud that I had now prepared a good talk,' he recalled. 'Just half an hour before the colloquium Bohr told me, "Well, it's obvious that we talk in English." I tried the best I could, but I think it was extremely poor' (Heisenberg, Conversations). However, with the help of Mrs. Maar, who was very kind to him, he learned after a while both Danish and English fairly well. Thus, for example, when he returned in spring 1926 to replace Hendrik Kramers as lecturer in Copenhagen, he gave his courses in Danish.

Heisenberg had arranged his arrival in Copenhagen in September 1924 so as to be available to Bohr during the period of Kramers' absence.[171] Unlike a German university's institute, which would be deserted during the summer vacation, Heisenberg did not find Bohr's Institute empty—though some of the visitors whom he had met in spring, e.g., John Slater, Harold Urey and Frank Hoyt, had left. On the other hand, Svein Rosseland was still around; so was Christian Møller, the only Danish theoretical physicist other than Bohr at the Institute. Among others at the Institute were the Danish experimentalists H. Marius Hansen and Sven Werner, who made important spectroscopic measurements; the Hungarian George de Hevesy, an old friend of Bohr's from the Manchester days, was more concerned with atomic chemistry. Ehrenfest's student G. H. Dieke came from Leyden for a short period in fall 1924. The American Ralph de Laer Kronig arrived in February 1925, at about the same time as Ralph Fowler from Cambridge, England. A little later David Dennison spent part of his International Education Board Fellowship in Copenhagen. Thus, Heisenberg could not really complain of being lonely at Bohr's Institute, as there were enough people to talk to. However, during the period from September 1924 to April 1925 he had the most contact with Niels Bohr and his closest collaborator Hendrik Kramers.

Hendrik Anthony Kramers had joined Bohr in Copenhagen already in fall 1916, while still a doctoral student of Ehrenfest's in Leyden. He had helped Bohr with his fundamental treatise 'On the Quantum Theory of Line Spectra' (Bohr, 1918a, b; 1922d). In particular, he had carried out detailed calculations on the intensities of spectral lines by applying Bohr's principles, which he submitted as his doctoral dissertation at the University of Leyden (Kramers, 1919). In 1916, together with Bohr, he had developed a theory of the helium atom; he had continued to perform calculations on this problem until he submitted a paper on it, containing negative results, in late 1922 (Kramers, 1923a). Just recently he had been involved in developing, with Bohr and the American visitor John Slater, the new radiation theory. Finally, he had derived the dispersion formula from this theory. Besides working with Bohr, Kramers had carried out some research of his own, either alone or in collaboration with others. For example, he had written papers on general relativity (Kramers, 1920a), on the quantum theory of mole-

[171] Kramers attended the 88th Congress of Scientists and Physicians (*Versammlung der Gesellschaft Deutscher Naturforscher und Ärzte*) in Innsbruck, 21–27 September 1924. He returned soon thereafter to Copenhagen, as Heisenberg mentioned his presence to Pauli in a letter, dated 30 September 1924.

cules (Kramers, 1923b; Kramers and Pauli, 1923), and on the theory of X-ray absorption and continuous X-ray spectra (Kramers, 1923d). He had acquired great expertise by propagating Bohr's work on atomic structure in many articles and talks in Denmark and Scandinavia, Holland and Germany. Kramers was indispensable not only with respect to the scientific work at Bohr's Institute; he had become Bohr's first assistant when the Institute for Theoretical Physics was founded in 1920 and had remained as his principal helper in all administrative matters. In addition, he had taken over most of Bohr's teaching duties, having been appointed '*Lector*' in 1922. Thus, next to Bohr, Kramers was by far the most important person at the Institute.

Though Kramers was his senior by only five years, Heisenberg was rather daunted by him when he first met him in 1924. He had expected that Kramers would play roughly the same role at Bohr's Institute, which Gregor Wentzel played in Munich, but he found that Kramers was much more. Heisenberg quickly discovered that, in many respects, Kramers was much superior to him. 'He knew, in general, things in life so very much more than I could ever hope to know,' Heisenberg admitted later (Heisenberg, Conversations, p. 216). Kramers appeared to him as having a very unusual combination of talents. On the one hand, he was a good sportsman and he could walk for many hours; on the other, he was intellectually most versatile. 'He spoke Dutch, Danish, German, French and English,' Heisenberg recalled. 'All these languages came out from him just as if they had been his mother tongue' (Heisenberg, Conversations, p. 215). Then Kramers played piano and cello; when Heisenberg and Kramers played together, Kramers would take over the cello part. 'How can a man know so much,' Heisenberg used to wonder. 'I had great difficulty in learning poor English and poor Danish, and before that time I just knew German. I could just play one instrument fairly well—that was the piano—nothing else. And I said to myself, "How can a man learn all these things?"' (Heisenberg, Conversations, p. 215). In addition, Kramers always seemed to be gay, charming and amusing. 'He could entertain the whole party when we were invited to the Bohrs' house . . . . He was always a perfect gentleman in every way,' noticed Heisenberg. 'He was the type of man which was far above my own reach and at the same time a type which was a bit strange to me' (Heisenberg, Conversations, p. 215). Thus, at first, Heisenberg had an 'enormous admiration' for Kramers, and at the same time he had difficulties in getting along with him, though Kramers treated him very nicely. Heisenberg even used to be angry with Kramers in the beginning because he did not take things so seriously as he thought they should be. 'He used to make jokes,' he recalled, 'when I didn't want to have any jokes made, just very nice jokes, certainly never any indecent ones, that was quite out of the question' (Heisenberg, Conversations, p. 215). Thus, it took him much longer to get accustomed to Kramers than to Bohr, and it was only after some time that he became friends with him.

Heisenberg was, of course, fully aware of what Kramers meant to Bohr's Institute. While Bohr was frequently occupied and did not have much time for people, Kramers always seemed to have time. Everybody first talked to Kramers before he talked to Bohr. And Kramers would listen to the ideas of younger people and try to help them by his friendly criticism. As Heisenberg remarked: 'Kramers contributed enormously to the whole development [of atomic theory at Bohr's Institute] just because of his physical strength. He was a man with an absolutely inexhaustible energy. He could work three days without sleep or anything. If Bohr wanted Kramers to calculate something, he would do it, and he certainly would do it very well' (Heisenberg, Conversations, pp. 217–218). Heisenberg knew that Kramers had done a great amount of work, and that his contributions to physics 'were very large, and perhaps it had not been seen so clearly only because he had always been together with Bohr and, so far, was a bit dependent on him' (Heisenberg, Conversations, p. 218). He was fully aware of the importance of Kramers' dispersion formula for the future quantum theory, but at the same time he was repelled by the fact that Kramers did not take the difficulties occurring in atomic theory as seriously as Bohr did.

It was evident that Heisenberg's feelings towards Kramers alternated between admiration and envious criticism. And his envy and criticism had, to some extent, a psychological background. In Munich and Göttingen, Heisenberg had had a privileged position; he had been the favourite student of both Sommerfeld and Born. In Copenhagen this position was occupied by Kramers; hence, subconsciously, he felt envious. This envy was enhanced by the superior qualities of Kramers, which Heisenberg noticed only too clearly. Although Kramers was considerably senior and held a deservedly established position, Heisenberg thought he had to compete with him.[172] Kramers, on the other hand, had no particular reason to take the young man too seriously; he did not regard him at all as a competitor, but more like a young visitor whom he could instruct. Certainly, in Heisenberg's opinion, he did not always estimate his (Heisenberg's) role enough. After all, Kramers had better things to do than to occupy himself with the psychological drives of an ambitious youngster; moreover, in 1924 he was quite preoccupied with the promotion of his own career. In any case, Heisenberg had difficulty in understanding him, and it took some time before he felt warm towards Kramers. 'It was certainly entirely my fault,' he remarked afterwards, 'because later on I liked him very much and thought that he was a very nice fellow, and everything went on excellently between us' (Heisenberg, Conversations, p. 215).[173]At that later time, however, Heisenberg had already found a firm place among the Copenhagen physicists.

Physics at Copenhagen was completely dominated by the personality of Niels Bohr. Heisenberg soon realized that the style of working was very different from

[172]A slight indication of Heisenberg's sense of competition with Kramers may be discerned from the fact that he proposed to arrive in Copenhagen in September 1924 just when Kramers would be away for a few days. Thus, Kramers' absence would give Heisenberg the opportunity of being alone with Bohr, thereby getting into closer contact with him without being disturbed.

[173]The difficulties, which Heisenberg had with Kramers in the beginning of his stay in Copenhagen, did not prevent him from immediately pursuing the discussion of scientific topics, for as he reported to Pauli: 'Kramers says just now that I should write to you, that one can discuss things much better with me than with you' (Heisenberg to Pauli, 30 September 1924).

the places he had known. 'For me, it was more exciting than Munich or Göttingen,' he remarked (Heisenberg, Conversations, p. 223). The young theoreticians could learn a lot from Kramers and, in spite of his preoccupation with many things, especially from Bohr. Heisenberg attributed Bohr's influence to the latter's great concern about the situation of quantum theory. He noticed that

> Bohr was more worried than anybody else about the inconsistencies of quantum theory. Neither Sommerfeld nor Born had been so much worried about things. Sommerfeld was quite happy when he could apply nice complex integrals [to solve problems], and he did not worry too much whether [his] approach was consistent or not. And Born, in a different way, was also interested mostly in mathematical problems. Inconsistencies were realized, but, after all, neither Born nor Sommerfeld really suffered [because of them], while Bohr couldn't talk anything else. (Heisenberg, Conversations, p. 207; also AHQP Interview)

This attitude was always with him, for as Heisenberg recalled:

> Whenever one went out for a walk with Bohr—sometimes I was invited to his country house [in Tisvilde] and went for long walks with him—he would always discuss these difficulties and what one could do about them, etc. In some way he suffered from the impossibility of penetrating into this very "*unanschaulich*" [unvisualizable], unreasonable behaviour of nature. The strongest impression which I obtained during the first few months in Copenhagen was just this: in discussions with Bohr I came to realize how terrible the situation was [in quantum theory] and how unavoidable the contradictions seemed to be. I realized how difficult it was to reconcile the results of one experiment with those of another. (Heisenberg, Conversations, p. 207; also AHQP Interview)

At that time in Copenhagen one spoke often about *Gedankenexperimente* (thought experiments), in which the particular contradictions of quantum theory would become obvious. One such *Gedankenexperiment* was the application of crossed electric and magnetic fields to the hydrogen atom, the difficulties arising from which in atomic theory had been discussed earlier by Oskar Klein (1924a, b) and Wilhelm Lenz (1924). However, one would seek to invent many more *Gedankenexperimente*, with the hope of being able to translate them into real experiments and then let nature decide about the disagreeable paradoxes of the theory.

The fact that someone should suffer physically and mentally from the problems confronting quantum theory was a new experience for Heisenberg; he himself had not suffered from such an affliction. Of course, like other people, he had had personal problems, and some of them had even caused him anguish, but his work in physics, even in the face of serious difficulties, had always challenged and exhilarated him. But in Munich and Göttingen he had not experienced such a sense of 'crisis' in quantum theory.

> Therefore, to get into the spirit of quantum theory was only possible in Copenhagen at that time. Of course, in other places one spoke about these things, too, but Bohr's

> attitude was that he would never stop before achieving utmost clarity. Bohr would follow things to the very end, to the point where he hit the wall. I very soon realized that there was nobody who had thought so deeply about the problems of quantum theory as Niels Bohr. That made a very strong impression on me. (Heisenberg, Conversations, p. 208; also AHQP Interview)

This impression was the greater, for Heisenberg, in fall 1924, was completely prepared and willing to accept Bohr's serious and rigorous attitude towards the problems of quantum theory, something which he would not have been ready to do a year or two earlier. The unsatisfactory results which he had obtained on the helium problem up to now (e.g., Born and Heisenberg, 1923b), and on various other questions of atomic theory which he had tackled, had made him receptive to Bohr's insistence upon facing the fundamental problems of quantum theory *all at once*. Moreover, he found Bohr very pleasant personally. 'I liked him very much,' he said later. 'We became very good friends, and we did many things together out in the country. It was an entirely new life' (Heisenberg, Conversations, pp. 208–209).

In many ways Bohr behaved differently from Born or Sommerfeld. He did not talk as much to the students as Sommerfeld did. 'It was only occasionally that he would come into the library to ask about something or appear to be excited. But when the discussions took place they dealt with special points that had to be solved and were mostly very critical' (Heisenberg, Conversations, p. 208). Then, for example in the discussions on resonance fluorescence or on the extension of Kramers' dispersion formula, one had to go through all the points with extreme care. Bohr applied the same care in writing papers or letters, which he would usually dictate to somebody. Heisenberg observed the following procedure:

> He would walk around in the room and dictate, and I would try to put it down on paper. In composing a paper Bohr would always change sentences again and again. Thus half a page could have been filled with a few sentences, and then everything was crossed out and changed again. And even when the whole paper was almost finished—say ten pages or so—the next day everything would be changed over again. So it was a continuous process of improvement, of change, and of discussions with others. (Heisenberg, Conversations, p. 209)[174]

This extreme care in formulating a paper was quite new to Heisenberg, and it was different from how Born and Sommerfeld wrote their papers. Sommerfeld,

[174]Bohr had developed the habit of dictating his papers to others very early; thus, he dictated his first (master's) thesis to his mother, his doctoral dissertation to his fiancée, and then his papers and letters to a succession of collaborators. Once Paul Dirac happened to be there when Bohr was composing a paper. Bohr paced as he dictated, went back and corrected, and so it went on for a long time. Then Dirac, during an impressive pause, said: 'Professor Bohr, when I was at school, my teacher taught me not to begin a sentence until I knew how to finish it.'

Dirac once said to Heisenberg: 'Bohr should have been a poet.' Heisenberg asked: 'Why a poet?' Dirac answered: 'He just takes too much trouble with the language, and he always improves the language. He should have written poetry' (Heisenberg, Conversations, p. 210).

for example, would first discuss all the essential parts of the paper, then he would just sit down and write it up. 'Of course, I would be allowed to make remarks and perhaps suggest improvements,' Heisenberg recalled, 'but it was written once and that was all' (Heisenberg, Conversations, p. 210).

Bohr's procedure did not necessarily contribute to making his papers easy to read. As Heisenberg remarked:

> The final text of Bohr's papers was so subtle, as he would think about half an hour whether in a certain case he would use the indicative or the subjunctive, and so on. The reader would just read over it and would not realize how much work was put into it. But for Bohr himself, that was quite different, because the writing of a paper was a process in which he would clarify his own mind. For him it was extremely important to make these changes and thereby get deeper and deeper into the problem. (Heisenberg, Conversations, p. 210; also AHQP Interview)

Heisenberg believed that Bohr's enormous influence on several generations of physicists arose from this habit of doing things, even composing his papers, together with others. Naturally, people to whom he dictated were obliged to spend a lot of time with Bohr, but Heisenberg never felt that this was a waste of time for him. 'On the contrary, I felt it was extremely agreeable to have this conversation and thereby clarify one's own mind. After I had written a paper or letter for Bohr, I always had the impression that I had learned something which I could use for my own work. And somehow I never felt that I had too little time for my own work. I always found time' (Heisenberg, Conversations, p. 212). Assisting Bohr in the writing of his letters and papers gave Heisenberg the opportunity of being frequently together with him during winter 1924–1925, and again in 1926 when he replaced Kramers as '*Lector*' in Copenhagen.[175]

Bohr was often occupied with the administration of the Institute, and this cut down the time available for talking with his research collaborators. However, in administrative matters also he exercised his usual care. For instance, once two mechanics in the workshop could not agree with each other, and it took him several days to smooth over the situation. The construction of the new parts of

[175] At the time of intense collaboration with Heisenberg, especially in 1926, Bohr would often come to Heisenberg's office in the morning and ask: 'Couldn't we write a few letters?' After he had written the letters, he would perhaps become involved in a paper and say: 'Could we not try to get a bit further in the paper?' Thus, Bohr and Heisenberg would sit together for two or three hours. But then, in the afternoon, Heisenberg would be mostly free. He would go to have his meal with Mrs. Maar, or just make some small repast for himself in his room upstairs in the Institute. In the evening, after 8 or 9 o'clock, all of a sudden Bohr might come up and say: 'Heisenberg, what *do* you think about this problem?' And then he would start talking; frequently they would go on up to 12 or 1 o'clock at night. 'Or sometimes,' recalled Heisenberg, 'he would call me to his flat near the Institute, and finally at 1 o'clock at night we would feel that we were tired and would take a glass of port wine and then go to bed' (Heisenberg, Conversations, p. 213). There was, however, no regularity in these meetings, for Bohr was not a man of fixed habits. In addition to these encounters, lectures had to be given, visitors had to be attended to, etc.

the Institute also occupied him very much. Bohr carried on an extensive scientific correspondence and spent much time on it. Writing scientific letters represented a way of doing physics. As Heisenberg recalled: 'He would ask me to help him in writing a letter, say to Fowler in Cambridge. And then, in the course of writing the letter he would discuss the problem with me and the whole problem would be clarified. So it [the writing of letters] was a mixture of physics and administration' (Heisenberg, Conversations, p. 217).[176] But sometimes Heisenberg found that Bohr was wasting his time in answering letters, such as to the man who claimed to have invented a perpetual motion machine.[177]

Heisenberg soon felt at home in Bohr's Institute. 'Here in Copenhagen I am very well, as you might imagine,' he wrote to Sommerfeld. 'However, I had to learn two languages in the first two months, English and Danish, and this is a little too much at once, but I manage somehow; there is, of course, no question of doing "well" [in this regard]' (Heisenberg to Sommerfeld, 18 November 1924). Though he was for the first time in a foreign country for a long period, he had little occasion to feel lonely, as the scientific work with Bohr and Kramers absorbed him completely.[178] The discussions of the problems of atomic theory really excited him; he especially liked to discuss with Bohr, 'because with him one got into the really desperate problems, into the discussion of those dreadful difficulties, and he would get to their bottom while Kramers would not take these difficulties so seriously as Bohr did' (Heisenberg, Conversations, p. 216).[179] In Copenhagen, Heisenberg became familiar with the full range of atomic problems —not just the specific problems like the anomalous Zeeman effects or the helium question. Bohr had already explained to him during his spring visit that all these problems were internally connected and had to be treated together. Thus, for instance, one could not separate the problem of the nature of radiation from the problem of describing many-electron atoms. Hence the Bohr–Kramers–Slater theory would also have something to do with such questions as the calculation of the states of helium; indeed, Kramers' dispersion theory was a first step in this direction. Heisenberg was aware that serious objections had been raised to the Copenhagen radiation theory by very competent critics, especially Einstein and

[176] Bohr made use of Heisenberg's help in writing letters in his first visit to Copenhagen. (See Bohr to Pauli, 24 March 1924.)

[177] This incident happened in 1926. A man had invented a perpetual motion machine, which violated the second law of thermodynamics. He sent a paper to Bohr, and Bohr explained in a letter why it was wrong. Klein and Heisenberg suggested to Bohr not to explain too much, otherwise the other would rewrite his paper and send it again. This is what he did, and it took Bohr a longer time to find the mistake. Bohr did not follow Klein and Heisenberg's advice not to answer this time. Thus, the story went on until Bohr finally refrained from discussing the seventh version of the perpetual motion machine (Heisenberg, Conversations, pp. 224–225; also AHQP Interview).

[178] Heisenberg wrote to Pauli on 8 October 1924: 'Otherwise I am quite well, though sometimes I have a little longing for Germany.'

[179] It is, of course, not so easy to decide whether Kramers did not feel the seriousness of the difficulties in atomic theory as much as Bohr. In any case, he did not reveal such feelings to Heisenberg.

Pauli.[180] Still he continued to share the general feeling that prevailed in Bohr's circle: that this theory offered the only way out of the inconsistencies between the classical wave description and the light-quantum description of light. Therefore, even the nonconservation of energy and momentum in individual atomic processes of the interaction of radiation and matter seemed to be unavoidable, though in Copenhagen also one doubted whether it was true. One had the impression that 'these things are perhaps still deeper than we imagined, and perhaps nature can even manage to conserve energy in spite of all that we have understood' (Heisenberg, Conversations, p. 219). Indeed, the objections of Einstein and Pauli were taken very seriously in Copenhagen, and the possibility was not excluded that in the radiation theory one had reached a point which, perhaps, one should not seek to describe at all. However, in spite of these considerations, people were convinced that the ideas of Bohr, Kramers and Slater were somehow on the right track. After all, the only way to deal with the problems of atomic theory was to apply the correspondence principle, and the Bohr–Kramers–Slater theory was exactly an attempt to describe radiation from atoms in agreement with the latter. Therefore, in Bohr's Institute, everyone, including Heisenberg, used this theory, or some aspects of it, whenever one had to treat radiation from atoms.

Heisenberg had decided in spring 1924 to follow Niels Bohr. Now he behaved in the same way as he had done in Munich and Göttingen: he followed his new professor completely and without reservation. He announced his turnabout to Bohr's views in a letter to Sommerfeld. He wrote:

> Besides I believe increasingly that the question of "light-quanta or correspondence principle" is merely a question of words. All effects in quantum theory must, of course, have an analogy in classical theory, since the classical theory is *almost* correct; hence [physical] effects always have two names, a classical and a quantum-theoretical one, and it is a matter of taste which one prefers. Perhaps Bohr's radiation theory is a very felicitous description of this dualism; I am expecting with great excitement the result of the Bothe–Geiger experiment. (Heisenberg to Sommerfeld, 18 November 1924)

With these words Heisenberg declared openly that from now on he was marching entirely in the footsteps of Niels Bohr.

III.4 Ad Majorem Correspondentiae Principii Gloriam

In summer 1924 the discussion about the theory of atomic structure attained a new level. On one hand, a new class of experimental data on the intensities of

[180] Heisenberg had had the opportunity of listening to Einstein's criticism from Einstein himself, during the latter's visit to Göttingen on 7 June 1924. On the other hand, Pauli wrote a long letter to Bohr on 2 October 1924, in which he systematically outlined both Einstein's arguments—which he came to know at the *Naturforscherversammlung* in Innsbruck—and his own against the Bohr–Kramers–Slater theory of radiation. Heisenberg learned the content of this letter, as Pauli's letters were openly discussed in Copenhagen.

spectral lines, especially of line multiplets, became available through the work of Leonard Salomon Ornstein and his students at the Physics Institute of the University of Utrecht (Dorgelo, 1924; Burger and Dorgelo, 1924; Ornstein and Burger, 1924d, f, g); on the other hand, the leading personalities of spectroscopic theory, Niels Bohr and Arnold Sommerfeld, felt the necessity of trying out a new, improved approach to deal with the new material: Sommerfeld was preparing a new edition of his *Atombau und Spektrallinien*, which would replace the third edition of 1922, and Bohr thought of rewriting the second part of his comprehensive essay on the application of quantum theory to atomic structure. In trying to elaborate on the new description, the two senior leaders of atomic theory pursued quite different approaches; they even disagreed about matters of principle, and this disagreement could not be removed in spite of a detailed exchange of opinions between Copenhagen and Munich.

During his visit to Leyden in spring 1924, Sommerfeld met Kramers, and the latter learned that Sommerfeld would give the principal talk on the foundations of quantum theory and Bohr's atomic model at the Innsbruck meeting of scientists and physicians (*Versammlung der Gesellschaft Deutscher Naturforscher und Ärzte*) the following September. On his return to Copenhagen, Kramers wrote a letter to Sommerfeld on 4 June 1924, emphasizing certain aspects of the theory of atomic structure that were important from the Copenhagen point of view. In particular, he argued in favour of an approach to atomic systems similar to the one that Bohr had taken in the case of the hydrogen atom, that is, first to develop an approximate theory in which only one quantum number entered, and then to treat the finer details as small perturbations. In his reply to Kramers, Sommerfeld stressed the alternative point of view—the same which he had taken in his first work on atomic structure in 1915; namely, that a system of several degrees of freedom had to be treated by using the phase integrals, and that in solving the equations of motion one should prefer the method of the separation of variables over the perturbation method. In the same letter, Sommerfeld also mentioned another point of disagreement that appeared to him to be more serious. He claimed that in Copenhagen the role of the correspondence principle in solving the problems of atomic structure had been greatly overestimated. 'The final view should be,' he wrote to Kramers, 'that the correspondence principle is a (highly valuable) limiting *theorem* of quantum theory, but not its *foundation* [fundamental principle]' (Sommerfeld to Kramers, 5 July 1924). He added that the intensity measurements at Utrecht supported his view. A month later Sommerfeld sent a preliminary manuscript, prepared for his talk at Innsbruck, to Copenhagen, and invited Kramers' criticism. Kramers, in his answer, tried to explain and defend the Copenhagen use of the correspondence principle as one of the principal features of quantum theory, which had proved to be valuable in the description of data. He said:

> Bohr is far from considering the correspondence principle as a foundation for an axiomatic formulation of quantum theory. Bohr's formulation of the principle is, of course, everywhere tentative and cautious, and it would, to say the least, be too

> early to conclude a "failure" or "inadequacy" of the correspondence principle from the beautiful intensity measurements at Utrecht. Rather, it is like this: as far as we have been able to discern until now, the correspondence principle does not enable us to give preference to any conjecture concerning the intensities of multiplets. We are confronted here with a problem in which the theory for determining the stationary states essentially fails; however, one may perhaps hope that the general correspondence point of view might offer, with the help of experiments, a hint for removing this difficulty. (Kramers to Sommerfeld, 6 September 1924)

Kramers' letter defending the correspondence principle did not reach Sommerfeld before he went to Innsbruck. Nevertheless, in his talk he somewhat softened his disagreement with Copenhagen; he declared: 'A correspondence-like treatment of the intensity problems provides only approximate values in a manner, which seems to be only little suited to the arithmetical simplicity of the data' (Sommerfeld, 1924c, p. 1048). And he took the same point of view in the fourth edition of *Atombau und Spektrallinien*, a copy of which he presented soon afterwards to Bohr.[181] In spite of the diplomatic manner in which Sommerfeld expressed his criticism of the correspondence principle in 1924, it became evident that the Copenhagen and Munich paths had diverged.

It is not that Bohr did not recognize the importance and implications of the Utrecht intensity measurements. Thus, he expressed his great interest in the results already in a letter to Ornstein, dated 5 July 1924. Having been occupied for several months with the question of statistical weights in atomic theory, he expected to derive certain helpful hints from the new empirical data.[182] The question of statistical weights was indeed a major problem to be solved, especially because certain difficulties, which had arisen in the discussion of the anomalous Zeeman effect data in light of the building-up principle (*Aufbauprinzip*), had to be removed. Bohr hoped to receive assistance in this matter from Heisenberg, an expert on the theory of anomalous Zeeman effects, for he wrote to Heisenberg: 'I intend to postpone the publication of my major review article [containing also a discussion of statistical weights] until you come here in September' (Bohr to Heisenberg, 18 June 1924). Sommerfeld's arguments against the correspondence principle in connection with the intensity measurements of Ornstein, Burger and Dorgelo had to be checked first. Hence Heisenberg, upon arrival in Copenhagen on 17 September 1924, plunged immediately into discussions with Bohr. A couple of weeks of intensive work followed, then Heisenberg wrote to Pauli: 'Together with Bohr I have again examined the problem carefully, and we arrived at the conclusion that it is not—as Sommerfeld says—that the sum-rules cannot be understood with the help of the correspondence principle; on the contrary they are a *necessary consequence* of the correspondence principle; they are indeed the most beautiful example of the fact that the

[181] See *Atombau und Spektrallinien* (1924d), Chapter Eight, pp. 657–658.

[182] That Bohr was working on the problem of statistical weights in spring 1924 is evident from a letter of Heisenberg. The latter wrote to Bohr: 'I am very curious to hear . . . how far your work on the quantum theory of statistical weights, etc., has progressed' (Heisenberg to Bohr, 15 May 1924).

correspondence principle occasionally yields unique conclusions' (Heisenberg to Pauli, 30 September 1924). Heisenberg quickly outlined the proof to Pauli. The main point was that the total intensities of all spectral lines emitted from the one-electron atom did not depend on the angle of the electron's angular momentum with the total angular momentum of the atom, *both* in classical and quantum theory; only the distribution of the total intensity among the multiplet's components would depend on the angle in question. Thus, it was possible to describe the intensity rules of complex multiplets by means of appropriate assumptions. 'We are very happy about this interpretation,' Heisenberg commented, 'for now the attacks against the correspondence principle are completely refuted.' He knew that Pauli had arrived at a similar conclusion, hence he added:

> But I am convinced that you have since long thought about all this in the same way, according to what you have written. If this is so, then you would do me a great favour by publishing your considerations, say in a short note in *Naturwissenschaften*. It is really simple, almost trivial. However, since recently the correspondence principle has been blamed so much, it would be good to publish it [i.e., the conclusion confirming the correspondence principle] "*ad majorem correspondentiae principii gloriam*" ["to the greater glory of the correspondence principle"]. (Heisenberg to Pauli, 30 September 1924)

Pauli had indeed come to very similar conclusions at about the same time and repeated them to Sommerfeld in Munich and Heisenberg in Copenhagen.[183] He was, however, not as completely satisfied with the results of the application of the correspondence principle as Bohr and Heisenberg were. Thus, he wrote to Sommerfeld: 'There is thus *very little* that can be concluded about the line intensities from the correspondence principle. This little, however, I must hold on to as certain' (Pauli to Sommerfeld, 29 September 1924). In particular, he had concluded that Ornstein's sum-rules satisfied the correspondence principle, but could not be derived from it. 'The correspondence principle demands only,' Pauli argued, 'that these intensity sums [i.e., the classical and quantum-theoretical ones] must become asymptotically equal for large k [where k denotes the angular momentum of the series electron], and one recognizes easily that this is also really true according to Ornstein' (Pauli to Sommerfeld, 29 September 1924). Then he pointed out a difficulty for the correspondence principle, presented by Ornstein's results for the weak line components arising from transitions with $\Delta j = -1$ (j being the inner quantum number or the quantum number of the total angular momentum of the atom): the data indicated that the intensities behaved as k^{-2} for large k, while the correspondence result was k^{-4}, k being the

[183] Pauli's letter of September 1924 to Heisenberg was lost, like almost all other letters to Heisenberg of this period. However, in his letter of 30 September, Heisenberg thanked Pauli for his letter that he had just received. In this letter Pauli also proposed detailed intensity formulae, which deviated slightly from the data of Ornstein and collaborators—and Heisenberg took note of these in his reply (Heisenberg to Pauli, 30 September 1924). All the results, which Pauli reported to Heisenberg, were also contained in Pauli's letter to Sommerfeld, dated 29 September 1924.

series electron's angular momentum. He also suggested formulae for the intensity ratios, which followed from the correspondence principle, and which differed slightly from the ones given by Ornstein and collaborators.

Although the Copenhagen physicists were happy about Pauli's procedure and results, they noted that he had not entirely followed Bohr's line. Heisenberg tried to convert Pauli by explaining the Copenhagen point of view in detail:

> We mean the following. The physical significance of Ornstein's rule in classical theory is evidently this: the total intensity $I_{+1} + I_0 + I_{-1}$ [where $I_{\pm 1}, I_0$, are the intensities of the components arising from the transitions $\Delta j = \pm 1, 0$] of the line depends, of course, only on the electron's orbit, but not its position with respect to the [atomic] core . . . In classical theory this is automatically satisfied, because the energy radiated away per second is determined *uniquely* by the orbit of the electron (apart from corrections having the order of magnitude of the coupling energy). The theorem is valid, therefore, already for the *single* atom. If one now goes to quantum theory, then it agrees totally with the spirit of every radiation theory if one says that the total quantum-theoretical probability for the transition [jump] is uniquely determined by the (virtual) electronic orbit and the corresponding radiated (virtual) energy [loss]; if one does not say that, then one cannot give any definition of the lifetime [of atoms], etc. If, however, one makes this assumption [of correspondence]—this assumption is, of course, not absolutely necessary, though it is a rather obvious analogy to the classical theory (and I could not imagine any theory that does not include this assumption, i.e., the correspondence principle)—then one arrives at Ornstein's rule . . . Therefore I mean: if one understands, as you [stated], the correspondence principle to signify the wrong assertion that one may derive the quantum-theoretical intensity by averaging over the classical one, then *you* are right, and in that case one *cannot* obtain Ornstein's rule through correspondence principle; if, however, one understands it as signifying an analogous logical connection to the classical theory, then *I* am right. (Heisenberg to Pauli, 8 October 1924)

Heisenberg hoped that Pauli would agree with his conclusions, and he proceeded to make further applications of the correspondence principle in the above-mentioned sense.

For one who, less than two years earlier, had accused Pauli of being '*verbohrt*' ('Bohrian'), Heisenberg's conversion to the Copenhagen faith came about amazingly quickly. The hard work for the greater glory of the correspondence principle probably caused him some feeling of uneasiness in relation to his erstwhile teacher, Sommerfeld, but he disposed of it by writing to him diplomatically that the differences between Copenhagen and Munich approaches could be considered merely as a 'quarrel about words' (Heisenberg to Sommerfeld, 18 November 1924). Heisenberg's new attitude was quite remarkable in view of his earlier close association with Sommerfeld. But in 1924 his attitude had changed fundamentally, for Heisenberg had become convinced that one had to follow Bohr's guidance in its entirety, both with respect to the methods as with respect to the spirit in which they had to be applied. It was a part of this spirit that the question was not so much whether the correspondence principle was right or wrong, or whether the known facts were entirely in agreement with it or sometimes contradicted it; the question for Heisenberg, in particular, became whether one could transform Bohr's method of 'careful groping,' characterized by the correspondence principle, into a precise mathematical device for attacking directly the problems of atomic physics and obtaining unique quantitative results. In this endeavour, the training which he had received from Sommerfeld in solving many specific problems would prove to be of inestimable help. He already considered the demonstration, arrived at in discussion with Bohr, that the intensity sum-rules of the Utrecht spectroscopists followed from the correspondence principle as 'the simplest example of [the latter's] sharpening' (Heisenberg, 1925a, p. 618). He would now use this sharpening of the correspondence principle as the new approach to many problems of atomic theory; he submitted the first results based on this method in a paper to *Zeitschrift für Physik* a little more than two months after his arrival in Copenhagen (Heisenberg, 1925a).

In this paper entitled '*Über eine Anwendung des Korrespondenzprinzips auf die Frage nach der Polarisation des Fluoreszenzlichtes*' ('On an Application of the Correspondence Principle to the Problem of the Polarization of Fluorescence Light'), Heisenberg treated several questions, the connecting link being the new procedure for handling the problems of atomic theory. He had already applied this procedure less consciously in his previous work on the anomalous Zeeman effects. The procedure consisted of the following: Heisenberg would first select an '*Ersatz*' (suitable 'substitute') model to describe the atomic system under consideration; then he would discuss the classical theory of the problem, and finally the quantum theory. The only difference between his earlier approach and this one was that now, under Bohr's influence in Copenhagen, he insisted meticulously on the correspondence and analogy of the classical and quantum-theoretical results. In the particular problem of explaining the intensity rules in multiplet spectra, for instance, he again used his favourite *Ersatz* model of the anomalous Zeeman effects. That is, he assumed that the multiplet spectra of complex atoms could be adequately described by referring to the following atomic model: the atom consists of an outer electron with angular momentum k, moving in an orbit with principal quantum number n; the orbit performs a precession with angular velocity ω_j around the axis of j, the total angular momentum of the atom; by this precession each emitted spectral line is split into three components (for doublet spectra the component corresponding to the transition $\Delta j = -1$ is missing), one being polarized parallel to the total angular momentum vector, the other two circularly polarized in opposite directions, with electric vectors (of the emitted radiation) remaining in planes perpendicular to the direction of j. In classical theory the total energy emitted by the electron is determined by the properties of the orbit. Hence, if one starts with the orbit in the absence of precession and slowly turns on the precession, the relative change in the motion of the electron is given by the ratio of ω_j to the precession frequency ω, which is a small quantity in general. It follows immediately that the total intensity of the classical line triplet deviates from the intensity of the single line in the absence of precession just by a term having the order of magnitude

ω_j/ω. However, in classical theory, the distribution of the total energy among the individual components depends strongly on the angle between the directions of the angular momenta of the electron (k) and the core (r), respectively.

In quantum theory the situation seemed to be completely different. In particular, one noticed that the system without the precession was degenerate; hence a difficulty arose in the application of the quantum conditions, a difficulty about the resolution of which opinions disagreed in Copenhagen and Munich. Heisenberg now proposed to adhere to a strict analogy with the classical result; that is, he assumed that in quantum theory also the total intensity should be determined only by the properties of the orbit, and the orbit should be influenced by the precession in roughly the same way as in the classical model. This procedure, which he called a 'logical sharpening of the correspondence principle' ('*sinngemässe Verschärfung des Korrespondenzprinzips*,' Heisenberg, 1925a, p. 617), allowed him to connect the degenerate and the nondegenerate system in quantum theory. Especially, he concluded that the total energy emitted in the line triplet (of the nondegenerate system) deviated from the energy of the single line (of the degenerate system) only by orders of magnitude of ω_j/ω. Thus he arrived at a complete explanation of the empirical observations: first, that the sum of the intensities of all spectral lines corresponding to transitions from a multiplet level is proportional to the statistical weight associated with j, the angular momentum of the atom; second, that in the presence of an external magnetic field the total intensity of one complex multiplet does not depend on the magnetic quantum number m, but deviates from the intensity of the multiplet in the case of no external field by orders of ω_L/ω_j (with ω_L the Larmor frequency) only. The latter consequence represents an adequate description of the observed spectroscopic stability, notably the fact that the properties of spectral lines do not change qualitatively when small external fields are applied to free atoms.

As a further example of the sharpening of the correspondence principle Heisenberg studied the polarization of the components arising from an unpolarized spectral line, emitted by an atom in the presence of external electric or magnetic fields. He argued that if the orbits of the series electron were not drastically changed, then the polarization of the components in classical theory would add up to zero, that is, the sum of the intensities of the components polarized perpendicularly (to the direction of the external field) would equal the sum of the intensities of the parallel polarized components. In quantum theory such a result appeared to be rather unlikely, for existence of the external fields provides a preferred direction to atomic systems, regardless of how small the field strength might be; and a preferred polarization would seem to follow automatically. 'Nevertheless, we have every reason to assume,' Heisenberg noted, 'that this polarization does not exist; on the contrary, the quantum-theoretical virtual oscillators that determine the radiation obey laws according to which the closest analogy between the classical and the quantum theory is retained' (Heisenberg, 1925a, p. 621). Heisenberg was perfectly aware that this last application of the correspondence principle was more hypothetical than the previous ones, for it implied the statistical averaging over the virtual oscillator fields of many atoms, but he quoted an observation in support of his conclusion: the absence of double refraction in gases which were exposed to external electric and magnetic fields.[184]

With these applications Heisenberg not only succeeded in explaining the empirical intensity sum-rules, his immediate goal, but also approached one of the principal problems of atomic theory of the day: namely, a consistent treatment of degenerate systems. The question of how to quantize such systems had been debated in summer 1924 by Sommerfeld and Kramers without reaching agreement. Heisenberg now showed that, at least in the problem of complex multiplets and their anomalous Zeeman effects, his application of the sharpened correspondence principle provided an answer to this fundamental question. And, in his paper, he proceeded to treat another such problem: the problem of fluorescent light emitted from atoms, both in the presence and absence of magnetic fields, the former case representing a nondegenerate system and the latter a degenerate one. Again he showed how the quantization rules for the degenerate system could be derived by starting from the nondegenerate case.

In fall 1924 a considerable amount of empirical information was available on the resonance fluorescence of atoms. In 1923 Robert W. Wood and Alexander Ellett had studied the resonance radiation from mercury and sodium atoms, stimulated by linearly polarized incident radiation; they had observed a strong polarization of the emitted light, which was produced by weak, but properly oriented, magnetic fields; they had also noted that this polarization of light could be destroyed by a magnetic field (in a certain direction), which, in the case of mercury vapour polarization, had a strength of less than 1 G (Wood and Ellett, 1923). Wilhelm Hanle in Göttingen had then suggested that the effect observed by Wood and Ellett was not a new magnetic phenomenon, but was related to the anomalous Zeeman effect (Hanle, 1923). During 1924 several physicists had tried to analyze the data on the basis of Hanle's suggestion, among them Peter Pringsheim (1924a, b), Georg Joos (1924), Gregory Breit (1924a) and John A. Eldridge (1924b). Fritz Weigert, on the other hand, had assumed a different origin of the polarization of fluorescence radiation, namely, due to the collisions of 'second' kind (Weigert, 1924). He argued that the incident radiation forces the atoms to take a preferred direction parallel to the direction of polarization of radiation, and the atoms oriented in parallel undergo collisions of the 'second' kind with the free electrons (in which they transfer to them a discrete amount of energy corresponding to the transitions between two stationary states; some free electrons must be present in each sample of atoms that is in thermal equilibrium); the free electrons, which 'remember' the polarization of the incident radiation, then collide with atoms (in collisions of the 'first' kind) and give rise to resonance radiation having the same polarization as the incident radiation. While the observations on resonance fluorescence were easily described by the classical theory of the Zeeman effect, i.e., by assuming isotropically bound series electrons, it seemed impossible to treat them on the basis of the Bohr–Sommerfeld

[184]The argument was as follows: If the intensities of the parallel and perpendicular components were different, they would also have different refractive indices and a double refraction would arise.

theory of atomic structure. This was due to the fact that in the latter the absorption of radiation and its reemission were quite different processes; the emitted radiation, in particular, depended on the structure of the atoms.

Since the treatment of the polarization of resonance fluorescence light in atomic theory caused difficulties, Bohr naturally became interested in the question. He hoped that the new radiation theory of Bohr, Kramers and Slater, in which the emission and absorption processes were treated more symmetrically, would allow one to remove the difficulties. Bohr had the advantage of receiving information prior to publication from James Franck's experimental institute in Göttingen, where the Wood–Ellett effect had been investigated. Thus, for instance, Heisenberg had already informed Bohr: 'Please tell Mr. Kramers that [Wilhelm] Hanle (a doctoral student here) has proved experimentally that his rotational effect is *not* the Faraday effect; hence one must, I believe, assume that even in the case of fluorescence there definitely must exist phase-relations [between the incident and the emitted radiation]' (Heisenberg to Bohr, 5 July 1924). Wilhelm Hanle, James Franck's student, had investigated in detail the Wood–Ellett effect in the resonance line of mercury at 2536 Å for his doctoral dissertation; in particular, he investigated the change in the plane of polarization for small magnetic fields. He found that his observation agreed with the prediction from the classical theory of the Zeeman effect; hence he could exclude Kramers' suggestion that the Wood–Ellett effect was a Faraday effect.[185] A few months later Hanle's work was completed, and Franck sent a copy of the manuscript of his paper to Copenhagen requesting Bohr to check the calculations (Franck to Bohr, 8 September 1924). The main difficulty that arose from Hanle's results for the quantum-theoretical explanation was clearly noted in the paper (Hanle, 1924). In quantum theory it had been assumed that each line component was emitted by a different atom; that is, it was due to the atom's internal structure, especially to the position (space quantization) of the orbit of the series electron. Now, in order to account for the effect observed by Hanle, i.e., in order to explain the turning of the polarization of resonance light in external magnetic fields, one had to assume that the radiation emitted by two distant atoms was in

[185] Hanle had presented certain results already on 22 June 1924 at the meeting of the Lower Saxony Section (*Gauverein Niedersachsen*) of the German Physical Society in Hamburg. (See *Verh. d. Deutsch. Phys. Ges.* (3) *5* (1924) p. 40.) Kramers, who attended this meeting, suggested that a test should be made whether Hanle's effect—i.e., the turning of the plane of polarization—had anything to do with the Faraday effect. Hanle proved afterwards that Kramers' suggestion was not correct, for the plane of polarization of the resonance fluorescence light was turned in the same sense as the Larmor precession in the Zeeman effect, opposite to the direction of the change in the Faraday effect. Moreover, the Faraday effect would increase with the length of the distance traversed by the light, while the Hanle effect occurred already with full strength in a small sample of fluorescent mercury vapour.

Wilhelm Hanle was born on 13 January 1901 at Mannheim and studied physics at the Universities of Heidelberg (under P. Lenard) and Göttingen, receiving his doctorate in 1924 (under J. Franck). He worked as an assistant at Göttingen (1924), Tübingen (1925) and Halle (1926). He became an Extraordinary Professor at the University of Jena in 1929; from there he moved to Leipzig (1935) and Göttingen (1937) in a similar position. In 1941 he was appointed Director of the Physics Institute of the University of Giessen.

phase and could superimpose; in classical theory this assumption presented no difficulty, but it did so in quantum theory. Hanle concluded in his paper:

> One cannot at all consider this [phase-relation]. Thus there remains only the assumption that for small amplitudes of the field there exists neither the space [directional]-quantization, nor the energy separation of the excited atoms into two terms. We rather arrive at the conclusion that one is forced to assume a special transition from the [originally] space-quantized, and therefore energetically different, atoms in strong magnetic fields to atoms which are not spatially quantized in a zero field. This result contradicts the usual assumption that even in a zero field the space-quantization remains completely intact, and just the separation [of the energy terms] becomes zero. (Hanle, 1924, p. 99)

The new experimental results put Bohr immediately to work. He thought about the explanation of the high polarization of resonance fluorescence light and sent a letter about it to James Franck. In the following month he wrote a small note on it for publication in *Naturwissenschaften*. Since he did not wish to interfere with the publication of Hanle's paper, he first sent his manuscript to Göttingen (Bohr to Franck, 11 October 1924). Franck agreed with the content of Bohr's note and began to test some of the consequences of his assertions; he even composed a little note himself on this question, but he did not publish it immediately.[186] In any case, Bohr's note on the polarization of fluorescence light was received by the editor of *Naturwissenschaften* on 1 November 1924—the same day as Hanle's paper arrived at *Zeitschrift für Physik*—and appeared in the issue of 5 December (Bohr, 1924).

The problem to which Bohr addressed himself was to determine how, in quantum theory, one could describe a polarization of resonance light from free atoms that was much stronger than the one obtained by averaging over arbitrary directions.[187] For this purpose he invoked the difference that existed between degenerate systems—such as the atom without an external magnetic field—and nondegenerate ones. He argued that in the former case there were more possibilities of transition from stationary states. He said:

> An immediate consequence of this is that the virtual oscillators corresponding to the transition probabilities from a given stationary state cannot be associated in a

[186] James Franck sent his note on the resonance fluorescence to Bohr on 25 October 1924 to Copenhagen. Bohr expressed reservations with respect to some of Franck's ideas. In particular, he thought that Franck might have misunderstood parts of his note. Thus, he wrote to Göttingen: 'My intention was only to emphasize [in the note] that a simple possibility exists to explain that the polarization without [a magnetic] field need not be the same, but is in general greater than the average value of polarization for arbitrary field directions' (Bohr to Franck, 1 November 1924). Franck, on the other hand, had concluded that for completely unperturbed atoms one would always obtain 100% polarization of the fluorescence light.

[187] The apparent incompatibility of the experimental results with the existing atomic theory was particularly reemphasized in the paper of J. Eldridge (1924b), which appeared in *Physical Review* in September 1924.

> unique way with the harmonic vibration components of the [electron's] motion. In contrast to the case of non-degenerate systems we must be prepared [for the possibility] that the behaviour of a degenerate atom is not determined, as far as the radiation is concerned, by the motion in the stationary state, but further specific assumptions must be made about the virtual oscillators. (Bohr, 1924, pp. 1115–1116)

Bohr then assumed, in agreement with what Heisenberg had written earlier (in his letter to Bohr, dated 5 July, quoted above), that one had to expect 'that phase-relations between the oscillator components, which act simultaneously in the case of degeneracy, may become important' (Bohr, 1924, p. 1116). Hence the problem of resonance fluorescence from atoms in the absence of external fields, like the problem of the helium atom, was a problem of degeneracy.

In order to approach this problem, Bohr began with the case of an atom in an external magnetic field. Then the polarization of the resonance radiation would be determined by the probability that the incident polarized radiation activates the corresponding virtual oscillator of the transition radiation. Bohr concluded: 'The result, which one thus obtains [in quantum theory], corresponds extensively to the consequences of the classical theory, if applied to the scattering of a non-isotropic harmonic oscillator. Hence it is obvious to assume that the state of polarization of the fluorescence light, which arises from the activated [excited] atoms, will be the same as that of the scattered radiation emerging from the atoms in the ground state' (Bohr, 1924, p. 1116). Even in classical theory, in the absence of a magnetic field—where the system is degenerate—a different behaviour followed: i.e., the influence of the incident radiation on the atom then depended not only on the harmonic component of the latter's electric moment, which was responsible for the resonance, but also on the magnitude of the total angular momentum of the atom. In quantum theory an analogous dependence must exist; it was connected with the fact that the direction of the total angular momentum of the atom can be changed by a finite angle when the atom is excited. Hence the assembly of virtual oscillators (describing the transitions under consideration) in degenerate systems must exhibit a degree of polarization similar to the one in classical theory, which could not be fully determined by the motion of electrons in the stationary states. Bohr thought that 'the vibrational state of these oscillators in the activated atoms may, therefore, depend in our case upon the type of excitation of the atoms, especially upon the direction of the light-vector of the exciting radiation' (Bohr, 1924, p. 1117). This suggestion restored the formal analogy to the classical theory without contradicting the principles of quantum theory, for Bohr believed that: 'It rather represents a feature which is characteristic of the tendency of quantum theory: here, in the case of polarization, there also shows up the absence of a direct relation between radiation and atomic motion—the same absence which one is used to assuming already for the frequency in the interpretation of spectra' (Bohr, 1924, p. 1117). Thus, while the correspondence principle was strictly satisfied in the quantum-theoretical discussion of resonance fluorescence, a quantitative treatment of the experimental observations of Hanle (1924) and the new observations of Wood and Ellett (1924) had still to be given. And this task was exactly to the taste of Werner Heisenberg, Bohr's new collaborator.

Heisenberg became interested in the problem of resonance fluorescence for several reasons. First, it was a problem of degenerate systems, similar to the two-electron problem (in the helium atom) which had concerned him recently. Second, he knew about the detailed experimental results long before publication, for his colleague Wilhelm Hanle had worked on it in Göttingen. Third, it was a problem of great interest to Niels Bohr, his new boss, who considered resonance fluorescence as a kind of test for the Bohr–Kramers–Slater radiation theory. Heisenberg, who was anxious to join Bohr's efforts as closely as possible, was eager to make a substantial contribution to the important problem of verifying the fundamental theory. Moreover, he was completely familiar with Bohr's ideas, as expressed in the latter's note to *Naturwissenschaften*.[188] And, coming from the schools of Sommerfeld and Born, he was especially prepared to derive quantitative results from Bohr's physical ideas.

In Copenhagen, Heisenberg grasped very well, perhaps better than anyone else at Bohr's Institute, the trend of the time: that the direction of physics in fall 1924 was 'to get clearer and clearer feelings about how nature worked' (Heisenberg, Conversations, p. 134). He knew that one had to exploit the correspondence principle for that purpose and, if necessary, to extend it. And he knew that, in spite of its weak points, he had to use the radiation theory of Bohr, Kramers and Slater. Many years later Heisenberg recalled:

> All this was part of the game to make the total number of linear oscillators to be the real picture of the atom. One felt that in the correspondence principle one should compare one of the linear oscillators with one Fourier component of the motion. Now keeping this in mind we said: "That's a bit vague. Can we put that a little more accurately? Could we not find places where we could do better?" So, wherever we could, we tried to push it a little bit further, and this way of broadening the correspondence principle meant that we had to find places where we could use the correspondence principle in a precise manner. So that it should not only give vague indications about intensities, but real laws about intensities. And, just as we have selection rules according to which no changes in angular momentum by more than one unit [in $h/2\pi$] are possible—that is a strict rule—so we should have the same strictness in the application of the correspondence principle. Thus I was very happy that in the case of polarization of the fluorescence light [I] could give a strict rule. (Heisenberg, Conversations, p. 134; also AHQP Interviews)

One should note that by his programme of making the content of correspondence principle more precise, Heisenberg tried to counter some of the criticism which Bohr's procedure had previously received from Sommerfeld—such as the fact that Bohr's arguments were too qualitative, and that exact numbers,

[188] It is most likely that Bohr dictated the resonance fluorescence paper in German to Heisenberg (Bohr, 1924); the reason for this supposition is that he was at that time the only visitor from Germany at Bohr's Institute.

especially the rational numbers, such as the ones observed by the Utrecht experimentalists, could not be derived from his methods. Heisenberg pushed the limits of his calculations, although, in the beginning at least, he did not always fully agree with Bohr and Kramers. He had, therefore, occasionally to defend his approach in Copenhagen.

With his note on resonance fluorescence Bohr had intended to propose certain ideas about how to explain the observed polarization in principle. He took proper care in discussing the empirical situation, but did not feel himself competent to make definite numerical statements. He emphasized the qualitative nature of his procedure, drawing attention to the possible effects that would modify his results. Thus, for example, he stated that an atom could never be exactly in a given state (of well-defined quantum number) because the virtual field of the atom is always perturbed by other atoms. While Bohr was overly cautious in making precise statements at this time, Heisenberg felt that one should march in the opposite direction: i.e., sharpen Bohr's qualitative hints to obtain quantitative results. In this attempt he met considerable resistance in Copenhagen, and a discussion developed, about which he reported later:

> I was just as happy about the result of a discussion with Bohr and Kramers on the question of the polarization of fluorescent light. Bohr had written a draft of a short note on this question in connection with some experiments at Franck's institute, while, disregarding all pictures and models, I used my more formal viewpoint on Bohr's problem and reached quantitative results that went somewhat further than Bohr's work. I succeeded in convincing Bohr and Kramers of the correctness of my formulae, but when I again returned to Bohr's office after lunch, Bohr and Kramers had agreed that my formulae were wrong and tried to explain their viewpoint to me. This developed into a long and heated discussion, during which, as I recall, the necessity for detachment from the intuitive models was for the first time stated emphatically and declared to be the guiding principle in all future work. Bohr's way of thinking, which in history is perhaps most clearly represented by such figures as Faraday and Gibbs, enabled him to expose the core of the problem with inimitable clarity, but he hesitated to take the step into mathematical abstraction, though he did not speak against it. We finally concluded that the formulae were correct, and I felt that we had come a good bit closer to the atomic theory of the future. (Heisenberg in Rozental, 1967, pp. 98–99)[189]

Although, as we shall presently explain, Heisenberg was by no means abandoning the use of models, he raised formal points, which at first met with the

[189] Bohr announced Heisenberg's results about resonance fluorescence in his letter to Franck in November. He wrote: 'I wish . . . to report that recently Heisenberg has succeeded in obtaining arguments from a very beautiful consideration that, for linear polarization of the incident light, the polarization phenomena without [magnetic] field are just the same as in the case in which a magnetic field parallel to the light-vector is present.' After giving the specific results in the case of mercury and sodium resonance lines, Bohr continued: 'Heisenberg intends to write a short note in the next days and he will, of course, send you immediately a copy of it, since you may perhaps be interested in his considerations before formulating finally your own note' (Bohr to Franck, 1 November 1924). It is most likely that Franck, on receiving Heisenberg's manuscript, did not submit his own note on resonance fluorescence for publication. Heisenberg's note was received by *Zeitschrift für Physik* on 30 November 1924.

opposition of Bohr and Kramers. In the above-mentioned discussion, for example, Heisenberg maintained Sommerfeld's point of view that, for some 'mystical reason,' an atom 'knows' about the direction in space under all circumstances. This belief contributed substantially to Heisenberg's assumption about spectroscopic stability (against the influence of external perturbations) in quantum theory, and it was responsible to a large extent for the success of his calculation. In any case, after careful consideration, Bohr arrived at the conclusion that he should not stop his young collaborator, and gave in. Heisenberg was very happy about this development. 'For the first time,' he recalled, 'I had the feeling that I had been able to convince Bohr about something concerning which we had disagreed before' (Heisenberg, Conversations, p. 132).

Evidently, Heisenberg based his treatment of resonance fluorescence on Bohr's previous note (Bohr, 1924). There, Bohr had argued, for instance, that the spatial quantization of atoms made sense only if the period of precession (in a magnetic field) was small compared to the average lifetime of the excited state; if it was large, one had a degenerate system, and the polarization had to be obtained from additional considerations, say, by starting from a nondegenerate system which approximated the degenerate one in some way. Heisenberg now performed this task explicitly in the following manner. First, he considered the analogy with the classical model. If polarized light falls on an atom, the frequencies of the incident radiation being very different from the atom's eigenfrequencies, then the scattered light will be polarized partially in the same plane as the incident light and partially in a perpendicular plane. By adding a weak external magnetic field, whose direction is parallel to the electric vector of the original radiation, one introduces small changes of the eigenfrequencies of the atoms, and the latter begin to perform a precession around the direction of the magnetic field. The intensities of the components of scattered light, whose planes of polarization are parallel and perpendicular to the direction of the external field, are not changed very much by turning on the external field; but the frequency of the perpendicularly polarized scattered light will be altered by amounts $\pm\omega_L$, where ω_L is the precession or Larmor frequency of the atom. If the Zeeman effect is normal, as it always is in classical theory, then a magnetic field does not change the type of motion of the electron, apart from the Larmor precession. Thus, in classical theory, one can reduce the degenerate system (the atom plus incident radiation without the external magnetic field), in a well-defined approximation, to a nondegenerate one; in this process, some frequencies would change by amounts ω_L, and the intensities of the scattered radiation might change by relative amounts of the order ω_L/ω, where ω is the frequency of the incident light. In a second step, Heisenberg transferred this situation to quantum theory, keeping in mind the correspondence point of view. He stated:

> Hence it is obvious, in the sense of the correspondence principle, to postulate the same in the case of quantum theory; that is, to assume that the degree of polarization of the scattered light will not be altered by applying a magnetic or electric field, whose direction is parallel to that of the light-vector. Although in quantum theory the degenerate problem is completely different from the problem of

the atom in an external field [which is not degenerate], we still believe that—as in the case of spectroscopic stability—the virtual oscillators obey laws, which permit such a close analogy between the quantum theory and the classical theory. Thereby we also obtain the possibility of computing in a quantitative way the desired polarization of the scattered radiation. (Heisenberg, 1925a, pp. 622–623)

From these considerations Heisenberg immediately drew conclusions for the cases in which radiation having frequencies very different from the eigenfrequencies of the atoms was scattered by the latter. In particular, he argued that the simple dispersion radiation—i.e., the radiation scattered by the atom without any shift in frequency—was, as in classical theory, completely polarized. He also noted the possibility of calculating the intensity of scattered light (which vibrates perpendicularly to the plane of incident radiation and is connected with the transitions $\Delta m = \pm 1$, m being the magnetic quantum number) with the help of the extended dispersion theory (which Kramers and Heisenberg were just working out). In the special case of the mercury resonance line at 2534.7 Å, arising from the mercury atom in the lowest state, only the magnetic level $m = 0$ exists; hence there were no scattered lines that were perpendicularly polarized, and the polarization of the corresponding resonance fluorescence light had always to be 100% complete. This would be different in the case of the sodium resonance lines; the sodium atom in the ground state possessed the magnetic quantum numbers $m = \pm \frac{1}{2}$, hence the resonance light could also be perpendicularly polarized. These results, which strictly held only for nondegenerate systems (i.e., for atoms in external fields), were also valid for degenerate systems (i.e., for free atoms). In other words, the polarization of the resonance fluorescence light remains the same as the one obtained for very small amplitudes of the external fields. And the only assumption, which one needed to arrive at this conclusion, was Heisenberg's assumption of the stability of atomic systems: i.e., the hypothesis that in quantum theory, as in classical theory, the polarization of radiation emitted by atoms is unaffected if one turns the small external field on or off.

Passing on to the problem of resonance fluorescence, i.e., to the cases where the frequencies of the incident light may excite the atoms to higher states, the analogy between classical and quantum theory appeared to be more difficult to maintain. There were two reasons: first, the resonance lines of atoms were connected with a mechanism which differed from the corresponding classical mechanism; second, one had to include the property of linewidths, which had not yet been properly understood in atomic theory. 'Nevertheless one might expect,' argued Heisenberg, 'that a close analogy between the quantum theory and the classical theory may also be carried through in the case of fluorescence; thus the stability assumption, which we have just used for scattering, remains valid' (Heisenberg, 1925a, p. 624). As a result, again the nondegenerate and degenerate cases could be related.

In order to treat specific examples, Heisenberg assumed—like his predecessors (Breit, 1924a; Gaviola and Pringsheim, 1924)—that the intensities of the resonance radiation were the same as the intensities of the spontaneous radiation from the excited states in question; in addition he used Landé's scheme of quantum numbers and the intensity rules of Ornstein and Burger. Then he turned to the case of sodium resonance fluorescence. For both lines, D_1 and D_2, only the levels with the magnetic quantum numbers $m = -\frac{1}{2}$ and $+\frac{1}{2}$ were excited by the incident light, which vibrated in a plane parallel to the direction of the external magnetic field. Since the relative intensities for transition radiation, without and with the change of the magnetic quantum number, were 4 and 1, respectively, in the case of the D_2-line, and 2 and 2, respectively, in the case of the D_1-line, the polarization for the D_2-resonance light was found to be $(4-1)/(4+1)$ or 60%, and zero for the D_1-light. In the case of mercury resonance fluorescence, only the radiation corresponding to a transition $\Delta m = 0$ was excited, which had to be completely polarized.[190]

Heisenberg compared these results with the experimental data. While the observations of Wood and Ellett (1923) and Hanle (1924) confirmed the result for mercury, there remained discrepancies between the theoretical and the observed sodium polarizations, which he ascribed to the unavoidable perturbations of the atoms in the experimental setup (e.g., the perturbations of atomic collisions, etc.). Indeed, Hanle would later on remove some of the discrepancy (Hanle, 1927), but by then the validity of Heisenberg's correspondence calculation was not doubted anymore. For the moment, Heisenberg was 'very happy' that in the case of polarization of fluorescence light he could give a 'strict rule' (Heisenberg, Conversations, p. 134). As for Bohr, the quantitative agreement of theoretical calculations with experimental results had less importance for him than the correctness of the underlying principles. He felt that he was on safer ground with his degeneracy argument alone, even though it did not yield quantitative results. He would not have pursued the question of resonance fluorescence as far as Heisenberg had done. He knew that the experimental data at that time were not beyond doubt, and the intensity observations of the spectroscopists at Utrecht involved tricky experiments. And yet, these results came from the best laboratories then around and had, somehow, to be reliable. But what convinced Bohr much more was the fact that Heisenberg had reproduced the data, or part of it, by using the correspondence principle. In particular, he had shown that the Utrecht intensity rules were completely consistent with the Copenhagen principles of quantum theory. In the light of this success Bohr finally withdrew his objections to the unsatisfactory assumption concerning a given space direction in the absence of the magnetic field, which alone was supposed to give that direction. In the future he would let Heisenberg pursue his path of 'sharpening' the correspondence principle; he would allow him to make even bolder assumptions, such as the ones he made in the problem of the anomalous Zeeman effects, to which he [Heisenberg] returned after completing a collaboration with Kramers on the extension of the latter's dispersion theory. After all, this work was '*ad majorem correspondentiae principii gloriam.*'

[190]The polarization given above should be compared to the ones obtained by averaging over all directions. This gives, e.g., 14% polarization for the sodium D_2-line, zero for the sodium D_1-line, and 27% for the mercury line.

III.5 The Kramers–Heisenberg Dispersion Theory

In spring 1924 Bohr and Kramers became interested in examining the problems of atomic structure from the point of view of dispersion theory. At first, this interest might appear strange because the problems of atomic structure, say, e.g., the calculation of the energy states of helium, would not seem to have any connection with the scattering of light by atoms, which was the principal concern of dispersion theory. However, Bohr and his collaborators had concluded that the problem of atomic structure could not be separated from the problem of the emission and absorption of radiation—and this could be considered as a problem of the dispersion of radiation. Moreover, in the context of the new radiation theory of Bohr, Kramers and Slater, Kramers succeeded in deriving a quantum formula describing the normal scattering of light having frequencies very different from the eigenfrequencies of the scattering atomic system. This formula represented an excellent example of a systematic application of Bohr's correspondence principle to atomic physics, and it indicated how further applications could be made. When Heisenberg arrived in Copenhagen in fall 1924, the possible extensions of Kramers' dispersion theory were at the centre of interest at Bohr's Institute, and he became thoroughly acquainted with them.

In the classical electron theory of dispersion of Hendrik Lorentz, Paul Drude and others, the dispersion of light was described in the following manner. Incident radiation of frequency ν and electric field strength E excites the quasi-elastically bound electrons, having electric charge $-e$, mass m_e, and eigenfrequency ν_0. As a result, an electric moment P ($= e$ times the distance of the displaced electron from its position in the absence of incident radiation) is induced, which is given by

$$P = \frac{e^2 E}{4\pi^2 m_e(\nu_0^2 - \nu^2)} \,. \tag{46}$$

From this moment one can calculate, e.g., the polarizability, α, of atoms having several different kinds of dispersion electrons, say f_l ($l = 1, 2, \ldots$), and one finds

$$\alpha = \frac{1}{E} \sum_l f_l P_l = \frac{e^2}{4\pi^2 m_e} \sum \frac{f_l}{\nu_{0l}^2 - \nu^2} \,. \tag{47}$$

Similarly, other quantities of interest also follow.[191]

[191] For example, if light of frequency ν is scattered by a sample containing N_l electrons of type l ($l = 1, 2, \ldots$), then the refractive index is given by

$$n = 1 + \frac{e^2}{m_e \pi} \sum_l \frac{N_l e^2}{\nu_{l0}^2 - \nu^2} \,,$$

as long as ν is very different from the eigenfrequencies ν_{l0}.

It was known that classical dispersion theory had a wide range of application; but after the successes of Bohr's theory of atomic structure it became obvious that the classical description of the dispersion of light would have to be replaced by an appropriate quantum-theoretical description. In spite of several early attempts to do so, the first real progress was obtained by Rudolf Ladenburg in 1921, who connected the classical theory with Albert Einstein's statistical theory of the emission and absorption of radiation by atoms (Einstein, 1916d). Thus, Ladenburg, in his pioneering paper, did not employ any specific model of atomic structure, but basically retained the classical description of the dispersion process (Ladenburg, 1921a). He introduced the quantum features by means of the following device: he put the classically calculated energy, emitted by a dispersion electron per second, equal to the corresponding expression in Einstein's statistical theory of emission and absorption of radiation. He was thus able to derive an equation for N, the number of dispersion electrons, connected with the transitions from a stationary state i to a state k of an atom, in an assembly of atoms in which N_i atoms are in the ith state; he obtained

$$N = N_i \frac{m_e c^2}{8\pi^2 e^2 \nu_{ik}^2} a_{ki}, \tag{48}$$

where ν_{ik} $(=(E_k - E_i)/h)$ is the absorption frequency and a_{ki} the Einstein coefficient of spontaneous emission from the higher state k.[192] Hence the number of dispersion electrons per atom in the state i became N/N_i, and the quantum-theoretical polarizability α assumed the value

$$\alpha = \frac{c^3}{32\pi^2} \sum_k \frac{a_{ki}}{\nu_{ik}^2(\nu_{ik}^2 - \nu^2)}, \tag{49}$$

with the sum running over all states k denoting the excitations of the state i. This expression was tested by Ladenburg and others and found to be in agreement with the data on various vapours and gases (see, e.g., Ladenburg and Reiche, 1923).

Bohr and his collaborators had not been actively involved in dispersion theory until the beginning of 1924, when the new radiation theory (of Bohr, Kramers and Slater) offered the possibility of discussing Einstein's 1916 treatment of radiation from the undulatory point of view. The central idea of the Bohr–Kramers–Slater theory consisted in the introduction of virtual fields of radiation that were associated with all atoms. The virtual field of an atom was assumed to be provided by an assembly of virtual oscillators, each of them having the frequency of an atomic transition. Hence it now seemed possible for the first time to treat the quantum-theoretical dispersion in a close analogy with classical

[192] In Eq. (48) it is assumed that both states i and k of the atom have equal weights, $g_i = g_k$; otherwise the right-hand side must be multiplied by g_k/g_i.

dispersion without giving up the principles of atomic theory. Only a few substitutions were needed in classical theory: instead of the eigenfrequencies of the quasi-elastically bound electrons, one had to put in the atomic transition frequencies, i.e., the frequencies of the virtual oscillators; the classical oscillator strength had to be related to another property of the virtual field, the transition amplitudes under consideration. In this context, it should be noted that in their new radiation theory, Bohr, Kramers and Slater had referred to the fact that Ladenburg, in his dispersion theory, had already made use of the idea of the virtual oscillator field. Ladenburg's 1921 paper, and the subsequent work of Ladenburg and his collaborators on the experimental determination of the number of dispersion electrons in atoms, interested Bohr very much at the time, for he was particularly concerned with the question of the statistical weights connected with atomic states.

Hendrik Kramers' theory of the intensities of the hydrogen lines (Kramers, 1919) had been quoted in Ladenburg and Reiche's paper (1923, p. 587) that gave a review of the status of the quantum theory of dispersion, as justifying the main steps in Ladenburg's (1921a) work—i.e., equating the classical and quantum-theoretical rates of emission. Now, in early 1924, Kramers relied on Ladenburg's work when he applied the Bohr–Kramers–Slater theory to derive his dispersion formula (Kramers, 1924a). From the Copenhagen point of view it seemed obvious that most of the relations in classical theory remained true, with appropriate modifications, in atomic theory. The problem then remained to compute, consistent with the quantum principles, the polarization vector of an atom, whose magnitude P was identical with the atom's induced electric moment. By assuming the directions of the polarization vector and the electric vector of the incident radiation to be parallel—an assumption which corresponded to that of isotropically bound electrons in classical theory—Kramers arrived at the following expression for α, the polarizability of the atom[193]:

$$\alpha = \frac{P}{E} = \frac{3c^2}{32\pi^2}\left\{\sum_l \frac{a_l^a}{(\nu_l^a)^2\left[(\nu_l^a)^2 - \nu^2\right]} - \sum_{l'} \frac{a_{l'}^e}{(\nu_{l'}^e)^2\left[(\nu_{l'}^e)^2 - \nu^2\right]}\right\}. \quad (50)$$

In Eq. (50), a_l^a ($l = 1, 2, \ldots$) and $a_{l'}^e$ denote the Einstein probability coefficients associated with the quantum transition frequencies ν_l^a and $\nu_{l'}^e$, respectively; as in Ladenburg's formula (49), these quantities should be written, say a_l^a as a_{ki} and ν_l^a as ν_{ik}, where i and k are two states of the atom. The upper indices a and e refer to the fact that the transitions in the first sum correspond to the absorption of radiation and the transitions in the second sum correspond to the emission of radiation from the state of the atom whose polarizability is considered.

[193] It should be noted that in Ladenburg's formula, Eq. (49), a factor 3 had been left out because Ladenburg had assumed that the oscillators were three-dimensional, in agreement with an investigation of Fritz Reiche (1919b). Kramers did not do so and retained the factor 3.

Evidently, Kramers' formula (50) extended Ladenburg's earlier formula (49) by introducing the second sum. With this sum Kramers took into account the fact that the atom under investigation may be in a state from which it can emit radiation spontaneously. Ladenburg's formula, then, would apply strictly to those atoms which are in the lowest state and can only absorb radiation. A particular point of Kramers' formula was that quantities $-a_{l'}^{e}(\nu_{l'}^{e})^{-2}$, which could be considered as representing (up to constant factors) what one called oscillator strengths in classical theory, were *negative* numbers. Kramers explained this apparent puzzle by referring to Einstein's concept of negative absorption, a concept necessary to describe the quantum nature of radiation. Although Kramers could not cite any empirical data that would favour Eq. (50) over Eq. (49), he argued that only his formula passed into the classical one, Eq. (47), in the limit of high quantum numbers, where the motion in successive stationary states differed by small amounts from each other.

In his first note on dispersion theory of March 1924, Kramers presented formula (50) as just an extension of Ladenburg's, which followed automatically if one applied correctly the conception of virtual oscillators in atoms. Several months later, in July 1924, in his second note to *Nature* on the dispersion formula (Kramers, 1924b), he sought to elucidate some points which were only briefly touched upon in his first letter. The central idea was that one had to write the classical expression for the polarizability on the right-hand side of Eq. (47) as a differential quotient, say $\partial\phi/\partial J$, where $\partial/\partial J$ stood for the derivative with respect to the action variables of the atomic system, and then to replace the differential quotient by the difference of two terms. He also pointed out that the dispersion formula 'contains only such quantities as allow of a direct physical interpretation' (Kramers, 1924b, p. 311). These ideas had been expounded earlier by Max Born in a paper entitled '*Über Quantenmechanik*' ('On Quantum Mechanics'), submitted to *Zeitschrift für Physik* from Göttingen in June 1924 (Born, 1924b). Born had considered general multiply periodic systems, described by the action variables $J_1, \ldots, J_f$ (and their conjugate angles), and had claimed that quantum formulae could be obtained from classical formulae by means of the following substitution:

$$\sum_k \tau_k \frac{\partial\phi}{\partial J_k} \rightarrow \int_0^1 \sum_k \tau_k \frac{\partial\phi}{\partial J_k}\, d\mu = \frac{1}{h}\left[\phi(n+\tau) - \phi(\tau)\right]. \tag{51}$$

On the classical left-hand side, ϕ is a quantity that can be expressed as a function of the action variables J_k, $k = 1, \ldots, f$; these action variables assume values that are integral multiples of Planck's constant h, i.e., the values n_k and $n_k + \tau_k$ (where n_k and τ_k are integers $\geqslant 0$). The μ-integration takes into account the fact that a quantum expression is obtained by averaging the classical expression over the values of the action variable, $J_k = h(n_k + \mu\tau_k)$, $0 \leqslant \mu \leqslant 1$; that is, any property of the quantum system, e.g., the frequency of the radiation, is not associated with one quantum state, but rather with two, denoted by the values of the action variables $n + \tau$ and n, respectively, where $n + \tau$ and n stand for a set of f pairs, $n_1 + \tau_1, \ldots, n_f + \tau_f$ and $n_1, \ldots, n_f$. With this translation prescription Born had been able to establish Kramers' dispersion formula, Eq. (50), for the scattering of light from arbitrary atomic systems.

The quantum-theoretical dispersion formula, like the corresponding classical one, applied strictly only to those cases where ν, the frequency of the incident radiation, was very different from the absorption and emission frequencies of the scattering atom or molecule. Only then—so Kramers had argued already in his first note—would the phase of the induced electric moment of the atom nearly coincide with the phase of the incoming radiation (Kramers, 1924a, p. 674). He had intended from the beginning to remove this restriction. Thus, he spoke on the scattering and absorption of light by atoms at Hamburg in June 1924 and announced that a detailed publication of his results would be forthcoming.[194] However, he was kept busy by many duties and did not find time to write his paper on dispersion until fall 1924, only after his return from the *Naturforscherversammlung* in Innsbruck at the end of September. By then Heisenberg was already in Copenhagen, and he soon participated in the discussions on dispersion theory.

Heisenberg had been aware of the problem of quantum-theoretical dispersion for several years. For example, at a discussion during Bohr's Göttingen lectures in June 1922, he had used an argument from classical dispersion theory in order to criticize Kramers' calculation of the quadratic Stark effect (Kramers, 1920b). The difficulty which Heisenberg pointed out, was well known among the quantum physicists at Munich ever since Sommerfeld had discussed the theory of dispersion in the light of Bohr's theory of atomic structure (Sommerfeld, 1915d, 1918a), and it could be formulated in the question: What are the frequencies that play a role in the dispersion of light by atoms and, especially, how do the transition frequencies of atoms enter into the dispersion formula? This question was related to the puzzling properties of radiation. In his atomic theory, Bohr had more or less assumed the emission and absorption of light-quanta in the transitions between discrete states. On the other hand, it was evident from the success of classical electromagnetic theory that light must be emitted and absorbed in spherical waves. No difficulty arose in classical theory, for the frequencies emitted from systems containing moving charged particles always had to be integral multiples of the mechanical frequencies. Hence, in the dispersion of light, the mechanical frequencies interacted with the frequency of the incident light, and the scattered radiation consisted of the interference of secondary waves with the incoming waves. This procedure could not be used in quantum theory because a given transition frequency corresponded neither to the

[194] See the reference to Kramers' contribution, entitled '*Über Streuung und Absorption von Licht an Atomen*' ('On the Scattering and Absorption of Light by Atoms'), presented at Lower Saxony Section (*Gauverein Niedersachsen*) of the German Physical Society, Hamburg on 21 June 1924, in *Verh. d. Deutsch. Phys. Ges.* (3) *5*, p. 37 (issue no. 3 of 15 August 1924).

mechanical frequency of the initial nor of the final orbit, but rather to a numerical average of these frequencies. Heisenberg had frequently discussed such questions with many people, including Pauli. That was also the reason why he participated in the discussion of Kramers' treatment of the Stark effect at the Bohr lectures in Göttingen. At that time, in June 1922, Bohr had been unable to give a satisfactory answer to the objection Heisenberg had raised. In Göttingen, where Heisenberg had gone in fall 1922, the dispersion problem was often discussed. As Pascual Jordan recalled:

> The problem of dispersion was discussed in Göttingen in this way between 1922 and 1923: whether it was so that the critical frequency in dispersion was the same as the one which gives rise to the spectrum of the discrete [atomic] lines, or whether they [i.e., the critical frequencies in the dispersion formula] had to be calculated from the eigenoscillations of the atoms due to the classically calculated electron orbits. Born had no definite opinion in this matter. I remember a discussion in which the three of us, Born, Heisenberg and I, were involved. I still tended to believe at that time that everything depended on the eigenfrequencies of the electron orbits . . . But Heisenberg said that he did not believe that; that is, if the dispersion theory was valid at all, then [Heisenberg thought] it was valid with the frequencies that do really occur in the case of spectral lines. And that persuaded me immediately . . . Then Born also agreed with Heisenberg's opinion, but I do not believe that he had thought much about this question earlier or had developed a definite opinion about it. (Jordan, AHQP Interview, First Session, p. 24)[195]

It seems to be quite remarkable that Ladenburg's paper was not considered as giving the right answer, although the Breslau physicist was well known to Born and Franck. Thus Jordan remarked:

> Ladenburg's work is an example of those cases in which somebody developed something important, but did not find a real appreciation of it for a long time. I cannot remember at all that Ladenburg's formula was ever mentioned in the many discussions on quantum theory in Göttingen during the first few semesters I was there. It somehow had not made its way, in general, into the quantum-theoretical thinking. (Jordan, AHQP Interview, First Session, p. 25)

Heisenberg also did not attribute the necessary importance to Ladenburg's paper when he first read it in Munich, for, as he recalled: 'My first impression of Ladenburg's paper, when we discussed it, was that it was too far away from the whole Bohr theory. [His] paper was very closely attached to the Einstein paper

[195] Heisenberg also recalled that he became occupied with the problem of the dispersion of light in quantum theory after the *Bohr Festival* in Göttingen:

> I had thought about the problem [he said]. I had turned around the problem in my mind quite frequently but without getting the grip on something. I didn't get hold of anything that I found interesting enough to really do something with . . . But I do remember that when we spoke about the difficulties of quantum theory, the problem came up again and again. But then, during that period, there also came out the paper of Compton [1923b], which occupied the minds of many people. This paper showed the strong reality of the picture of light-quanta. So all this contributed to the interest in the problem of dispersion. (Heisenberg, Conversations, p. 124)

and insofar, of course, was good physics. That everybody saw. But one did not see the connection between Ladenburg's paper and the actual calculation of intensities' (Heisenberg, Conversations, p. 123). Indeed, both in Munich and later in Göttingen, the theoreticians had not been alert to the arguments based on Einstein's statistical radiation theory of 1916; the latter was too far removed from the theory of multiply periodic systems, which were treated there in the early 1920s. But the situation changed altogether when Kramers' first note on dispersion theory came out. As Heisenberg recalled:

> It was only the first paper of Kramers [1924a] that established the connection between Einstein's ideas and the idea of the atom as a collection of linear oscillators which somehow had to do with the electron and this 'somehow,' of course, had to be explained. It was quite clear in Kramers' paper how one should connect the idea of the linear oscillators in the atom with Einstein's idea of probability. That had not been so clear in Ladenburg's paper. (Heisenberg, Conversations, p. 123)[196]

Though Heisenberg could not do anything with Ladenburg's paper, for it was based on Einstein's theory and did not appeal to him, it was different with Kramers' paper. When the latter appeared, he immediately found that 'this idea of somehow using the harmonic oscillator in connection with the atomic models appealed to me' (Heisenberg, Conversations, pp. 123–124). In fact, he had already employed it in the paper on the intensities of multiplet lines and their Zeeman components (Heisenberg and Sommerfeld, 1922b). In some way, he thought, he could now pick up the earlier treatment and improve upon it since Kramers had hit upon the key idea of the quantum-theoretical treatment of dispersion. 'So I don't know,' Heisenberg remarked later, 'whether I would have come back to dispersion if it was not with Kramers' (Heisenberg, Conversations, p. 124). As soon as Kramers came out with his formula, Born in Göttingen saw a way to work on the dispersion problem, and he immediately incorporated it into his paper '*Über Quantenmechanik*' (Born, 1924b). Though Heisenberg was occupied at that time with the problem of the anomalous Zeeman effects, he was extremely interested in Born's work. On one hand, he helped Born in carrying out the calculations[197]; on the other hand, he was aware of the fact that Born's general treatment of quantum systems, of which Kramers' dispersion formula was only one result, might help him in his own work on many-electron systems.[198] His renewed attack on the helium problem failed to give the expected

[196] Heisenberg also remarked: 'The second paper of Ladenburg and Reiche [1923] I didn't know at all' (Heisenberg, Conversations, p. 123).

[197] Born acknowledged Heisenberg's help in the following words: 'I am also greatly indebted to Mr. W. Heisenberg for much advice and help with the calculation' (Born, 1924b, p. 380, footnote 1).

[198] About this connection of Kramers' dispersion theory with the problem of many-electron atoms, Heisenberg reported to Pauli: 'Born has thought about the following: If one considers Kramers' dispersion formula in terms of the usual perturbation procedure, then one finds that it may be translated as describing the coupling of *two electrons*' (Heisenberg to Pauli, 8 June 1924). Heisenberg, after the completion of his paper on the anomalous Zeeman effect (Heisenberg, 1924e), turned to a new discussion of the helium problem with the new methods of Born.

success; in turn, he got into the task of extending Kramers' dispersion theory in Copenhagen, in close collaboration with Kramers himself.

Up to fall 1924 Kramers had published only the two notes in *Nature* on the dispersion formula. The original argument gave the formula, but said almost nothing as to how it had been obtained (Kramers, 1924a). The second note expanded the argument a little more, but still a thorough discussion was lacking (Kramers, 1924b). Hence, Kramers' first goal was to formulate the dispersion theory systematically, providing all the theoretical motivations and derivations. His second goal was to remove the limitations of the dispersion formula, Eq. (50), and to incorporate those dispersion phenomena in which the scattered radiation did not have the same frequency as the incident one. Again, this was Kramers' idea, for as Heisenberg confirmed later: 'Kramers had more or less the idea that one should extend his paper on the dispersion, and also the paper on the lines of the different frequencies' (Heisenberg, Conversations, p. 136).[199] With the last remark, Heisenberg had in mind a note of the Viennese physicist Adolf Smekal, entitled '*Zur Quantentheorie der Dispersion*' ('On the Quantum Theory of Dispersion') and published as a letter to *Naturwissenschaften* in October 1923 (Smekal, 1923c). In his note, Smekal had discussed all dispersion phenomena by assuming radiation to consist of light-quanta, and studying their energy and momentum exchanges with atoms in detail. In particular, he had predicted the existence of anomalous dispersion connected with the change of the frequency of the scattered radiation, i.e., the occurrence of the frequencies

$$\nu + \nu_l \qquad \text{and} \qquad \nu - \nu_{l'}, \tag{52}$$

where ν is the frequency of the incident radiation and ν_l and $\nu_{l'}$ correspond to the transition frequencies of the scattering atom[200]; the reason being that any atom in a given state, which is struck by incident radiation, should be able, with a certain probability, to pass on to a lower state by losing the energy $h\nu_l$ or to a higher state by acquiring the energy $h\nu_{l'}$. The effect predicted by Smekal was, however, only one particular example of a much more general class of effects in atomic theory. Another such example was an effect observed in connection with the resonance radiation of atoms in a magnetic field; this was noted explicitly by Heisenberg in his paper on resonance fluorescence. After discussing the resonance radiation, which is polarized parallel to the direction of the incident

[199]This recollection is confirmed by a statement in the paper published by Kramers and Heisenberg. They noted: 'The idea of relating the scattering action of atoms in the presence of external radiation to the scattering action of the atomic system expected from the classical theory, in accordance with the correspondence principle,—given by Smekal—first occurred to Kramers in connection with his work on dispersion theory' (Kramers and Heisenberg, 1925, p. 687).

[200]The frequencies given by Eq. (52) are only special cases of the ones obtained by A. Smekal. In his note, he also took into account the possibility that the momentum of the atom may change in the dispersion process (see Smekal, 1923c, p. 874, Eq. (1)); we have neglected this possibility here.

radiation, he had remarked:

> In contrast to it the nature of the perpendicularly-polarized light is no longer that of usual scattering. But, because of the rotation of the atom around the axis of the external field, the perpendicularly-polarized scattered light will be radiated in two frequencies, which deviate from the frequency of the activating light by the (positive or negative) rotational frequency of the atom around the axis of the field. We do not wish to discuss here more closely the special properties of this type of scattered light, which possesses a quantum-theoretical analogue in an effect predicted by Smekal, who was motivated by the light-quantum theory. (Heisenberg, 1925a, p. 622)

Besides the fact that his latest work on resonance fluorescence had a connection with Kramers' programme to extend dispersion theory, Heisenberg was eager to participate in the latter for other reasons. In particular, he quickly realized the fundamental importance which Bohr attributed to the entire dispersion-theoretic approach and, ambitious as he was, he wished to get on this programme immediately. He was very happy when Bohr and Kramers drew him into their discussions on the subject. At Bohr's Institute, unlike Göttingen, all the people used to discuss together the most interesting physical problems of the day; specific problems were not just worked out by one or two people, who then published the results in a paper, but everybody had a share in them. 'It was like in Sommerfeld's Institute,' Heisenberg recalled later. 'There were continuous discussions between all people who took these things seriously' (Heisenberg, Conversations, p. 140). Therefore, often certain people contributed ideas to the solution of a problem, whose name would not, and could not, be mentioned in the final publication.[201] In the case of Heisenberg's collaboration with Kramers on dispersion theory the situation was less involved. And yet, a difficulty arose that he had not encountered before. As he recalled: 'For some time it was not clear whether we should write a joint paper, or whether Kramers should write it alone. And actually, it was suggested—I don't know whether by Kramers himself or by somebody else—that Kramers should publish it alone, because most of the work was due to him' (Heisenberg, Conversations, p. 139). Heisenberg did not like this suggestion at all, for in matters of publication he was accustomed to a generous attitude of the senior people—especially Sommerfeld's, who was always anxious to give his students every credit for their contributions. Looking back on

[201] This style was very much due to Niels Bohr, who used to discuss all questions of atomic theory with his scientific collaborators and visitors, but often did not put his own name among the authors of a paper. Still, it is not difficult to trace Bohr's contributions to the physics done at his Institute. Apart from the nominal acknowledgement of Bohr's help, his arguments and ideas in other people's papers were in most cases quite obvious, for they were known from his own publications. The situation was different for those who spent only a short period at the Institute. In the atmosphere of Bohr's Institute, the ideas were somehow squeezed out of people; these ideas fell, as it were, into a big pot, where they melted and joined with other ingredients, and afterwards reappeared in a more or less completely modified form. Thus, it was always difficult to find out who had originally contributed what part of the final solution of a problem.

the work on dispersion theory, Heisenberg remarked:

> I was a bit hurt by this idea [of Kramers' publishing the paper alone] because I felt that I had made an actual contribution. But I simply said, "Well, I leave that entirely to Bohr (as I had left it always to Sommerfeld). Bohr shall decide. I don't care which way his decision goes." Bohr then probably thought, "After all, the young man has contributed a bit, why not put his name also on it? Everybody knows that it is really Kramers' theory because Kramers had written the first notes." This must have been Bohr's idea more or less. I think that was quite right. (Heisenberg, Conversations, p. 140; also AHQP Interview)[202]

While Kramers' leading role in the work on dispersion theory in fall 1924 was evident, one has to determine what specific contribution did Heisenberg make. In the joint paper it was simply stated that, 'the further elaboration of this idea [i.e., to extend Kramers' dispersion formula], which is given in the present paper, resulted from discussions between the authors' (Kramers and Heisenberg, 1925, p. 687). Thus, the main aid which Kramers obtained from his collaborator was in formulating the physical ideas in mathematical equations and calculating the results. Heisenberg was well prepared for this task, as he could make use of his previous knowledge, which he had acquired in helping Born with his paper entitled 'On Quantum Mechanics' ('*Über Quantenmechanik*,' Born, 1924b). The collaboration turned out to be quite exciting, especially since Kramers and Heisenberg had first to establish a common basis for understanding. As Heisenberg recalled later: 'We did not agree to begin with, and Kramers did not take the problem quite as seriously as I did. But certainly he also tried in the same direction as I did; both Kramers and Bohr worked in the same direction. Bohr did not go so much into the details of the [description of] classical motion, and the mathematics of the classical motion was not at the centre of his interest. Of course, Kramers knew that very well' (Heisenberg, Conversations, p. 135). Heisenberg, trained in the mathematical approach of Göttingen, found that Kramers did not insist as much, as he himself was used to, on carrying through the formal calculations to the very end; Kramers would rather introduce additional physical arguments. Evidently, mathematics was not taken so seriously in Copenhagen, and Heisenberg was not willing to tolerate this attitude. He thought that it was necessary to discuss the dispersion phenomena as new examples of a

[202] The question, whether or not Kramers and Heisenberg should publish together on dispersion theory, was settled quite soon. In his note on the polarization of fluorescence light, which was signed 1 November 1924, Bohr already mentioned in a footnote: 'This topic [i.e., the extension of dispersion theory] will be treated in greater detail in a paper of [Kramers and Heisenberg], which will soon be published' (Bohr, 1924, p. 1116, footnote 5). In the same spirit Heisenberg reported to Sommerfeld a little later: 'I shall write a note with Kramers on "Smekal's Compton Jumps" (Smekal, *Naturwiss*. **11**, 873, 1923 [Smekal, 1923c]). These jumps have a rather nice analogue in classical theory and are, in addition, necessary in order to explain the polarization effects in question [in resonance radiation]' (Heisenberg to Sommerfeld, 18 November 1924).

It appears that the original plan was to write a short note on the extension of Kramers' dispersion theory to include the effect predicted by Smekal; only later was it decided that Kramers and Heisenberg would be co-authors of the long paper on dispersion of light by atoms.

sharpened application of the correspondence principle. This approach implied that mathematics had to be employed very rigorously. Heisenberg remarked: 'I did not try to discuss the problem primarily from the side of Einstein's laws [of absorption and emission of radiation] and physics, but I simply said, "Let us see what classical mechanics does. Let us just assume that we have a classical atom like hydrogen and we put an electric wave on it, and then see what happens to the Fourier component"' (Heisenberg, Conversations, p. 136). These classical terms had then to be worked out individually and finally to be replaced by proper quantum-theoretical terms, following the prescriptions used earlier by Kramers (1924a) and Born (1924b). Now, in going through this procedure in detail, Kramers and Heisenberg hit upon a difficulty, towards which they assumed completely different attitudes.

The point of disagreement between Kramers and Heisenberg was the following. In the quantum-theoretical formula describing the induced electric moment of an atom—which now generalized Kramers' dispersion formula, Eq. (50)—the terms had always the same structure: the numerator consisted of a product of two transition amplitudes multiplied by an exponent containing the sum or difference of two frequencies; while in the denominators, sums or differences of two frequencies—not the same as in the exponent—occurred.[203] The physical interpretation was that each term corresponded to two successive transitions of the scattering atom: first, from the state P to the state Q, and then from the state Q to the final state R. By evaluating all the terms carefully, Heisenberg arrived at the following result: the electric moment of an atom in a given state P should become very large when the incident frequency ν approaches a frequency of transition, ν_{QR}, from a level Q, higher than P, to a level R, lower than P; this was due to the occurrence of the denominator $\nu_{QR} - \nu$. This result seemed to contradict physical intuition, which indicated that resonance effects should arise only when the transitions from the state P to the state Q (or R) were possible. Heisenberg and Kramers were puzzled; theoretically, the unphysical term was absolutely necessary, but from the physical point of view it should not be there. About this, they disagreed sharply. 'Kramers was willing to drop the term, but I did not want to drop it,' recalled Heisenberg later (Heisenberg, Conversations, p. 138). Heisenberg believed in the validity of the quantum-theoretical perturbation theory, which had been developed by Born and now used in the extension of Kramers' dispersion theory. Kramers, however, insisted that the whole thing had to be understood from the physical point of view; he took seriously what actually happened in nature, and argued that if the experiment did not show a resonance

[203] In the case of the coherent dispersion of light, which is covered by Kramers' formula, Eq. (50), the induced electric moment has the same structure: in each quantity a_l^a (or $a_{l'}^e$) the product of two Fourier amplitudes, A_l^a and $\overline{A_{l'}^a}$ is contained; any term in the l- or l'-sum with quadratic denominator, say $(\nu_l^a)^2 - \nu^2$, can be split into two terms with linear denominators, $\nu_l^a + \nu$ and $\nu_l^a - \nu$, respectively; moreover, the only frequency that appears in the exponent is ν, the frequency of the incident radiation.

line there should not be such a term in the perturbation calculations. They took their case to Bohr for a decision. As Heisenberg recalled later: 'We would both stand at the blackboard, and Kramers would defend his scheme and I would defend mine. Bohr listened to the arguments. But we did not reach a conclusion then' (Heisenberg, Conversations, p. 140). Only after several weeks did Kramers find the solution: the term, whose existence was advocated by Heisenberg, was actually there; but there was also an interference between this term and another, which was due to spontaneous radiation with the same frequency, and this interference removed the apparent resonance effect arising in the dispersion scheme. Thus, from his formal point of view, Heisenberg was right; so was Kramers, with his physical argument. And Bohr was happy about the resolution of the Kramers–Heisenberg disagreement. The two collaborators then worked out all the terms in the quantum-theoretical dispersion formula, a task which they completed by mid-December 1924.

Since Kramers regarded dispersion theory as his own problem, and also because he was the senior and more experienced, he undertook to write the paper. He completed the paper, entitled '*Über die Streuung von Strahlung durch Atome*' ('On the Scattering of Radiation by Atoms') later in December and submitted it to *Zeitschrift für Physik*, where it was received on 5 January 1925 and published about two months later (Kramers and Heisenberg, 1925).[204] This paper contained all the results of Kramers and Heisenberg's calculations, not just a discussion of the incoherent scattering predicted earlier by Smekal; it gave the derivation of the general quantum-theoretical dispersion formulae from the corresponding classical ones with the help of the prescription for discretization, Eq. (51). The style of the paper reflected the fact that it had originated in Copenhagen: there were only a few formulae, introduced or followed by long and careful physical argumentation, including discussions of the relevant available observations. Thus, the authors not only succeeded in presenting the most complete results based on the correspondence approach to the problem of the scattering of light by atoms, they wrote in such a way that every physicist, theoretician or experimentalist, interested in the subject could understand. Kramers, the senior author—an experienced expositor—made full use of his talent to illuminate difficult theoretical questions. In contrast to the style prevalent at that time in papers on quantum theory—whether they originated in Göttingen, Copenhagen, or other places—the Kramers–Heisenberg paper stood out for its

[204] Heisenberg left Copenhagen a few days before 22 December 1924 to spend his Christmas vacation in Munich and the Bavarian mountains as Bohr reported to Pauli (Bohr to Pauli, 22 December 1924). Kramers must have completed the paper by himself by the end of the year. Heisenberg, on receiving the final manuscript of the paper from Copenhagen during his stay in Munich, wrote his comments to Kramers in a letter, dated 8 January 1925. In it he said: 'Many thanks for sending me the final manuscript; I have gone through it very carefully and find it very nice.' He then added a few remarks concerning certain details of the calculation, which led Kramers to correct some of his statements. And he promised: 'I shall remove the little unevenness in the German while correcting the proofs' (Heisenberg to Kramers, 8 January 1925).

clarity and straightforwardness, in which the results were obtained from a few intelligible assumptions.[205]

The clarity of the style was evident from the organization of the paper. In the introduction were stated the relations of the new description of atomic structures, with special reference to the Bohr–Kramers–Slater theory of radiation; then was stated the main problem to be treated, namely, to extend the previous formulae describing the dispersion of light by atoms to cases of incoherent scattering, resulting in radiation of frequencies different from those of the incident radiation; finally, the method of approach was mentioned, which was to follow closely the analogy to the classical description of dispersion phenomena. In Section 2, the theory of the scattering of light by a classical system was worked out in detail. The translation of the expressions for the induced electric moment of the system, which is created by the incident radiation, into the corresponding quantum-theoretical expressions was done in the following two sections. Section 3 was devoted to coherent scattering, i.e., to cases in which the frequency of radiation was not changed by the scattering process; as a special case, then, Kramers' formula, Eq. (50), was obtained. In the long Section 4, the authors analyzed various types of incoherent scattering, finding among others the types predicted earlier by Smekal on the basis of light-quantum considerations. The consistency of all these results with the accepted principles of quantum theory, such as the existence of thermal equilibrium between atoms and radiation and the assumption of virtual fields in atoms, was pointed out in the conclusion. The main theoretical results were thus obtained in Sections 2 to 4, and we shall outline them in the following.

Since Kramers and Heisenberg knew the method of obtaining quantum-theoretical formulae describing multiply periodic systems from the corresponding classical ones, they put particular emphasis on getting the complete classical equation for that quantity of a multiply periodic system which was responsible for the scattering of light, i.e., its electric moment $P(t)$. From a perturbation calculation they found that for a system of f periodic degrees of freedom, on which radiation characterized by the electric field vector $E(t) = \mathrm{Re}(E \cdot \{\exp(2\pi i\nu t)\})$—with E the (vector) amplitude of the electric field and Re referring to the real part of the expression within brackets—falls, the electric moment was given by[206]

$$P(t) = P_0(t) + P_1(t). \tag{53}$$

[205] Of course, Kramers followed the style he was used to for many years, a style which had its origin in Copenhagen. Thus, in contrast to the papers of Born and the Göttingen school, he always used fewer formulae. Unlike Born, he did not insist on formulating a problem in the greatest possible generality or look for the most general solution of a given problem. In Copenhagen, one followed Bohr's method of arguing on the basis of the empirical information at hand. However, because of the definite and unambiguous results of dispersion theory all vague and obscure speculations, which made many papers on quantum theory between 1923 and 1925 difficult reading, could be avoided.

[206] The symbols $E(t)$, E, $P(t)$, $P_0(t)$ and $P_1(t)$ denote vectors in the following equations. The same is true of the Fourier coefficients C_τ and the corresponding quantum-theoretical scattering amplitudes, A^a and A^e, below. We have dropped the vector notation here for the sake of simplicity.

The first term on the right-hand side of Eq. (53) represents the electric moment of the unperturbed system, that is,

$$P_0(t) = \sum_\tau C_\tau \exp(2\pi i \omega t), \tag{53a}$$

where C_τ ($= C_{\tau_1,\ldots,\tau_f}$) stands for the Fourier coefficient of the electric moment associated with the frequency ω ($= \tau_1\omega_1 + \cdots + \tau_f\omega_f$), and $\sum_\tau$ is an f-fold sum, in which $\tau_1, \ldots, \tau_f$ assume integral values; $\omega_1, \ldots, \omega_f$ are the frequencies of the unperturbed system. The second term on the right-hand side of Eq. (53) yields, in first approximation, the electric moment of the system induced by the perturbation through the incident radiation. This term can be expressed as the double sum[207]

$$P_1(t) = \mathrm{Re}\left\{ \sum_\tau \sum_{\tau'} \frac{1}{4}\left[\frac{\partial C_\tau}{\partial J'} \cdot \frac{(E \cdot C_{\tau'})}{\omega' + \nu} - C_\tau \cdot \frac{\partial}{\partial J}\left(\frac{(E \cdot C_{\tau'})}{\omega' + \nu} \right) \right] \right.$$
$$\left. \cdot \exp\left[2\pi i(\omega + \omega' + \nu)t \right] \right\}. \tag{53b}$$

The τ- and τ'-sums (each of which is again an f-fold sum) run over integral numbers, both positive and negative, while the differential symbols are defined as $\partial/\partial J = \tau_1 \partial/\partial J_1 + \cdots + \tau_f \partial/\partial J_f$ and $\partial/\partial J' = \tau'_1 \partial/\partial J_1 + \cdots + \tau'_f \partial/\partial J_f$, respectively, and $\omega' = \tau'_1\omega_1 + \cdots + \tau'_f\omega_f$. $P_1(t)$ is the total induced electric moment of the system (of f periodic degrees of freedom), and it accounts for all possible scattering effects in classical theory. In particular, Eq. (53b) states that

> under the influence of the incident light, the system will emit scattered radiation of an intensity which is proportional to the intensity of the incident light; when separated into its harmonic components, it contains not only the frequency ν of the incident light but also frequencies which can be represented as the sum or difference of ν and a frequency ω^0 . . . . The frequency ω^0 itself need not appear in the motion of the unperturbed system. Rather it can be seen from [Eq. (53b)] that ω^0 is always of the form $\pm|\omega| \pm |\omega'|$, where ω and ω' are two frequencies that actually occur in the unperturbed motion. (Kramers and Heisenberg, 1925, p. 689)

In the case of coherent scattering, the frequency ω^0 is always zero; hence the double sum in Eq. (53b) simplifies into a single one, since τ' always assumes the negative values of τ, and one obtains for the induced electric moment

$$P(\nu, t) = \mathrm{Re}\left\{ \sum_\tau{}' \frac{1}{4} \frac{\partial}{\partial J}\left[\frac{C_\tau(E \cdot \bar{C}_\tau)}{\omega - \nu} + \frac{\bar{C}_\tau(E \cdot C_\tau)}{\omega + \nu} \right] \cdot \exp\left[2\pi i \nu t \right] \right\}, \tag{54}$$

[207] The product $(E \cdot C_{\tau'})$ denotes the scalar product of the two vectors E and $C_{\tau'}$.

where the prime on the summation symbol indicates that the τ-sum has to be restricted to those values for which the resulting ω is positive, and $\bar{C}_\tau$ $(= C_{-\tau})$ is the complex conjugate of the coefficient C_τ.

Now the authors had to adapt Eq. (53b) to quantum theory. Thus, the classical frequencies of the multiply periodic system (and their higher harmonics) had to be replaced by the transition frequencies of the atom, ν_{qu} (i.e., $\omega \to \nu_{\text{qu}}$), which coincided with ω only in the limit of high quantum numbers. Similarly, the classical amplitudes C_τ had to be changed into A_{qu}, the transition amplitudes of the atom, which could be either absorption amplitudes, A^a, or emission amplitudes, A^e. Finally, one had to substitute differences in place of differential coefficients in accordance with the rule (51). In the case of coherent scattering, this procedure yielded—corresponding to the classical expression (54)—the quantum-theoretical-induced electric moment,

$$P_{\text{qu}}(\nu,t) = \text{Re}\left\{\sum_a \frac{1}{4h}\left[\frac{A^a(E\cdot\bar{A}^a)}{\nu_a-\nu} + \frac{\bar{A}^a(E\cdot A^a)}{\nu_a+\nu}\right]\right.$$

$$\left. - \sum_e \frac{1}{4h}\left[\frac{A^e(E\cdot\bar{A}^e)}{\nu_e-\nu} + \frac{\bar{A}^e(E\cdot A^e)}{\nu_e+\nu}\right]\right\}. \tag{55}$$

The a-sum goes over all frequencies ν_a, for which the atom shows selective absorption; the e-sum goes over all discrete emission frequencies. Equation (55) is identical to Kramers' Eq. (50) if all vectors, A^a, $\bar{A}^a$, A^e and $\bar{A}^e$, are parallel to the vector E; in that case the identity of the two expressions is quickly realized if one introduces the relation between the Einstein coefficients and the quantum-theoretical amplitudes (e.g., $a_{ik} = a^e = (16\pi^4(\nu^e)^3/3c^3)(A^e\bar{A}^e)$, etc.). For atomic systems, whose atoms are not bound isotropically, the Kramers formula (50) will fail to hold. The authors noted that the range of validity of the quantum-theoretical equation for coherent scattering was similar to that of the corresponding classical formula; it would apply basically to those frequencies ν of the incident radiation that were very different from the absorption and emission frequencies. Equation (55) should, however, still be useful in the vicinity of a resonance line, as it predicts strong absorption in that region, which is also actually seen. If, on the other hand, ν approaches an emission line, the conclusions from Eq. (55) may have to be modified, for then spontaneous emission comes into play.

For incoherent scattering, the double sum in Eq. (53b) has to be taken into account in classical theory, and that also remains true in quantum theory. The induced electric moment corresponding to $P_1(t)$, Eq. (53b), is obtained by transcribing all terms on the right-hand side according to the rules stated above; thus one finds a physical interpretation for the terms in the expression for $P_1(t)$. Suppose that the electric moment of an atom has to be calculated when it is in a stationary state P, and that under the influence of incident radiation it goes over into a state Q. Then the classical formula tells us that this process occurs in two steps: first, by exciting the harmonic denoted by the indices τ, then by adding further excitations given by the indices τ'. In quantum theory, there corresponds to these steps a composite transition as follows: first, from the state P to an intermediate state R; second, from R to Q, the transition frequencies being ν_p and ν_q, respectively. With this interpretation, the quantum-theoretical moment corresponding to $P_1(t)$, Eq. (53b), becomes

$$P_1^{\mathrm{qu}}(t) = \mathrm{Re}\left\{ \sum_Q \sum_R M(Q,P:R) + \sum_Q \sum_R M(P,Q:R) \right\}, \tag{56}$$

where

$$M(P,Q:R) = \frac{1}{4h}\left[\frac{A_q(E\cdot A_p)}{\nu_p+\nu} - \frac{A_p(E\cdot A_q)}{\nu_q+\nu} \right]\cdot \exp\left[2\pi i(\nu_p+\nu_q+\nu)t\right]. \tag{56a}$$

That is, to the dispersive electric moment of an atom a sum of terms contributes, each term representing a transition from the given state P to another, possible state Q of the atom, via a third state R. Moreover, the frequency sum, $\nu_p + \nu_q + \nu$, associated with $M(P, Q:R)$, may be either positive or negative, and these two possibilities give rise to different physical interpretations: first, if it is positive, an atom, while going from the state P into the state Q, will absorb the energy-quantum $h\nu$ and emit the energy-quantum $h(\nu_p + \nu_q + \nu)$; second, if it is negative, an atom will make the same transition by first losing the energy-quantum $h\nu$ and then the energy-quantum $h[-(\nu_p + \nu_q + \nu)]$. While the first process had been predicted by Smekal (1923c), the second one did not seem to agree with the light-quantum concept, as it involved the emission of two light-quanta.[208] In the R-sum of the expression for $P_1^{\mathrm{qu}}(t)$, all intermediate states R have to be taken into account, via which one can reach a given state Q from the state P. Finally, the Q-sum signifies that, in order to calculate the quantum-theoretical moment of an atom in the state P, one has to consider the transition to all states Q that are compatible with energy conservation. As a consequence, the Q-sum in the first term in Eq. (56) includes all states Q, whose energy is smaller than the energy of the state P plus the energy-quantum $h\nu$; however, in the second sum, one is restricted to those states Q whose energy is smaller than the difference between the P-state energy and $h\nu$. Hence, if one considers *only the contribution of a state Q, whose energy is higher* than the energy of the state P,

[208] Such a process—if all particles were massive—would correspond to a collision of three bodies and seemed to be a rare event.

one finds from Eq. (56) the following result:

$$P_1^{\mathrm{qu}}(\nu - \nu^*) = \mathrm{Re}\left\{ \frac{1}{4h} \left[\sum_{R_a} \left(\frac{A_2(E \cdot \bar{A}_1)}{\nu_1 - \nu} + \frac{\bar{A}_1(E \cdot A_2)}{\nu_2 + \nu} \right) \right.\right.$$
$$+ \sum_{R_b} \left(\frac{\bar{A}_4(E \cdot \bar{A}_3)}{\nu_3 - \nu} - \frac{\bar{A}_3(E \cdot \bar{A}_4)}{\nu_4 - \nu} \right)$$
$$\left.\left. + \sum_{R_c} \left(- \frac{\bar{A}_6(E \cdot A_5)}{\nu_5 - \nu} - \frac{A_5(E \cdot \bar{A}_6)}{\nu_6 - \nu} \right) \right] \exp\left[2\pi i(\nu - \nu^*)t\right] \right\}, \tag{57}$$

where $h\nu^*$ is the (positive) energy difference between two states Q and P. In Eq. (57), the R_a-sum includes all states, whose energy is higher than the energy of the state Q; the R_b-sum includes those states whose energy lies between the states Q and P; and the R_c-sum includes the states lower than P. The frequencies ν_1 and ν_2 correspond to the transitions from P to R_a and Q to R_a, respectively; ν_3 and ν_4 are the frequencies of the transitions from P to R_b and from Q to R_b, respectively; ν_5 and ν_6 are the frequencies of the transition from P to R_c and from Q to R_c, respectively. All frequencies $\nu_1, \ldots, \nu_6$, are taken to be positive.

The discussion of the quantum-theoretical formula describing the incoherent scattering is now straightforward. Let us turn to the situation described by Eq. (57).[209] The R_a-sum yields large values of the induced moment in those cases, in which ν, the frequency of the incident light, comes close to the absorption frequency ν_1 (corresponding to the transition from the state P to the states R_a), or in those cases, in which the scattered frequency, $\nu - \nu^*$, approaches the frequency ν_2. This behaviour has to be expected for physical reasons. But, by investigating the R_c-sum, a strange situation is observed. The last term states, in particular, that the atom in the state P should exhibit a large induced electric moment also under the action of incident light having a frequency close to the transition frequency ν_6. This frequency does not refer to any transition from the state P at all, but to the transitions from Q to R_c. Now it is evident on physical grounds that only those frequencies are critical which approach the transition frequencies from that state whose scattering moment is considered. The term with the unphysical denominator (i.e., the second term in the sum over R_c in Eq. (57)) is an example of the terms which Kramers originally wanted to drop in writing down the dispersion formulae. His resolution of the puzzle was the

[209] Kramers and Heisenberg also discussed other cases: e.g., contributions to $P_1^{\mathrm{qu}}(t)$, in which either the states Q have lower energy than the state P or the same energy. In the latter case (the states P and Q identical) one has either coherent scattering or a degenerate system, in which the phase of the scattered light becomes undetermined (Kramers and Heisenberg, 1925, pp. 700–701).

following. For small denominators, $\nu_6 - \nu = \delta$, the frequency of the scattered radiation, $\nu - \nu^*$, assumes the value $\nu_5 - \delta$; i.e., it comes close to the absorption frequency ν_5 (corresponding to the transition from P to R_c). Now the frequency ν_5 is already contained in the virtual field of the atom, namely, in $P_0(t)$, Eq. (53a), which gives the quantum-theoretical expression replacing the classical unperturbed electric moment; and it gives rise to spontaneous radiation from the atom, described by the term $\mathrm{Re}\{A_5 \exp[2\pi i \nu_5 t]\}$. If the frequency $\nu - \nu^*$ approaches the frequency ν_5, the term $P_1^{\mathrm{qu}}(\nu - \nu^*)$—which is actually $\mathrm{Re}\{-(A_5(E \cdot \bar{A}_6)/4h\delta)\exp[2\pi i(\nu_5 - \delta)]$—and the term with frequency ν_5 in $P_0(t)$ begin to interfere destructively because they carry opposite signs. For a more detailed comparison of the two terms, one has to take into account the finite lifetime of the stationary states. It was possible to do so only in a crude model; however, the result was that the terms describing the scattered radiation and the spontaneous emission had indeed the same order of magnitude. Hence, in the special case of incident radiation with frequency close to ν_6, the action of the last term on the right-hand side of Eq. (57) was neutralized, and Kramers was thus able to save the physical interpretation of the quantum-theoretical dispersion formula, Eq. (56).[210]

The paper of Kramers and Heisenberg was well received, especially by those physicists who tended to follow Niels Bohr. One reason for this was that the results were consistent with what were considered to be the indispensable tenets of atomic theory. For example, Kramers and Heisenberg explicitly stated in their concluding remarks (Section 5) that their dispersion formula, Eq. (56), would leave the energy distribution in blackbody radiation invariant, and it would also lead to the usually assumed (and observed) statistical equilibrium of atomic states in samples of atoms. In particular, their entire reasoning was based on the correspondence principle, which was the most important tool of quantum theory in Copenhagen. Thus they remarked:

> The discussion in this paper shows that there can hardly be any other way than through a formula of the type [(56)], where P is of the characteristic form [(56a)], of satisfying the "correspondence" requirement that the scattered radiation induced in an atomic system by external radiation should coincide in the limit of large quantum numbers with the scattered radiation demanded by classical theory. Even if the description of the true scattering moment should require a less simple expression for P, it would be difficult to escape the conclusion that the moment of the scattered radiation will assume a particularly large value whenever a condition of the form $\nu + \nu_p$ or $\nu + \nu_q = 0$ is satisfied. (Kramers and Heisenberg, 1925, p. 707)

And, after referring to the removal of unphysical terms in the dispersion formula

[210]Similar difficulties, as in the case described by Eq. (57), arise in other cases. The situation is always the following: among the terms, which contribute to the induced moment, there is one which becomes large if the frequency ν approaches a frequency that is not associated with the state of the atom under investigation. But in all cases there is a term in the spontaneous emission that cancels the contribution from the unphysical term.

due to interference with spontaneous radiation, they remarked:

> From this one may conclude that an interpretation, according to which the scattering action of an atom in the presence of incident (external) radiation must be considered as the action of an atom in a definite stationary state, necessarily implies the view that also the spontaneous radiation of an atom must be described as the radiation in a definite stationary state and not, say, as an action which the atom exerts only in a transition between stationary states. This, however, is exactly the hypothesis introduced by Slater, which led to the views developed in detail by Bohr, Kramers and Slater. (Kramers and Heisenberg, 1925, p. 708)

The feeling that the work on dispersion theory had confirmed the radiation theory of Bohr, Kramers and Slater was particularly strong in Kramers. He therefore thought that the task for the immediate future was to seek to describe other atomic phenomena in the language of dispersion theory.[211] Kramers' view of pursuing such a programme did not change even when in April 1925 the experiment of Walther Bothe and Hans Geiger on the Compton effect disproved one of the fundamental assumptions of the Copenhagen radiation theory, namely, that energy and momentum were conserved only statistically in atomic processes (Bothe and Geiger, 1925a). As Max Born had noted already a year earlier, the dispersion formula was independent of this assumption. Thus, even those who were critical of the Bohr–Kramers–Slater radiation theory agreed with the results of the new dispersion theory. One of them was Pauli; since fall 1924 he had assumed a completely negative attitude towards the Copenhagen radiation theory, yet a year later he referred to 'the beautiful dispersion formula' of Kramers.[212] Still, the positive reception of the paper of Kramers and Heisenberg by most experts in 1925 was surprising for several reasons. First, their theory departed not only from classical theory, but also from the old quantum theory. Yet many people seemed to have confidence in the dispersion formula 'because it looked all right.' It seemed to represent a reasonable compromise for describing the different aspects of nature that one knew about: it took care of the undulatory properties of radiation as well as of many light-quantum aspects; in particular, it seemed to account for several features of the Compton effect. The second point was that, in spite of all the theoretical arguments supporting the Kramers–Heisenberg formula, it could not be based on direct empirical evidence; nothing had been observed that could be compared with it. The effect predicted by Smekal, and included in Eq. (56), would not be seen until several years later, when the Indian physicist Chandrasekhara Venkata Raman (1928) and the Russians Grigorii Landsberg and Leonid Mandelstam (1928) independently discovered the discrete displacement of the frequency of scattered light.

[211] In 1925 Kramers published several papers on the dispersion-theoretical approach in atomic physics (see, e.g., Kramers, 1925a).

[212] In a letter to Ralph Kronig he wrote: 'Many greetings to Kramers, whom I basically love very much, especially when I think of his very beautiful dispersion formula' (Pauli to Kronig, 9 October 1925).

All one knew was that Ladenburg had obtained very good results with his formula, Eq. (49); Kramers' formula, Eq. (50), appeared to be better because of theoretical arguments, and Eq. (56) just occurred, again on theoretical grounds, as a proper generalization of Kramers' formula. Heisenberg characterized the feelings of the physicists close to Bohr by saying:

> One felt that one had now come a step further in getting into the spirit of the new mechanics. Everybody knew that there must be some new kind of mechanics behind it. Nobody had a clear idea of it, but still one felt that the dispersion formula was a good step in the right direction. The feeling about what the new mechanics should look like was rather widespread, at least among ten or fifteen people at different places, in Copenhagen, Göttingen, Cambridge, Leyden, etc. (Heisenberg, Conversations, p. 144)

For Heisenberg, the collaboration with Kramers on the dispersion formula had a quite different effect. As we have mentioned before, he had not planned to work on this problem when he arrived in Copenhagen; he had been drawn into it through discussions at Bohr's Institute, and also because of his ambition to contribute to a theory that was considered to be important and of fundamental interest. However, once he got into it, he participated in it with full force and contributed his own ideas to it. Afterwards he drew his own consequences from this work. He noticed, in particular, that the dispersion formula provided the first example of a problem in which the use of the correspondence principle yielded an unambiguous result—a result, in which all the terms could be explained and had a definite meaning, including the amplitudes and the phases. He recalled later:

> That made a strong impression on me at that time. Then I saw that the analogy [of the quantum-theoretical quantities] to the Fourier components in classical physics was really very close, because not only the absolute value [of the transition amplitudes], but even the phase must somehow be well defined. Of course, only later on did one learn that the phase was well defined except for an arbitrary phase factor in every stationary state. But when one had an intermediate state then this arbitrary phase went out. The important thing was that one could actually have interference. Therefore all this interference argument of Kramers' was important; one could see that there the phase comes in and is important. (Heisenberg, Conversations, p. 141; also AHQP Interview)

Thus, the collaboration with Kramers strengthened in Heisenberg's mind the consequences, which already had occurred to him in connection with the treatment of resonance fluorescence. The main lesson was that in atomic theory the electron orbits had somehow to be replaced to get a better idea of what was happening. Soon a programme emerged, which he described as follows: 'One could see that the Fourier components were the reality, and not the orbits. So one had to look for those connections between the Fourier components, which were true in classical mechanics, and to see whether or not similar connections were true also in quantum mechanics—i.e., if one took, instead of the Fourier components, the transition amplitudes of the real lines of atoms' (Heisenberg, Conversations, p. 135).

This programme, to which Heisenberg would turn several months later, led to the first steps in the formulation of quantum mechanics. However, in winter 1924–1925 he did not yet go so far; he rather became concerned again with the most difficult problem of atomic theory: the problem of complex spectra and the anomalous Zeeman effects. With this work he fulfilled the purpose of his going to Copenhagen—though, in a way, it delayed the progress towards the discovery of quantum mechanics.

III.6 Atomic Models and Complex Spectra

During the year 1924 there occurred a polarization of views among those who worked at the frontiers of atomic theory. This polarization resulted from the fact that people gave different answers to the question as to how the crisis in the explanation of crucial quantum phenomena, such as the Compton effect or the spectra of complex atoms, had to be overcome. There was general agreement that most of the difficulties of the usual theoretical description pointed to one fundamental difficulty, but physicists disagreed sharply about the methods by which this difficulty might be overcome. The two principal competing approaches were represented, respectively, by Niels Bohr and Arnold Sommerfeld, who were also the authors of the theory of atomic structure that had been found to be wanting. While Bohr and his co-workers relied in their research on the correspondence principle and a new radiation theory, in which the discrete, nonclassical features of light were accounted for by virtual fields of the atoms and one could dispense with the light-quantum, Sommerfeld considered the latter as one of the basic cornerstones of a future theory. In addition, as we have noted earlier, Bohr and Sommerfeld followed quite different methods of explaining the properties of complex spectra: Bohr proceeded by treating simple systems and calculating the more complicated ones with the help of perturbation theory and correspondence arguments; Sommerfeld still regarded the phase integral method as the proper tool for arriving at the correct description of the data, especially since it allowed one directly to find ratios of integral numbers, which seemed to be demanded not only by the observed patterns of spectral lines but also by the ratios of their intensities.

By studying the papers published since early 1923 one could recognize how the respective approaches of Bohr and Sommerfeld diverged more and more. At the same time, hidden from the scientific public, another unexpected opposition to the Copenhagen view developed slowly. This opposition came from Wolfgang Pauli, formerly a student and collaborator of both Sommerfeld's and Bohr's. After the unsuccessful attempts to arrive at a consistent theory of anomalous Zeeman effects on the basis of Bohr's principles of atomic theory in 1923, Pauli had retired from contributing to the problems of atomic structure. Since he believed that those questions were too difficult to be answered at the time, he had assumed in 1924 the role of a watchman, observing critically the experimental and theoretical progress brought about by others, while working on a completely different problem himself—the kinetic theory of heat conduction in solids. Pauli followed two developments most critically: one was the theory of the anomalous Zeeman effects, which Heisenberg had worked on since fall 1923; the other was the radiation theory of Bohr, Kramers and Slater of early 1924. It is true that, as a result of his visit to Copenhagen in April 1924, Pauli had been persuaded by Bohr to accept the new theory, but later in the year he changed his attitude entirely. Following the Innsbruck meeting of scientists and physicians (*Versammlung der Gesellschaft Deutscher Naturforscher und Ärzte*), where he met and discussed with Albert Einstein, Pauli wrote a long letter to Bohr, dated 2 October 1924, in which he summarized both Einstein's and his own arguments against the Bohr–Kramers–Slater theory. Pauli partly shared Einstein's objections, such as the insistence on strict conservation of energy and momentum in the interaction between light and atoms, although he admitted that experiments had so far not decided either for or against it. His main arguments, however, he directed against another consequence that had been derived from the Copenhagen radiation theory: namely, that two different types of resonance radiation occurred in nature, one emitted from excited atoms and the other from unexcited atoms.[213] 'I am convinced,' wrote Pauli, 'that to this distinction there corresponds nothing the least real in actual fact, and that no experiment exists which allows one to determine the contributions of the excited and the unexcited atoms to the scattered radiation separately' (Pauli to Bohr, 2 October 1924). He then went on to discuss the *Gedankenexperiment*, by means of which Bohr had tried to support the separate existence of the two kinds of resonance radiation and showed that it did not work. Moreover, Pauli argued that the different types of radiation, predicted on the basis of the Bohr–Kramers–Slater theory, contradicted the correspondence principle. He concluded: 'Hence (i.e., because of the connection to the classical theory) I do not believe it is possible in a reasonable way to distinguish within a [spectral] line Compton jumps and real quantum jumps' (Pauli to Bohr, 2 October 1924).[214]

Pauli's arguments against certain consequences of the Bohr–Kramers–Slater theory of radiation had an immediate impact on the work in Copenhagen. These arguments were examined, in particular, in connection with the problem of resonance fluorescence. After discussing the details of the dispersion of light,

[213]Bohr, Kramers and Slater had stated in their paper that one part of resonance radiation occurred through the decrease of the intensity of incident light, due to the destructive interference with the secondary wavelets arising from the illuminated, unexcited atoms; the other part, which was connected with the induced transition of excited atoms, served 'only as an accompanying effect by which a statistical conservation of energy is ensured' (Bohrs, Kramers and Slater, 1924, p. 798).

[214]The argument in the paper of Bohr, Kramers and Slater was that the Compton jumps were due to a coherent process, i.e., the interference of the incident light with the secondary wavelets emitted from illuminated electrons bound in atoms.

including the incoherent scattering (in Kramers and Heisenberg's extension of the quantum-theoretical dispersion formula), Bohr and his co-workers arrived at the following conclusion: 'Hence one is allowed to assume that also in the general case of nondegenerate atomic systems that part of the fluorescence light, which is emitted by atoms in the ground state, has the same structure as the part of this [fluorescence] light arising from activated [i.e., excited] atoms' (Bohr, 1924, p. 1116, footnote 5). Thus, the Copenhagen physicists agreed that, from the point of view of the correspondence principle, the separation of fluorescence light into two types did not make sense. They admitted frankly that the consequences from their radiation theory should not be taken literally. 'Perhaps I should have a bad conscience because of the radiation problems,' wrote Bohr to Pauli in December 1924. 'However, even if it might be a crime from a logical point of view, I must confess that I am nevertheless convinced that the trick ("*Schwindel*") of mixing the classical theory and the quantum theory will turn out to be fruitful in tracing the secrets of nature in many situations yet' (Bohr to Pauli, 11 December 1924).

In the same letter, in which Bohr answered Pauli's objections concerning radiaton theory, he inquired about Pauli's new ideas about atomic theory that had been brought to his attention in a very indirect way. Pauli had indeed been stimulated by his discussions with Einstein and others at the Innsbruck meeting (*Naturforscherversammlung*) to concern himself again with the quantum theory of atomic structure. One result was soon available and reported to Sommerfeld in a letter, dated 29 September 1924, as we have mentioned in Section III.4: Pauli derived the intensity rules of Ornstein and his collaborators in agreement with the correspondence principle, thus supporting the Copenhagen point of view. But then he turned to other questions and found that the empirical data demanded a description that would contradict many of the sacrosanct assumptions of Copenhagen physicists. First, Pauli reconsidered his old problem, the anomalous Zeeman effect, and arrived at an interesting conclusion: if, in the usual atomic-core model (of Landé and Heisenberg), one calculates the relativistic correction to the g-factor of the core, then it is given by a quantity minus δ ($= \frac{1}{2}\alpha^2 Z^2/n^2$, where α is the fine-structure constant, Z is the effective electric charge in units of the absolute value of the electron's charge and is close to the atomic number, n is the principal quantum number), and this correction is large for heavier atoms—e.g., for mercury δ should be about 18%. As Pauli had confirmed from Alfred Landé and Ernst Back of Tübingen, such relativistic corrections had not been observed, and one had to assume an unknown compensating effect if one wanted to stick to the atomic-core model of the anomalous Zeeman effects. Although a compensating effect could not be excluded, Pauli, in a paper submitted to *Zeitschrift für Physik* on 2 December 1924, argued in favour of doubting the foundation of the atomic-core model (Pauli, 1925a). In this paper he suggested in particular that the closed shells (representing the core of alkali atoms) should not contribute to the magnetic moment and the angular momentum of the atom, hence the relativistic corrections would also not show up in the anomalous

Zeeman effects. To explain the latter, he proposed the following alternative view: The magnetic moment and the angular moment of an atom, as well as the splittings of its energy terms in external magnetic fields, have their origin solely in the properties of the series electron; in particular, the doublet structure of alkaline spectra and the breakdown of Larmor's theorem (in the anomalous Zeeman effects) '*arises, according to this view, by a peculiar, classically not describable kind of duplicity of the quantum-theoretical properties of the series electron*' (Pauli, 1925a, p. 385).

As the principal fact in support of his new point of view concerning the origin of complex spectra and anomalous Zeeman effects, Pauli quoted the observations of Millikan and Bowen (1924b) and of Landé (1924a, b), pointing to the common origin of relativistic doublets and optical doublets. Now, using his new point of view, he immediately derived a most important consequence for the theory of atomic structure: he combined the assumption that closed shells in atoms do not contribute to the complex structure of atomic spectra with Edmund C. Stoner's proposal (Stoner, 1924) to reorganize the quantum numbers of electrons in atoms to give a new description of atomic structure, different from Niels Bohr's—which had served for more than three years as the fundamental basis for all calculations. In Bohr's theory, two quantum numbers—the principal quantum number n and the quantum number k, obtained from the relativistic fine structure—described the electron orbits in atoms; several electrons could have identical quantum numbers (e.g., in the ground state of helium two elecrons were on the $n_k = 1_1$ orbit, in beryllium two electrons occupied the 2_1-orbit, and in neon four electrons occupied both the 2_1- and 2_2-orbits). Second, the building-up principle (*Aufbauprinzip*) was valid; it guaranteed the permanence of given quantum numbers associated with each electron, when further electrons were added to an atom. Third, empirically there were only $2n^2$ electrons in all orbits whose principal quantum number was n (i.e., in the nth shell); this followed from the length of the periods in the system of elements. Although the latter fact had not yet emerged from any theoretical reasoning, Bohr had hoped to explain it ultimately with the help of the correspondence principle; in particular, he thought that the selection rules might work in this direction. Bohr's atomic theory had been very successful, its greatest success being the prediction of the properties of the element with the atomic number $Z = 72$, which was discovered in 1922 at Bohr's Institute in Copenhagen and christened 'hafnium.' However, by 1924 the failures of the theory had become evident. For example, serious difficulties had arisen with the number and organization of X-ray levels, apart from the principal theoretical problems in many-electron systems. Moreover, not the slightest progress had been made in the problem of explaining the completion of electron shells. Then Stoner had an idea; he found that 'the number of electrons in each completed shell is double the sum of the inner quantum number as assigned, there being in the K, L, M, N levels when completed, 2, 8 ($= 2 + 2 + 4$), 18 ($= 2 + 2 + 4 + 4 + 6$), 32, . . . electrons' (Stoner, 1924, p. 722). He also suggested that the number of electrons associated with each sublevel was equal to twice the value of the inner quantum number. Stoner thus characterized the atomic levels by three instead of two quantum numbers: the principal quantum number n; the azimuthal quantum number k_1; the inner quantum number k_2 $(= j + \frac{1}{2}$, j being the inner quantum number of Sommerfeld), which was connected with the relativistic correction in X-ray spectra. Thus, he rearranged the 2_1- and 2_2-levels of Bohr, each of which was occupied maximally with four electrons, in the following way: he introduced a level denoted by the quantum numbers $n = 1$, $k_1 = 1$ and $k_2 = 1$ and occupied it maximally with $2k_2 = 2$ electrons; a level denoted by the quantum numbers $n = 2$, $k_1 = 2$, $k_2 = 1$ and occupied it again with maximally two electrons; and finally a level denoted by the quantum numbers $n = 2$, $k_1 = 2$ and $k_2 = 2$, which may be occupied by maximally $2k_2 = 4$ electrons. The sum of maximum occupation numbers was eight, as in Bohr's previous assignment. Stoner noticed immediately that his assignment yielded 'the observed term multiplet as revealed by the spectra in a weak magnetic field' (Stoner, 1924, p. 725).

The advantage of Stoner's proposal was obvious. It provided a better way of counting the atomic levels than Bohr's, which had not even been unique; it solved the problems of X-ray spectra and their respective weights (see also Sommerfeld, 1925a). Pauli thought that Stoner's suggestion had solved Bohr's problem of completing the atomic shells. He even argued that this suggestion should be sharpened in the following sense: the states of an electron are completely determined by the four quantum numbers n, k_1, k_2 and m_1 (m_1 being the magnetic quantum number associated with the splitting by external fields; in strong magnetic fields one may replace k_2 by a second magnetic quantum number m_2); two or more equivalent electrons never existed in atoms, for which all four quantum numbers, n, k_1, k_2 and m_1, were the same. Pauli's sharpened formulation or rule applied strictly only to atoms in strong magnetic fields, but it could also be transferred to the case of weak fields with the help of the adiabatic principle.

Pauli, whose attention had been attracted to Stoner's paper by a complimentary reference to it in Sommerfeld's new, fourth edition of *Atombau und Spektrallinien* (a copy of which he received from his former professor in November 1924), immediately drew some consequences. For example, he noticed that the last statement of his rule, which was later called the 'exclusion principle,' automatically took care of the hitherto unexplained fact that in the spectra of alkaline-earth elements, the triplet s-term having the same principal quantum number as the singlet s-term—the ground state—did not exist.[215] While he was still collecting further confirmation of his new concepts, Pauli informed Alfred Landé in Tübingen and Arnold Sommerfeld in Munich about his results; and via Landé

[215] Pauli's exclusion principle states that two electrons with identical quantum numbers, $n = k_1 = k_2 = 1$, must have magnetic quantum numbers $m_1 = \frac{1}{2}$ and $m_1 = -\frac{1}{2}$, respectively. Then the internal quantum number of the atom must be $j = 0$, and only a singlet s-term exists.

and Leyden the news ultimately reached Bohr in Copenhagen.[216] Bohr was, of course, very interested in Pauli's ideas, and he immediately wrote to Pauli, explaining his curiosity in the following words: 'Not the least to follow your advice that I should publish my considerations on the Rydberg–Ritz formula, I am occupied these days in analyzing more closely in a short note the extent of the fraud ("*Schwindel*") on which the theory of atomic structure is founded. I would therefore very much like to hear from you some more details about the direction of your new thoughts' (Bohr to Pauli, 11 December 1924). Pauli, in turn, was happy about Bohr's interest and sent him instantly the manuscript of a paper, entitled '*Über den Zusammenhang des Abschlusses der Elektronengruppen im Atom mit der Komplexstruktur der Spektren*' ('On the Connection between the Completion of Electron Groups in an Atom with the Complex Structure of Spectra'), which he had just completed. In the accompanying letter he wrote: 'Just today I was about to send the enclosed manuscript of an as yet unpublished paper to Heisenberg. (Especially to Heisenberg, for I believe that of all physicists he will be the last to agree with it.) Just now your kind letter arrived, about which I am enormously happy. I see from it that you always think of me very kindly, and that at the moment you are very interested in what occupies me. Therefore I am sending the manuscript directly to you' (Pauli to Bohr, 12 December 1924).

It may sound a little surprising that Pauli expressed the belief that *Heisenberg* would be the last to agree with the paper he had written. Actually he had contradicted in his paper one of *Bohr's* basic arguments, namely, that the completion of electron shells in atoms had something to do with the correspondence principle. And he wrote to Bohr:

> I have told you repeatedly that in my opinion the correspondence principle has in fact nothing to do with the problem of the closure of electron goups in the atom. You have always replied to me that you believed that I was *too* critical in this regard. Now, however, I believe that I am quite certain of my stand. The exclusion of certain stationary states (*not* of transitions), which one is dealing with here, is rather similar in principle to the exclusion of the states $m = 0$ or $k = 0$ (in the case of the hydrogen atom) than, say, to the selection rule $\Delta k = \pm 1$. Do you still stick to your alleged application of the correspondence principle in this situation? I believe that in reality everything is much simpler; one need not talk at all, in this connection, about some harmonic exchange. (Pauli to Bohr, 12 December 1924)

This was a direct reproof of Bohr's programme. And yet, Pauli pointed out that

[216] The first person, who learned about Pauli's renewed interest in complex spectra, was Landé. To him Pauli wrote two letters, dated 10 and 14 November 1924, dealing with the proposal to consider the relativistic correction to the g-factors; and in the letter of 24 November 1924 he gave a detailed report concerning the new description of electrons with four quantum numbers. The main reason for writing to Landé was to obtain from Tübingen experimental data that might confirm or disprove his ideas. Pauli authorized Landé explicitly to mention his ideas privately to Paul Ehrenfest on a visit to Leyden (Pauli to Landé, 14 and 24 November 1924). At Leyden, Dirk Coster heard the news; he took it to Kramers, who, in turn, informed Bohr.

his new point of view was not destructive, for he wrote:

> Now, finally let me turn to a fundamental point. The view, from which I start, is—I am quite sure—nonsense. For one does not succeed in this way in reconciling the type of motion of the precessing central orbit, as required by the correspondence principle, with the relativistic doublet formula. This type of motion appears to me to be inconsistent at the moment even with my assumptions. But I believe that what I do here is *no bigger* nonsense than the present interpretation of the complex structure. My nonsense is conjugate to the hitherto usual nonsense. Just because of that I believe that this nonsense must necessarily be tried in the present state of the problem. The physicist, who once will succeed in compounding both of these nonsenses, will obtain the truth! (Pauli to Bohr, 12 December 1924)

Thus Pauli considered the main virtue of his alternative view of complex structure to be that it might help physicists to overcome the previously insurmountable difficulties of atomic theory and to restart working with a fresh spirit. Indeed, he expressed considerable optimism; perhaps, he thought, it was possible to extend the correspondence principle to apply also to atoms having several electrons by taking into account both views, the old (Bohr's) and the new (his own). On the way towards a future, satisfactory theory, Pauli added, one had to introduce further changes in the concepts of classical theory. Hence he wrote to Bohr:

> It seems to me that the relativistic doublet formula shows without doubt that not only the dynamical concept of force, but also the kinematic concept of motion of the classical theory, will have to experience a profound modification. (Therefore I have also consistently avoided using the notion of an "orbit" in my paper.) Since this concept of motion also lies at the basis of the correspondence principle, the efforts of the theoreticians must concentrate on its clarification. I believe that the energy and momentum values of stationary states are much more real than "orbits". (Pauli to Bohr, 12 December 1924)

He concluded this consideration of principles by proposing the following programme:

> The goal (not yet achieved) must be to deduce these [i.e., the energy and momentum values] and all other physically real, observable properties of the stationary states from the (integral) quantum numbers and quantum-theoretical laws. We must not bind the atoms in the chains of our prejudices ("*nicht die Atome in die Fesseln unserer Vorurteile schlagen*")—to which, in my opinion, also belongs the assumption that electron orbits exist in the sense of ordinary mechanics—but we must, on the contrary, adapt our concepts to experience. (Pauli to Bohr, 12 December 1924)

Pauli's letter and the enclosed manuscript were studied in Copenhagen with great attention and care. 'We are all quite enraptured by the many new, beautiful things which you have discovered,' replied Bohr ten days later (Bohr to Pauli, 22 December 1924). The Copenhagen physicists were happy that Pauli, after a long abstinence, had again begun to participate in work at the frontiers of atomic theory. Although Pauli's new beginning spelled a major criticism of the earlier endeavours on the theory of atomic structure—it was directed both against Bohr's theory of the periodic system of elements and the core theory of multiplet structure—Bohr especially welcomed Pauli's optimistic attitude. Therefore, he did not quarrel about specific points in Pauli's proposal, but answered frankly:

> I need not utter any general criticism, since you have yourself characterized, better than anyone else could do, the whole thing as complete insanity in your letter. On the other hand, I also do not believe that you can emphasize the insanity of the usual momentum considerations more strongly than we have done here already in our discussions before we learned about your latest results . . . Altogether, both kinds of insanity [i.e., the old views of Bohr and others, and Pauli's new views on atomic theory] may be connected too closely to the truth so that one cannot criticize them as isolated aspects. (Bohr to Pauli, 22 December 1924)

Bohr then argued that, since Pauli had not given up the (classical) concepts of space altogether, he did not take the verdict against the explanation of the completion of electron groups from the point of view of the correspondence principle as the final word. He emphasized his conviction that the specific coupling between electrons having identical quantum numbers might eventually work in the same direction as the exclusion principle. 'You see, dear Pauli,' Bohr remarked, 'that you have achieved everything to set our thinking into motion. I have a feeling that we stand at a turning point, since now the extent of the entire swindle has been characterized so exhaustively' (Bohr to Pauli, 22 December 1924).

In his letter to Pauli, Bohr also reported about the status of work at his Institute. Since Heisenberg's return to Copenhagen in September, Bohr, Kramers and Heisenberg had been involved in continuous discussions on the problem of many-electron atoms and complex structure. However, no progress had been achieved so far. Only one specific point had become clear in connection with the investigations on resonance fluorescence: the electron motion in atoms was largely independent of the reaction of atoms to external actions. Bohr and Heisenberg hoped to use this result profitably in their future treatment of many-electron atoms, especially in considering the mutual interactions of the bound electrons. But in general the Copenhagen theoreticians encountered great difficulties, and it soon became evident that the prevailing conceptions of atomic structure did not provide appropriate models that could be treated with the sharpened correspondence approach. At this very moment Pauli's new ideas were extremely welcome; and Bohr and Heisenberg had a great desire to meet Pauli as soon as possible in order to discuss the entire matter in detail and draw further consequences, which could form the basis of an improved theory of atomic structure. Bohr invited Pauli to visit Copenhagen on his return to Hamburg from Vienna (after the Christmas vacation) in January 1925, while Heisenberg hoped to meet his friend earlier in Munich.

Pauli was, of course, right in assuming that Heisenberg's previous views on the structure of many-electron atoms were opposed to his recent findings. For example, the hypothesis, that completed shells of electrons in atoms would not contribute to their magnetic moment and angular momentum, clearly contradicted the core model, which Heisenberg had considered as his precious achievement and had used repeatedly, especially to explain the anomalous Zeeman effects. It is true that in the course of time the core model had adopted many features that could not be interpreted in the corresponding classical theory; Heisenberg had therefore emphasized the formal nature of the rules that followed from the improved core model of summer 1924 (Heisenberg, 1924e). Similarly, Pauli's investigations on the quantum numbers of electrons and their organization in atoms appeared to be as formal as the most recent quantum-theoretical description of the anomalous Zeeman effects. Hence Heisenberg, after a quick look at Pauli's new manuscript, wrote to him:

> Today I read your new paper, and it is certain that I am the person who is most happy about it: not only because you have pushed the swindle to a hitherto not anticipated height of trickery and have thus easily beaten all records, of which you have accused me (by introducing *single* electrons with *4* degrees of freedom); but altogether, I triumph because also you (*et tu, Brute*) have returned with lowered head ("*gesenktem Haupt*") to the land of the Philistines of formalism ("*Formalismusphilister*"). But don't be sad, you will be received there with open arms. And if you believe that you have yourself written anything against the present types of swindle, then you are wrong: because swindle times swindle will not give anything right, hence two swindles can *never* contradict each other. Therefore I congratulate you!!!!!!!! (Heisenberg to Pauli, 15 December 1924)

Since Heisenberg, in Copenhagen, followed the programme of providing adequate classical models for his earlier formal rules for complex structure, it might have escaped him that Pauli was about to abolish the model-dependent description of atomic phenomena altogether. But, after some reflection, he soon grasped the main point, and wrote to Landé: 'The most important in Pauli's paper appears to me really the negative aspect: that all hitherto prevalent points of view must be considered in a very symbolic and formalistic way (especially, Pauli's own)' (Heisenberg to Landé, 4 January 1925). This recognition of the formal nature of the model-dependent consideration had been familiar to Heisenberg for some time; hence he welcomed the fact that Pauli, who had always severely criticized his ad hoc assumptions in atomic theory as not being consistent with (classical) dynamics, had now converted to the same attitude. As long as the real model of the atom was not known, Heisenberg did not mind having an alternative to the insufficient core model. By the end of 1924, Heisenberg was no longer sticking stubbornly to his earlier views, and his general attitude may be construed from Bohr's remark to Pauli: 'Although we are not blind to the possibility that the limited validity of Larmor's theorem in classical theory perhaps assumes a central place in the entire swindle, still Heisenberg's conscience has not really become worse because of your remarks on this point; this is something which also could be borne with difficulty' (Bohr to Pauli, 22 December 1924). Where Pauli had expected strong hostility towards his new ideas, he found an open mind: in December 1924 Bohr and Heisenberg were fully aware of how unsatisfactory the entire model-dependent description of atoms was.

Heisenberg planned to devote the second part of his stay in Copenhagen in early 1925 to investigating the problem of complex spectra. In January 1925, on returning from an extended Christmas vacation in Munich and its vicinity, he began to work intensively on his programme.[217] From the very beginning he took into account, together with the core model, Pauli's new description of many-electron atoms, following his suggestion that 'perhaps the final solution of the problems discussed here will lie in the direction of a compromise between the two points of view' (Pauli, 1925a, p. 385).[218] In this sense Heisenberg remarked to Landé: "What do you think about Pauli's new paper [on the completion of electron groups in atoms]? Altogether I find it to be very beautiful, and if once one succeeds in combining the different formalisms of the Zeeman effect together into a unified description, then one would have arrived at a new theory' (Heisenberg to Landé, 4 January 1925). With this guiding principle in mind, he still made only slow progress, partly because in Copenhagen one first needed an understanding—based on the correspondence principle—of the characteristic two-valuedness of the electron, which had arisen from Pauli's new assignment of four quantum numbers to the electron. It was Kramers, who by his study of resonance fluorescence, seemed to have found an appropriate answer. As Heisenberg reported to Landé in a letter, dated 18 February 1925, it followed from Kramers' calculations, 'that the external radiation leads, through its proper "stress" on the atom, to a duplicity of the radiation reaction in the atom, such that with one energy value of the atom two types of radiation reaction, and with one type of radiation reaction two energy values, are associated.' This result appeared to Heisenberg to be the key to the previously undefined concept of '*Zwang*' (stress or constraint); it was also related to his earlier findings in the theory of complex structure. Hence he thought that 'this duplicity, which follows with near certainty from all assumptions of Bohr's radiation theory, now possesses a strange similarity with the "$\int_j^{j+1} H_{kl}\, dj$-mould ['*Schimmel*,' referring to

[217] Heisenberg left Copenhagen shortly after 15 December 1924 (as Bohr indicated in his letter to Pauli on 22 December) to go to Munich. He was not yet back in Copenhagen on 10 January 1925, when Bohr (with Kramers' help) wrote a letter to Pauli. Heisenberg mentioned in a letter to Landé: 'I myself have to stay on here [in Munich] another couple of weeks, since I have hurt myself in skiing' (Heisenberg to Landé, 4 January 1925).

[218] Heisenberg most probably met Pauli at Munich in early January 1925 and discussed with him the problems of atomic structure, for Pauli wrote to Bohr: 'I shall also go to Munich and definitely hope to meet Heisenberg at the "Institute for Number Mysticism" there. I have already taken up contact with him' (Pauli to Bohr, 31 December 1924). Heisenberg informed Landé: 'Next week Pauli comes here [to Munich]; I hope to talk to him a little. Then he will travel, so far as I know, through Tübingen to Hamburg' (Heisenberg to Landé, 4 January 1925).

prescription]"' (Heisenberg to Landé, 18 February 1925). That is, his old formal prescription (for obtaining the quantum-theoretical Hamiltonian of a multiply periodic system from the classical one) came out simply as a consequence of the Bohr–Kramers–Slater theory of radiation. With respect to his own problem, the treatment of many-electron atoms, he now generalized Kramers' result to the following hypothesis: If two atomic systems A and B (say, an outer electron and an atomic core) interact, then the 'constraint of their interaction' ('*Zwang ihrer Wechselwirkung*') creates a 'duplicity' ('*Zweideutigkeit*') in such a way, that to *one* pair of energy values of the two system (i.e., the energies of the two systems neglecting the interaction energy) there correspond *two* mechanical interaction energies, and *vice versa*. Heisenberg further argued: 'This statement may be evidently illustrated intuitively in two ways: 1. A acts on B, such that in B there arise for *one* energy two mechanical models responsible for the interaction. 2. B acts on A, such that in A there arise for *one* energy two mechanical models responsible for the interaction' (Heisenberg to Landé, 18 February 1925). From there Heisenberg immediately concluded: 'If one now applies the above hypothesis to the ion and the [series] electron, and further expresses (what is plausible for several reasons) the duplicity by two values of the angular momentum quantum number, then one recognizes that the schemes 1 and 2 become *qualitatively* identical with Pauli's new and the old $\int_j^{j+1} H\,dj$-prescription, respectively. Indeed, [the idea in] Pauli's new paper is completely equivalent to the duplicity of the angular momenta in the $\int_j^{j+1} H\,dj$-mould, if the electron and the ion are exchanged' (Heisenberg to Landé, 18 February 1925).

The result that his own scheme describing the anomalous Zeeman effects and Pauli's new scheme accounted for the same physical situation made Heisenberg quite euphoric, and he mentioned it in a short letter to Pauli, dated 26 February 1925. Pauli replied that, although he still disagreed in principle with the Copenhagen radiation theory, he would probably come to terms with Heisenberg about the theory of complex structure. He wrote:

> As to what you write about the k_1, k_2- [i.e., Pauli's new description] and the J, R-prescription [i.e., the core model of 1924] we shall, I believe, easily agree. Nothing do I want less than to bet on a certain mould [referring to these prescriptions as "mould" ("*Schimmel*")]. (They are all almost equally wrong.) I even think it is quite probable that one might describe the closure of the [electron] groups also by your prescription, if one excludes for the same principal quantum number the averaging over certain diagonals. (Pauli to Heisenberg, 28 February 1925)

Pauli also pointed out again the principal failures of the core model, especially the ad hoc assumption that the core should have a double magnetic moment (compared to the one in classical theory). He suggested, therefore, that one should attack the problem of complex structure on a deeper level and connect it with the problem of quantum theory of degenerate systems. Finally he expressed the hope that he would discuss these matters in detail with Heisenberg during his forthcoming visit to Copenhagen in March 1925.[219] Heisenberg took Pauli's letter as an encouragement to proceed with his programme of describing the complex structure from two different points of view. Already one week after he had received Pauli's letter, he announced to Landé that things 'now looked more reasonable,' and he asked the Tübingen theoretician some specific questions about the g-values of mercury in connection with his investigations (Heisenberg to Landé, 6 March 1925). On 10 April 1925, *Zeitschrift für Physik* received Heisenberg's new paper entitled '*Zur Quantentheorie der Multiplettstruktur und der anomalen Zeemaneffekte*' ('On the Quantum Theory of Multiplet Structure and the Anomalous Zeeman Effects,' Heisenberg, 1925b).

The paper on multiplet structure and anomalous Zeeman effects, which was the third work that Heisenberg submitted from Copenhagen, contained a complete discussion of the problem at that time. This paper had the blessing of Niels Bohr; also Pauli, who was in Copenhagen when the paper was completed, agreed with its overall content. The central goal of the paper was to formulate what Bohr had called 'the stress [constraint], which is not analogous to the action of external force fields' ('*der Zwang, der nicht mit der Wirkung äusserer Kraftfelder analog ist,*' Bohr, 1923c, p. 276), and to derive quantitative conclusions from it. Pauli had just before proposed to ascribe this nonmechanical stress to 'a peculiar, classically not describable type of duplicity of the quantum-theoretical properties of the series electron' (Pauli, 1925a, p. 385). However, in Copenhagen one sought the origin of the nonmechanical features exhibited by the complex structure of atoms in the fact that 'the coupling of the electrons in the atom implies a mechanically not-describable stress,' and then concluded: 'This leads to the result that, in spite of the unique determination of the entire atom in its stationary states, one needs a certain duplicity for the description of the behaviour of single electrons' (Heisenberg, 1925b, p. 841). That is, the duplicity of the properties of the electron, as proposed by Pauli in his previous paper (Pauli, 1925b), was regarded as a consequence rather than as the original fact. The reason was that only under this assumption did it seem possible to apply the correspondence approach. Heisenberg thus started his treatment of complex spectra with the following hypothesis:

> If an electron and an atomic core interact with each other, then the energy of this interaction possesses a reciprocal duplicity, such that for given well-defined stationary states of the atomic core and the outer [series] electron there always exist two values of the interaction energy and, therefore, also two stationary states of the entire atom; inversely, to a given value of the interaction energy there always belong two systems of stationary states of the electron and the atomic core. (Heisenberg, 1925b, pp. 841–842)

In order to use this hypothesis, which was a special case of the general one that

[219] In his letter to Heisenberg quoted above (28 February 1925), Pauli announced his arrival in Copenhagen around 15 March.

he had enunciated earlier in the letter to Landé, dated 18 February, which we have already mentioned, Heisenberg turned to discuss 'symbolic-type pictures' ('*symbolmässige Bilder*') corresponding to the classical description. The non-mechanical stress, in particular, brought about the use of two different symbolic models in connection with the theory of complex spectra: 'I. The electron acts on the atomic core through a non-mechanical stress, such that the stationary state of the atomic core seems to become doubled . . . II. The atomic core acts on the electron through a non-mechanical stress, such that the stationary state of the electron becomes doubled' (Heisenberg, 1925b, p. 842). However, the important point in applying the two different models was the following: all consequences derived from either of the models did not contradict each other. The reason for this conclusion was that, in spite of the nonuniqueness of the stress, the energy states of an atom were uniquely defined. Hence Heisenberg expected that

> rather the two pictures [i.e., symbolic models] will have to complement each other, just because of that uniqueness of their consequences, in such a way that the quantities which remain undetermined in one scheme will be determined in the other, and *vice versa*. Both schemes together will provide, so to say, a convergent method of calculation for determining the properties of the stationary states of the atom. If one considers the interaction of several electrons with one atomic core, then there may even be an entire sequence of equivalent pictures, which, if taken together, enable one to describe the stationary states of the atom. (Heisenberg, 1925b, p. 842)

As an example of the application of different models, Heisenberg first selected the atoms emitting doublet spectra, i.e., those consisting of a noble gas-like core and one series electron. In these cases the situation was very simple; there existed just the two models I and II. Model II had the advantage that it allowed one to calculate the energy states of the atom quantitatively; for example, in the absence of external fields one found that W_{int}, the interaction energy of the electron, assumed—for each set of quantum numbers n and k ($= k_1$ of Pauli)—two values, $W_{\text{int}}(n, k + \frac{1}{2})$ and $W_{\text{int}}(n, k - \frac{1}{2})$.[220] If, however, one wanted to apply the correspondence principle, one had to use model I, for in that case the mechanical orbits of the electrons had to be uniquely defined. Then one could calculate, for instance, the intensity of the transition of the series electron in the absence of an external field. For a strong external field, one could immediately obtain the energy states from model II by considering the magnetic quantum number m_k ($= m_1$ in Pauli's notation) instead of the quantum number k. By identifying the results of model II with those from model I, Heisenberg concluded, in particular, that the atomic core (in model I) must have the g-factor 2. The change of the magnetic components of doublet spectra in an external magnetic field of decreasing strength was obtained only with the help of model I;

[220] Heisenberg's result agreed with Landé's observation that the doublet splitting was accounted for by the relativistic energy difference (Landé, 1924a). However, it also allowed for deviations from Landé's result in those cases where the orbits of the series electron penetrated into the atomic core.

the reason again was that model II did not permit one to apply the correspondence and the adiabatic principles. However, the g-factors entering Landé's formula for the anomalous magnetic splitting of doublet lines in a magnetic field, Eq. (41), could most easily be determined within model II; one had just to take the ratio of k, the integral angular momentum of the series electron, and the values $k + \frac{1}{2}$ and $k - \frac{1}{2}$, appearing in the terms for the interaction energy. Thus, the g-factor must always be 2 $(= 1 : \frac{1}{2})$ for the s-term because the angular momentum of the electron in the lowest state is one unit (of $h/2\pi$), and the angular momentum in the interaction term is half a unit.

Models I and II also worked in principle for atoms having two outer electrons, such as those of the alkaline earths. In that case, one had just to consider the coupling between the core and the combined two electrons, or between their respective angular momenta. For example, if one of the outer electrons—say, the inner one—is in an s-state (angular momentum $\frac{1}{2}h/2\pi$), then i, the quantum number of the combined angular momenta of the electrons, takes on the two values $k + \frac{1}{2}$ and $k - \frac{1}{2}$, where k is the quantum number of the outer electron. As a result, two types of series spectra arise, denoted by the letters $s, p, d, \ldots,$ and $S, P, D, \ldots,$ respectively, which assume the same series limit. If the inner electron is on a p-orbit, four types of series spectra arise, etc. These results looked fine, but Heisenberg also noticed that neither model I nor model II indicated the types of series spectra; especially, they did not indicate that in the first example only singlet and triplet spectra exist. However, he found a means to handle this problem, too. The double-valuedness of the atom allowed one, in the case of atoms with two outer electrons, to choose a third model (model III). For this purpose, one associated the double-valuedness of *both* electrons with the atomic core; the latter angular momentum then assumed the magnitudes $r = \frac{1}{2}$ and $\frac{3}{2}$ (in units of $h/2\pi$), respectively. The momenta of the two outer electrons then combined (according to the vector addition) to yield a vector whose magnitude was denoted by i; the interaction energy of the atom then depended on an angular momentum quantum number l, which assumed the values $i + \frac{1}{2}$ and $i - \frac{1}{2}$, respectively; hence for any given arrangement of the core and the two electron orbits the interaction energy again possessed two values. Now, by assuming in this model that the mutual interaction of the two outer electrons was large compared to the interaction between the atomic core and the system of outer electrons, Heisenberg proved the singlet and triplet nature of the series spectra of two-valence-electron atoms, in which the inner electron was on an s-orbit. He further showed that the results derived from the three models did not contradict each other. However, an application of the correspondence and adiabatic principles was possible only to model III. This model also yielded the g-factor in weak magnetic fields. Provided the mutual interaction of the two electrons was much stronger than their interaction with the core, the g-factors were given by Landé's formula, Eq. (41), in which the quantum number k was replaced by the quantum number l. If, however, the interactions were comparable—physically, this meant that the relativistic corrections in model III assumed the same order of magnitude as the electron–electron interaction—Heisenberg expected irrational g-values.[221] With the help of model III, Heisenberg determined the relative splittings between the components of the complex multiplets of free atoms; however, their absolute size had again to be found from model II.

After briefly sketching how the treatment might be extended to atoms with more than two outer electrons, Heisenberg gave the principal steps of the application of the correspondence principle to obtain the selection rules and the intensities of complex spectra. Since the mechanical model describing the electron orbits had to be unique, he had either to use model I (for one-valence-electron atoms) or model III. This procedure yielded results in agreement with the intensity rules and the sum-rules discovered by Ornstein and his collaborators in all those cases, where the transition of just one electron was involved. But also the case of line multiplets, which were associated with the transitions of two or more electrons, could be treated. For example, by considering the coupling between two electrons—in a similar way as the coupling between atomic radiation and incident radiation in dispersion theory—Heisenberg derived combination frequencies. They corresponded to transitions in which one electron changed its quantum number k by 0 or 2 (in units of $h/2\pi$), while the other electron changed k by ± 1 unit; at the same time, j, the total angular momentum of the atom changed by 0 or ± 1 unit, and—provided the coupling between the outer electrons and the core could be neglected—also the quantum number l (of model III) changed in the same way. The only problem, which still remained to be solved, was to show that the core quantum number r obeyed the selection rule $\Delta r = 0, \pm 1$.

With the treatment of complex structure, which he completed in early April 1925, Heisenberg fully exhausted the possibilities of the Copenhagen approach based upon the correspondence principle. He succeeded in describing, at least in principle, all available data. For example, he explained the new observations of Henry Norris Russell and Frederick Albert Saunders on alkaline-earth lines (Russell and Saunders, 1925); these authors, incidentally, had proposed a model description, which was identical with Heisenberg's model III. Heisenberg was also able to discuss the latest results of Ernst Back on the Zeeman effects of neon lines (Back, 1925a).[222] Moreover, he reproduced Otto Laporte's empirical scheme spectra emitted by atoms having more than two outer or valence electrons (Laporte, 1923a). Altogether there seemed to remain hardly any empirical result,

[221] As an example of irrational g-values, Heisenberg cited the situation in the higher-order multiplets of neon. However, he assumed that when the coupling between the core and the outer electrons dominates all other interaction energies, then the g-values might be rational numbers again, although they would not follow Landé's formula.

[222] In his paper, Back argued that the g-values of the neon states were rational numbers within experimental errors (Back, 1925a, pp. 330–331, Tables 6 and 7). Heisenberg, who believed that these g-values had to be irrational (because of non-negligible relativistic corrections to the interaction energy of the outer electrons according to model II), claimed that Back's rational numbers did not agree with the ones obtained for other spectra (Heisenberg, 1925b, p. 853, footnote 2). Hence there remained the question as to how to represent the observations correctly.

which could not be accounted for by one of Heisenberg's models. As far as detailed applications were concerned, Heisenberg probably did not have the time, or take the time, to work them out. In any case, there were other theoreticians around who performed the necessary calculations and comparison with empirical data very well. One of them was Ralph de Laer Kronig, who went to Copenhagen in February 1925 with a more or less complete manuscript on the intensity of Zeeman lines of triplet spectra (Kronig, 1925a). His results were in accord with Heisenberg's sharpened application of the correspondence principle, as expressed in the latter's paper on resonance fluorescence.[223] Kronig continued to extend his results to describe also the intensities of what Landé and Heisenberg had earlier called the multiplets of higher order, as observed in the case of atoms having two or more electrons (Kronig, 1925b). For that purpose he made ample use of Heisenberg's model III; in the case of the neon spectra he especially achieved a nearly perfect agreement between theory and experiment. Another application of Heisenberg's schemes was made by Friedrich Hund, who in two papers—one submitted in June and the other in August 1925—organized the neon spectra as well as the spectra of most many-valence-electron atoms (Hund, 1925c, e). Heisenberg helped both Kronig in Copenhagen and Hund in Göttingen, but he did not get further involved in these applications. To him, the entire status of the theory seemed to be preliminary, as he stated in the conclusion of his paper:

> It should be hardly necessary to emphasize the preliminary and in many respects unsatisfactory character of the formulation of the problem of multiplet structure, which is attempted in this paper. Although the quantum-theoretical laws of the interaction of electrons within the atom are doubtlessly characterized by great simplicity, it seems that at the moment there exists no other way to interpret these laws than to employ model-dependent pictures of a symbolic nature in which this simplicity is hardly reflected in a satisfactory manner. (Heisenberg, 1925b, p. 860)

The simplification of the description of complex structure and complex spectra, of which Heisenberg dreamt in April 1925, would come in two separate steps: in the first step, the correct laws of quantum mechanics would have to be discovered and, second, the concept of electron spin would have to be properly assimilated. The latter would finally abolish the mystery connected with the earlier introduction of the nonmechanical stress or duplicity.

With all its virtues and shortcomings, Heisenberg's work on complex spectra represented a certain final state of the efforts in Copenhagen concerning the sharpened application of the correspondence principle. It threw some light on the particular difficulties of the quantum-theoretical description of atomic phenomena: that one did not possess unique mechanical models of a physical situation, and two or more models could be chosen, which represented different, even

[223]See Kronig's reference to Heisenberg's paper and to discussions with him (Kronig, 1925a, p. 888, footnote 2).

contradictory features. This fact was already part of Heisenberg's principal hypothesis, which he called 'reciprocal duplicity' ('*reziproke Zweideutigkeit*'), namely, that to each stationary state of a many-electron atom there corresponded two interaction energies, and vice versa. Naturally this hypothesis gave rise to a 'dualistic' description, for as Heisenberg remarked: 'However, due to the nature of the thing, the above hypothesis implies a characteristic duality in the selection of model-dependent pictures' (Heisenberg, 1925b, p. 842). The wording, which Heisenberg used here, was identical with Niels Bohr's at that time. In the early months of 1925 Bohr was concerned with a fundamental investigation of atomic processes and the interaction between atomic systems. Bohr arrived at the conclusion that the known processes of interaction would fall into two classes: to each process of the first class—to which, for instance, two collisions between atoms belonged—the inverse exists, and Bohr called them 'reciprocal interactions'; for the other class of processes—which includes the interaction between radiation and matter—such reciprocity does not exist. For example, according to the radiation theory of Bohr, Kramers and Slater, there was no strict coupling between the processes of emission and absorption of radiation, hence resonance absorption fell into the class of 'irreciprocal' interactions. An important property of reciprocal interactions was that the laws of (classical) mechanics applied to them, hence energy and momentum conservation were satisfied. For irreciprocal interactions, said Bohr, 'only a dualistic description appears to be possible at the moment, in which the mutual reactions of the systems involved are connected with each other only through probability laws, according to which the conservation laws assume at first the form of statistical laws' (Bohr, 1925a, p. 142). Bohr discussed these ideas in his analysis of the collisions of atoms with charged particles. He hoped that the separation into reciprocal and irreciprocal interactions might help to understand the nature of all atomic processes and provide the proper means of describing them successfully.

The attempt to develop a common language to speak about quantum phenomena was a significant feature of the environment at Bohr's Institute in 1924 and 1925. Although Niels Bohr, for many years, had paid special attention to the verbal formulation of physical phenomena, it was mainly the radiation theory of Bohr, Kramers and Slater, with its unusual description of atomic processes, which required a new language. This language was fashioned, in particular, by Bohr, Kramers and Heisenberg, and it proved to be helpful in deriving results even in those cases where no complete or consistent mathematical formulation could be found; Heisenberg's treatment of complex structure was a glowing example of this kind. Bohr and his co-workers considered their common, special language as an essential tool of quantum theory, for it corresponded perfectly to the paradoxical situations existing in atomic physics. Those, who did not closely follow the Copenhagen approach to atomic theory, often got the impression that Bohr and his inner circle had developed the code of a secret science or a kind of religion. This feeling was expressed, though jokingly, in letters exchanged between Heisenberg and Pauli. Already in summer 1924, Heisenberg had referred to Bohr as having the authority of a 'pope,' when he wrote that he would publish his formal theory of the Zeeman effects against Pauli's criticism, but with '*Päpstlicher Segen*' ['papal blessing' of Bohr (Heisenberg to Pauli, 8 June 1924)]. Pauli picked up the comparison and developed its details. He referred to Hendrik Kramers, Bohr's closest collaborator, as 'His Eminence' ('*Seine Eminenz*'), the cardinal in the pope's entourage, and called himself a 'non-believer' ('*Ungläubiger*') in the Copenhagen religion centered around the dogma of the Bohr–Kramers–Slater theory of radiation (Pauli to Heisenberg, 28 February 1925). The Copenhagen theoreticians, of course, tried to convert Pauli. Thus, Heisenberg wrote to him: 'We are enormously happy about your coming, and we look forward to the possibility of wrestling with you (intellectually, of course) and convincing you in all matters—in which there might still exist disagreement—of our point of view' (Heisenberg to Pauli, 26 February 1925). Heisenberg did have the opportunity of 'wrestling' with Pauli during his visit to Copenhagen in March–April 1925, but more so did Niels Bohr. And he might well have succeeded in converting Pauli, had new difficulties not arisen in April 1925.

About three weeks after Bohr had submitted his paper, entitled '*Über die Wirkung von Atomen bei Stößen*' ('On the Action of Atoms in Collisions,' Bohr, 1925a)—which, together with Heisenberg's paper on complex structure, represented the most recent and complete formulation of the Copenhagen approach to quantum theory—to *Zeitschrift für Physik*, he received a letter from Hans Geiger (Geiger to Bohr, 17 April 1925). In this letter, Geiger informed his former colleague from the Manchester days that the experiment on the Compton effect, which he had just carried out with Walther Bothe (see Bothe and Geiger, 1925a), gave a result in favour of the light-quantum theory, i.e., it contradicted the Copenhagen radiation theory. Bohr had meanwhile already developed certain doubts whether his conclusions concerning the collisions of atoms and Kramers' conclusion concerning radiation could be upheld (Bohr to Heisenberg, 18 April 1925); now Geiger's letter alarmed him completely. He still tried to rescue what he could of the radiation theory; for example, he argued that the usual space–time description of atomic phenomena had to be abandoned.[224] For Pauli, however, the situation was clear: the Bohr–Kramers–Slater theory, which he had never liked anyway, was definitely dead. And, just as dead were the unpleasant consequences, such as the possibility of having two different types of resonance radiation, the one from excited atoms and the other from unexcited atoms—which he had already criticized earlier (Pauli to Bohr, 2 October 1924). In a letter to Ralph Kronig, Pauli remarked: 'I always prefer to say that I have not yet found a complete picture of the processes than to accept, even temporarily, such awful conceptions which hurt my scientific instincts. This is my last word in this matter' (Pauli to Kronig, 21 May 1925). Pauli was convinced that the entire fabric of Copenhagen views, especially Bohr's philosophy, had collapsed. Since it

[224]See Bohr's reply to Hans Geiger, dated 21 April 1925, and his postscript to his article on atomic collisions (Bohr, 1925a, pp. 154–157, especially p. 157).

was the only available philosophy, which sought to give a complete description of quantum phenomena, he concluded with some despair:

> Physics at the moment is again very muddled; in any case, for me it is too complicated, and I wish I were a film comedian or something of that sort and had never heard anything about physics. Now I do hope nevertheless that Bohr will save us with a new idea. I beg him to do so urgently, and convey to him my greetings and many thanks for all his kindness and patience towards me. (Pauli to Kronig, 21 May 1925)

Heisenberg had left Copenhagen before the new difficulties arose. Thus, he was in a very optimistic frame of mind about the status of atomic theory in general, and dispersion theory in particular; he had just seen how Pauli had succeeded in giving a most skillful explanation of certain combination lines emitted by atoms (which had complex spectra) in the presence of external electric fields (Pauli, 1925c). This result signified, indeed, a brilliant application of the Kramers–Heisenberg dispersion theory. Altogether, when Heisenberg resumed his research work in Göttingen in the beginning of May, he did not see the slightest reason to change his approach to atomic physics, especially the philosophy he had learned from Niels Bohr.

Chapter IV

Sunrise in Helgoland

In early April 1925 Werner Heisenberg returned to Germany. At Bohr's Institute in Copenhagen he had completed his scientific education. There he had obtained his deepest insight into the problems and difficulties of quantum theory, especially in relation to its application to atomic structure. In Copenhagen, as in Munich and Göttingen earlier, he had been able to learn and absorb new, promising methods of approach to atomic problems. By the end of his stay in Denmark, Heisenberg had become as much attached to Bohr as he had been to his previous teachers Sommerfeld and Born; in fact, the Copenhagen experience meant even more to him. In the letter, in which he thanked Bohr and his wife for their warm hospitality, Heisenberg remarked: 'With respect to science the past half year was for me the most beautiful of my entire life as a student. I am almost sad about the fact that [scientifically] I must carry on wretchedly alone by myself in the future' (Heisenberg to Bohr, 21 April 1925). He knew perfectly well that in Göttingen, where he had to work during the following summer semester, he would miss the inspiration of Bohr's Institute, especially the wise counsel of Bohr himself. The occasional exchange of letters and news would not be enough to replace the intense, fruitful discussions, which had taken place during the past half year. Still, Heisenberg had to continue the work, even under the less favourable conditions that (he thought) existed in Göttingen. He soon adapted himself to his home university again and began to investigate a new problem, namely, the calculation of the intensities of the hydrogen lines, which he hoped to carry through without difficulties. However, in dealing with this problem, following his method of a sharpened application of the correspondence principle, Heisenberg met unexpected obstacles. He discovered that the quantum-theoretical formulation of this problem involved an ambiguity. In order to avoid it, he selected a simpler mechanical system, the anharmonic oscillator. After overcoming certain difficulties, he managed to integrate the equation of motion of the latter by referring only to strictly quantum-theoretical principles, which involved in particular the application of a new multiplication rule of physical quantities. Calculating the energy of the anharmonic oscillator according to the new rules, Heisenberg found that it was conserved, at least in the approximation which he had employed. This result indicated that an unwanted consequence of the radiation theory of Bohr, Kramers and Slater—i.e., only statistical conservation of the energy and momentum in processes involving the interaction of radiation and matter—might be circumvented in the final quantum theory of atomic systems. The idea of purely statistical conservation of energy and momentum in individual atomic processes had been proved to be incorrect by the outcome of the Bothe–Geiger experiment, which had become available in April 1925. Heisenberg's treatment of the anharmonic oscillator, completed about two months later, appeared as a promising step to a more final theory of atomic mechanics.

IV.1 Return to Göttingen and the Hydrogen Problem

Heisenberg left Copenhagen immediately after completing his paper on complex spectra (Heisenberg, 1925b).[225] Before going back to Göttingen he took a vacation for about three weeks in Southern Germany. From Munich he wrote to Bohr:

> During the past two weeks I have not thought anymore about physics even for a minute, but have once again visited a wonderful part of Germany. In the Württemberg country we hiked through old small towns with gateways and little towers and nooks and alleys, where the spirit of the middle ages surrounds one all over and —in the moonlight—lets one dream of the past, of mercenaries and nightwatchmen. Later we descended from Odenwald, near Heidelberg, down into the Rhine valley, an infinite sea of blossoming fruit trees, and by the bewildering splendour and fragrance we felt enchanted. (Heisenberg to Bohr, 21 April 1925)[226]

After telling Bohr about his vacation, he announced: 'In the beginning of next week I shall return to Göttingen' (Heisenberg to Bohr, 21 April 1925). Thus, he arrived at his university before the beginning of the semester and resumed his work.

In spring 1925 Heisenberg had to fulfill his teaching duties as a *Privatdozent* at the University of Göttingen.[227] At Born's Institute he was now a most valuable partner for discussions on atomic theory, especially having collaborated for half

[225]This paper was then typed up in Copenhagen and sent to the *Zeitschrift für Physik*, where it was received on 10 April 1925. (See Bohr to Heisenberg, 18 April 1925: 'As agreed, your paper was sent before Easter [Easter Sunday was on 12 April 1925] to [Karl] Scheel [the editor of *Zeitschrift für Physik*], who has confirmed its receipt.' (See also Heisenberg's letter to Bohr dated 21 April 1925, in which he referred to having received the typed version of his manuscript ('*Reinschrift*') sent from Copenhagen.)

[226]Heisenberg had planned the tour through Southwestern Germany earlier in Copenhagen. Thus, he had written to Landé: 'It is possible that at Easter I might pass again through Tübingen, but I don't know anything definite yet' (Heisenberg to Landé, 6 March 1925). It is not clear whether Heisenberg really went to Tübingen or met Landé; probably not, for he did not mention anything to Niels Bohr about it in his letter of 21 April 1925.

[227]There is no record of the subject on which Heisenberg lectured during the summer semester 1925. In the official catalogue of the University of Göttingen, which appeared before the beginning of the semester (*Amtliches Namensverzeichnis Winterhalbjahr 1924/25—Verzeichnis der Vorlesungen Sommerhalbjahr 1925, Dieterichsche Universitäts-Buchdruckerei*, Göttingen 1925), it is just stated on p. 24: '*Dr. Heisenberg kündigt später an*.' ('Dr. Heisenberg will announce [his subject] later on.') In the *Heisenberg Archives* in Munich there is a manuscript entitled '*Elektro- und Magnetooptik*' ('Electro- and Magneto-Optics') containing the material of his first lecture course.

a year with Niels Bohr, the greatest authority in that field. Indeed, Heisenberg was considered in Göttingen as the representative of the Copenhagen point of view and approach to quantum theory. As Born remarked: 'I think that Heisenberg had more connections with Copenhagen than I. I was there once for a few days only, and never really belonged to Copenhagen. But Heisenberg did' (Born, Conversations, p. 20).[228] With great interest the Göttingen theoreticians studied Heisenberg's most recent paper on the complex structure of many-electron atoms, the typed version of which had become available in late April. Pascual Jordan, in his note on atomic structure (Jordan, 1925c), and Friedrich Hund, in his interpretation of complex spectra (Hund, 1925c), made use of Heisenberg's results.[229] For the first time Born's Institute turned into a centre of theoretical spectroscopy through the work and interaction of Heisenberg, Hund and Jordan. And Born reported enthusiastically to Einstein:

> On the whole my young people, Heisenberg, Jordan and Hund, are brilliant. I find that merely to keep up with their thoughts demands at times a considerable effort on my part. Their mastery of the so-called "term zoology" is marvellous. Heisenberg's latest paper, soon to be published, appears rather mystifying but is certainly true and profound; it enabled Hund to bring into order the whole of the periodic system with all its complicated multiplets. This paper, too, is soon to be published. (Born to Einstein, 15 July 1925)

While Born praised Heisenberg's work (1925b) and its consequences (Hund, 1925c), Heisenberg himself became increasingly aware of its shortcomings. In Copenhagen he had been mainly discontent with one point: the simplicity exhibited by the atomic data had not come through in the theoretical description. But after his departure he learned that the theoretical basis (especially the dualistic view), on which he had built the approach to complex structure, was about to be given up, for Bohr wrote to him in Munich on 18 April 1925: 'I could tell you much, at least much that is negative, for I am now looking at many things with much greater doubt than at the time you were here. I think I will not publish at all my paper on the deceleration of α-particles, of which I have just received the galley proofs. The paper of Kramers, who is still in Holland, was not even sent away and will hardly be published in this form.' Thus, as early as mid-April 1925, Bohr considered his own work on reciprocal and irreciprocal atomic processes and Kramers' investigation on resonance fluorescence as being incorrect; with it much of the motivation of Heisenberg's many-model approach to complex spectra disappeared.[230] However, Bohr in-

[228] It should be noted, however, that both James Franck and Max Born had also direct connections with Niels Bohr; indeed, Born had just spent several days at the end of February 1925 in Copenhagen, and Franck exchanged several letters with Bohr during March and April 1925.

[229] Hund acknowledged: 'I am very much obliged to Mr. Heisenberg for his helpful interest in this work' (Hund, 1925c, p. 371).

[230] Kramers' paper on resonance fluorescence was actually withdrawn, but Bohr's paper on the collision of α-particles with matter was published in *Zeitschrift für Physik* with a later addendum (Bohr, 1925a).

formed him: 'In this situation, I have taken the liberty of shortening the introduction of your paper [Heisenberg, 1925b] a little bit and to mention, instead of the collision and radiation phenomena, just a few words about the "stress" ("*Zwang*"). Now you must see for yourself whether you are satisfied with it. I shall be glad, in any case, to hear from you after you have received the galley proofs' (Bohr to Heisenberg, 18 April 1925). Heisenberg, though agreeing to the changes made by Bohr in his paper, wondered about Bohr's new, pessimistic attitude. He guessed that Pauli, who was just then visiting Copenhagen, was behind it; in any case, he inquired: 'I would be interested to hear how your discussions with Pauli have gone on' (Heisenberg to Bohr, 21 April 1925). But the Copenhagen theoreticians discovered more and more difficulties with the previous views. For example, Kramers, in investigating the situation of electron orbits penetrating into the atomic core (*Tauchbahnen*), pointed to the fact that the azimuthal quantum number k as well as the principal quantum number n might become double-valued. This result not only contradicted the empirical data, but also the basic assumption in Heisenberg's treatment of many-electron atoms.[231]

The difficulties connected with the foundation of the description of complex structure constituted only one part of the troubles that confronted Bohr in spring 1925. Bohr wrote to Heisenberg:

> In spite of all the darkness, the secrets of the atoms are still better known at the moment than the general description of how the quantum processes take place in space and time. [He continued:] Stimulated in particular by conversations with Pauli, I agonize these days with all my strength to become familiar with the mystery of nature and I try to prepare myself for all eventualities, even for the acceptance of a coupling of quantum processes between different atoms. (Bohr to Heisenberg, 18 April 1925)

Bohr added that he regarded the consequences of such an assumption as extremely serious. Heisenberg, remembering the penetrating discussions of the fundamental issues of radiation theory in Copenhagen, knew that only important reasons could have driven Bohr to give up the Bohr–Kramers–Slater theory. He soon learned about the outcome of the Bothe–Geiger experiment on the Compton effect, which could provide just such reasons. 'I would be very interested,' he wrote to Bohr, 'to hear about your views, now that the Geiger–Bothe experiment has evidently decided in favour of a coupling [i.e., in favour of strict energy-momentum conservation in processes involving the interaction of radiation and matter]' (Heisenberg to Bohr, 16 May 1925). Moreover, he did not see the point why Bohr wanted to withdraw his paper on the passage of α-rays through matter. In any case, Heisenberg believed that without discussing with Bohr, without being in Copenhagen, he would not be able to make progress in the fundamental questions of radiation theory. Therefore, he did not join Max Born and Pascual

[231] Bohr mentioned his intention of carefully going over the proof-sheets of Heisenberg's paper (1925b) on complex structure during his forthcoming visit to England (see Bohr to Heisenberg, 10 May 1925).

Jordan in working on radiation theory, and wrote to Bohr: 'Born and Jordan want to reform radiation theory, but I have no great faith [in what they are doing]; the problem is just much more difficult in reality than people here tend to imagine' (Heisenberg to Bohr, 16 May 1925).[232]

The changed situation in atomic theory made it difficult for Heisenberg to select a proper topic for his work in Göttingen. Of course, he could have tried to improve on the foundation of the theory of many-electron atoms. Evidently he would have done that if he had still been the person of a year ago. But now he was aware of the difficulties connected with such a programme; and without constant discussions with Bohr he did not expect to make real progress. Still he wanted to treat a problem of atomic structure, remembering Bohr's view that these problems were less affected by the desperate status of radiation theory. Heisenberg felt that he had to treat the case of an atom in which the unexplained 'stress' ('*Zwang*') did not play a prominent role. And the only such atom he could think of was the hydrogen atom. Its mechanical description seemed to be unique; there was no complex core, from which any duplicity could arise, at least not as long as one neglected relativistic effects. Hence, in this simplest atomic system Heisenberg could hope to apply his standard procedure of the sharpened correspondence principle because the mechanical model was uniquely determined. By the middle of May he was working on the hydrogen problem. 'Recently I have been occupied with intensities [of spectral lines], notably in the case of hydrogen,' he reported to Bohr. 'So far not much has come out of it. The present conditions are still not entirely sufficient to obtain the intensities uniquely [from them], but I do wish to try to proceed further' (Heisenberg, 16 May 1925).

To take up the hydrogen problem at this stage might appear, at first sight, to be an audacious step, for the very basis of the success of the old Bohr–Sommerfeld theory had begun to appear doubtful to those who were most closely associated with it. Let us briefly recall the development since the inauguration of the quantum theory of atomic structure. The spectrum of the hydrogen atom had been the most prominent result of Bohr's theory (Bohr, 1913b). Sommerfeld had then treated the hydrogen atom as the prototype for his model of atoms, from which he derived the details of the structure of spectral lines (Sommerfeld, 1915b, c). Paul Epstein, a collaborator of Sommerfeld's, had calculated the Stark effect, i.e., the splitting of single Balmer lines in an electric field (Epstein, 1916a, c, e), while Peter Debye (1916a, b) and Sommerfeld himself (1916d) had derived the (normal) Zeeman splitting of a line into three components in a magnetic field. In addition, Sommerfeld had obtained the intensity rules for a

[232] In a letter to Ralph Kronig, Heisenberg expressed his opinion on the status of radiation theory in some detail. He wrote:

> According to your letter, there must exist a terrible confusion about the theory of radiation in Copenhagen. If I had been there, I would first plead—as with Zeeman effects—for a formal dualistic theory: everything must be describable both by wave theory and by the use of light-quanta . . . I myself don't dare to deal with such dangerous problems, of which one cannot be sure at all whether they are ready to be solved now.' (Heisenberg to Kronig, 8 May 1925)

given Balmer line by counting the number of possible transitions, excluding the forbidden ones, and assigning to the elliptical orbits weights which decreased with their eccentricity (Sommerfeld, 1915c; 1916b). Finally, Hendrik Kramers, in his thesis, had discussed the intensities of the Balmer lines and their Stark and Zeeman components from the point of view of the correspondence principle (Kramers, 1919). All these results agreed statisfactorily with empirical observations, and even as late as 1925 one could consider the hydrogen problem as completely solved, but for two particular points. One was very technical and intricate, for it concerned a special situation, namely, the simultaneous application of crossed electric and magnetic fields to a hydrogen atom. In December 1923 Oskar Klein, Bohr's former collaborator, then at Ann Arbor, Michigan, had shown that with the help of adiabatically changing crossed-fields one could turn an allowed Kepler orbit into a forbidden pendulum orbit, i.e., a degenerate path of an electron passing through the nucleus (Klein, 1924a, b). This fact clearly contradicted the accepted principles of atomic theory, especially Paul Ehrenfest's adiabatic principle. The second point was more obvious, namely, the fact that Bohr's frequency condition, which served to determine the transition frequencies, stood in complete contradiction to classical mechanics. Hence the successes of the theory of the hydrogen atom appeared to be rather accidental. Thus, Hendrik Kramers, in his review of Bohr's theory at the Innsbruck *Naturforscherversammlung* (Congress of German Scientists and Physicians) in September 1924, declared:

> The theory of the hydrogen spectrum owes its success to the fact that the motions in the stationary states can be described in this case with the help of classical electrodynamics, or—to put it slightly differently—with the help of mechanics based on Coulomb's law of attraction. That this is possible, is not at all self-evident, since the fundamental postulates of Bohr are in open contradiction to classical electrodynamics. Thus the strange transitions between stationary states, whether stimulated by radiation or by collisions, cannot be described in principle by classical methods, not even in the case of the hydrogen atom; and the same is true for the interaction [of atoms] with the radiation field. (Kramers, 1924c, p. 1050)

At the same time as the situation with the Bohr–Sommerfeld atomic model deteriorated irretrievably, one discovered an alternative way to describe the interaction between radiation and matter—the dispersion-theoretic approach of Kramers (1924a, b), Born (1924b), and Kramers and Heisenberg (1925)—and this method yielded satisfactory results. Indeed, by the beginning of 1925 the emphasis on the problems of atomic theory had shifted completely from what it had been, say in June 1922. Then, on the occasion of the *Bohr Festival* in Göttingen, the construction of mechanical models had reached its peak, its crowning success being represented by Bohr's explanation of the periodic system of elements (Bohr, 1921e). The outstanding problem of that time had been how to handle the details of the interaction between matter and radiation. Such problems could now be properly treated by means of the quantum-theoretic dispersion theory, but success in this domain had made suspect all the 'good' results of the previous atomic mechanics. In an address to the sixth Scandinavian Mathematical Congress in Copenhagen on 30 August 1925, Bohr summarized the situation as follows: 'While this description of optical phenomena was entirely in harmony with the fundamental ideas of quantum theory, i.e., the correspondence principle, it soon appeared that it stood in strange contradiction to the use of the mechanical pictures previously employed for the analysis of stationary states' (Bohr, 1925b, p. 851). In addition to the objection in principle, Bohr listed two major failures of the erstwhile quantum theory of multiply periodic systems: first, the impossibility of treating properly the action of very slowly varying fields, the outstanding difficulty showing up in the case of a hydrogen atom in crossed electric and magnetic fields; second, the fact 'that the theory of periodic systems was apparently helpless in the problem of the quantitative determination of the transition probabilities on the basis of the mechanical pictures of the stationary states' (Bohr, 1925b, p. 851). Bohr called the latter defect 'especially unsatisfactory.' The only recipe for obtaining the intensities of the radiation emitted from atoms was to use the correspondence principle approach. But the question had to be answered as to how one could then bring in the known information from atomic structure, which so far had been expressed only in terms of mechanical models.

In the light of the Copenhagen programme it was quite natural, therefore, that one of the most significant problems to be tackled should be the calculation of the intensities of the hydrogen spectrum. Questions concerning the intensities had indeed a prominent place in Bohr's Institute in spring 1925. Ralph Fowler, then visiting from Cambridge, treated the intensities of the band spectra (Fowler, 1925a), and Ralph Kronig—as we have mentioned above—worked out the intensities of the components of complex spectra of atoms with and without external magnetic fields (Kronig, 1925a, b). Both used the method of 'refining' (Fowler) or 'sharpening' (Heisenberg, Kronig) the correspondence principle, which led in particular to sum-rules for the intensities under consideration. In May 1925 Kronig also thought of treating the intensities of the hydrogen spectrum. He recalled later:

> I was aware that the success at which one had arrived with these intensity calculations of multiplets and Zeeman components led to the question whether one could sharpen or make precise the correspondence principle in other cases, too. And one of the problems where the correspondence principle had been studied semiquantitatively was the Stark effect of the atomic hydrogen spectrum. I spent quite some time on trying to get rigorous expressions . . . of these Stark components. (Kronig, AHQP Interview)

The study of the Stark effect of hydrogen lines, which Kronig called 'semiquantitative,' had been carried out by Hendrik Kramers in his thesis (Kramers, 1919). Kramers had there expanded the rectangular coordinates of the electron orbits in terms of the multiplet Fourier series. In the case that no external field was present, the planar motion of an electron in a relativistically precessing Kepler orbit could be described by a double Fourier series. For example, Kramers wrote a special combination of x and y as

$$x + iy = \sum_{\tau_1} \sum_{\tau_2} C_{\tau_1, \tau_2} \exp\{2\pi i(\tau_1 w_1 + \tau_2 w_2)\}, \tag{58}$$

where w_1 and w_2 are the two angular coordinates of the doubly periodic system and the summation indices τ_1 and τ_2 assume integral values. The coefficients of the harmonic expansion, C_{τ_1, τ_2}, determine the motion completely. In case that an external field was applied to the hydrogen atom, as in the Zeeman or Stark effects, the motion of the electron had to be considered as being spatial; hence the system would become triply periodic and three angular coordinates, w_1, w_2 and w_3, would enter into the harmonic expansion having the Fourier coefficients $C_{\tau_1, \tau_2, \tau_3}$. The hydrogen atom exhibited an advantage, which Kramers had made use of in his thesis: in the Bohr–Sommerfeld quantum theory of multiply periodic systems one had exact solutions of the equation of motion of the system with and without external fields. From these solutions one could obtain the frequencies of the emitted radiation by taking the differences of energies between two stationary states according to Bohr's frequency condition, in agreement with experiment. That is, the observed frequencies did not coincide with the orbital frequencies or their higher harmonics, which followed from classical electrodynamics. Only in the limit of high quantum numbers associated with the periodic degrees of the motion (via the phase integral) would the classical frequencies and the quantum transition frequencies coincide. For low quantum numbers, however, Kramers had realized that the quantum frequencies could be represented 'in a simple way as the *mean value*' of the classical frequencies in the two stationary states connected with the transition (Kramers, 1919, p. 330). This observation had permitted him to suggest a calculation of the intensities of the emitted lines with the help of the following procedure. In classical electrodynamics the intensity of an harmonic denoted by the indices τ—i.e., τ_1 and τ_2 for the unperturbed hydrogen atom and τ_1, τ_2 and τ_3 for the atom in an external electric field—is proportional to the absolute square of the Fourier coefficient C_τ. Because of the correspondence principle this result should remain true if, in the Bohr–Sommerfeld theory, one considers comparatively small orders of the harmonics (i.e., the τ's are small in comparison to the quantum numbers determining the stationary states involved in the transition) in the case of very high quantum numbers. Hence, in the correspondence limit the probabilities of spontaneous transitions, which in quantum theory account for the intensities of the radiation, are asymptotically proportional to $|C_\tau|^2$. For transitions between states having low quantum numbers Kramers had suggested obtaining the probability in question by replacing the classical Fourier coefficient C_τ by $\overline{C}$, where the latter quantity denotes a suitably chosen mean value of the Fourier coefficients, associated with the initial and final states of the transition, respectively. With such a prescription Kramers had succeeded in calculating values for the intensities of the Stark effect components, of the fine-structure components and of the Zeeman effect components of the hydrogen atom (Kramers, 1919, §§6–8). In general, the theoretical and experimental results had agreed satisfactorily.

Kramers, in his thesis of 1919, had used a specific averaging procedure to arrive at the Fourier coefficient $\bar{C}$ entering in the expression for the quantum-theoretical transition probability or intensity.[233] He had noticed, however, that in some extreme cases the averaging procedure led to poor results because the $\bar{C}$ thus obtained 'do not satisfy the fundamental condition that small external forces can only produce small changes in the intensity distribution of spectral lines' (Kramers, 1919, p. 330, footnote 1). Five years later a definite averaging procedure seemed to be available; this was Born's discretization prescription (51). This procedure had already worked very well, as it had yielded the quantum-theoretical dispersion formulae (Kramers and Heisenberg, 1925). It had also been possible to show in special examples, such as the polarization of resonance fluorescence light, that Kramers' above-mentioned difficulty with respect to the effect of weak external forces on the intensity of radiation from atoms might be satisfactorily handled. Under these premises Heisenberg and Kronig believed themselves to be prepared well enough to resume the problem of calculating the intensities of the hydrogen lines. In order to achieve their goal they had to rewrite Kramers' classical intensity formulae with the help of the discretization prescription (51). The task was not an easy one because the classical expressions were very complicated. For example, even the simplest electron motion in the hydrogen atom—a nonrelativistic Kepler orbit—had to be described with the help of Fourier coefficients, each of which was a combination of two Bessel functions; thus Eq. (58) reduced in that case to

$$x + iy = \text{const.} - \sum_{\tau} C_{\tau} \exp\{2\pi i(w\tau)\}, \tag{59}$$

with

$$C_{\tau} = \frac{a}{2\tau}\left[\left(1 + \frac{k}{n}\right)J_{\tau-1}(\tau\epsilon) - \left(1 - \frac{k}{n}\right)J_{\tau+1}(\tau\epsilon)\right], \tag{59a}$$

where k and n denote the azimuthal and principal quantum numbers, respectively, w is the angular variable (denoting the position of the electron in the Kepler orbit), a the semi-major axis, ϵ $(=\sqrt{1 - k^2/n^2}\,)$ the eccentricity of the ellipse, and $J_{\tau\pm1}(\tau\epsilon)$ are Bessel functions of order $(\tau \pm 1)$. In the case of the more complicated electron motions, as in the presence of external electric or magnetic fields, the Fourier coefficients (of the double and triple series) were combinations of products of Bessel functions. And these Fourier coefficients entered into the classical expressions for the intensities, which then had to be reformulated in quantum theory.

[233]See the prescription given in Kramers, 1919, p. 330, footnote (1).

Heisenberg started to work on the problem of calculating the hydrogen line intensities with the help of a promising new idea, which he immediately reported to Ralph Kronig. He wrote: 'In this I have hit upon a very amusing possibility. If one expands the Bessel functions, which occur in the intensities, in terms of powers (of the eccentricities)—this expansion always converges—, then in quantum theory one can rewrite the first term of this series according to the principle that the jumps should disappear at the boundaries' (Heisenberg to Kronig, 8 May 1925). Heisenberg then proposed to obtain the quantum-theoretical intensities from the classical ones by substituting binomial expansions for certain power expansions; for example, he proposed to change the factor $(n - k)^{\tau - 2}$ (where n and k are the principal and azimuthal quantum numbers, respectively, of the hydrogen orbit, and τ the index of the Fourier coefficient) occurring in the classical intensity formula into the product $(n - k)(n - k - 1) \ldots (n - k - \tau + 3)$ in the quantum formula.[234] He argued:

> If now one impudently converts this entire power series for I [i.e., the intensity] simply into a binomial series or polynomial series, respectively, then one obtains series, which break off after a certain number of terms (hence I^{qu} [the quantum-theoretical intensity] would always be rational [i.e., a rational function of the quantum numbers, as followed from the measurements]), which nevertheless pass over, in the limit $n \to \infty, k \to \infty$, into Bessel functions. I shall have faith in these formulae only if I can determine which difference equations they [i.e., the terms of the power series of the intensity, or rather the quantum-theoretically converted terms] satisfy. (That they satisfy a difference equation seems probable to me.) Hence I shall try further and see how things develop. (Heisenberg to Kronig, 8 May 1925)

In the beginning Heisenberg was very optimistic. He told Born: 'One has such an impression that one could now almost guess all the intensities, if only one did it well enough' (Heisenberg, Conversations, p. 205). Doing well enough implied, of course, the use of all the promising ideas which had been found earlier in the programme of discretization of atomic theory, such as the replacement of j^2, the square of the angular momentum variable, by $j(j + 1)$. Born agreed that Heisenberg should attack the hydrogen problem. Heisenberg recalled: 'Born said, "That's a good idea, you try it. Then you probably want to know something about the classical Fourier components; there you need Bessel functions." And he handed me a book in which I could learn Bessel functions' (Heisenberg, Conversations, p. 205).[235] Heisenberg found that he had indeed to acquaint

[234] Instead of the product $(n - k)(n - k - 1) \ldots (n - k - \tau + 3)$, Heisenberg, in his letter to Kronig of 8 May 1925, wrote the product $(n - k)(n - k - 1) \ldots (n - k - \tau + 2)$. This was a mistake, as can be seen by taking the correspondence limit, which is $(n - k)^{\tau - 1}$ with Heisenberg's product instead of $(n - k)^{\tau - 2}$ (as the classical formula demands).

[235] The book on Bessel functions, which Heisenberg borrowed from Born, was either *Funktionentafeln* by E. Jahnke and F. Emde (1909), or the first volume of *Courant-Hilbert* (1924). It is more likely that Heisenberg took the Jahnke–Emde book because Born had quoted it in connection with the treatment of the intensities of hydrogen Stark components in his recent book on *Atommechanik* (Born, 1925, p. 256, footnote 1).

himself with the mathematical tool of Bessel functions because in the intensity problems, which he had treated before, only trigonometric functions had entered.[236] Therefore, he took the book home and tried hard to solve his problem during the next couple of weeks. Many years later he described his efforts in the following words: 'In the hydrogen atom you have rather complicated multiplications, rather complicated mathematics, and still at the end you have a Bessel function as an amplitude. But then you see that this Bessel function comes out of many very complicated processes in the calculation. And I felt that I should try to imitate the process in the same way as I had imitated in the dispersion formula the process of the perturbation theory, that is by multiplying amplitudes and so on' (Heisenberg, Conversations, pp. 260–261). After a few weeks Heisenberg realized that this task was very hard, probably much too hard for him. 'I saw that I couldn't really imitate these processes,' he recalled, 'because the mathematics was so complicated. Then I started to think about the fundamental side . . . I gradually saw that the first thing I must know was: if I have all of the amplitudes of a coordinate x, and of another coordinate y, how can I get the amplitudes of the product xy?' (Heisenberg, Conversations, pp. 260–261)

Given the tremendous ambition with which Heisenberg had carried out his earlier tasks until he reached successful results, one might wonder why in May 1925 he gave up so early on the hydrogen problem. Certainly it is no explanation that Kronig did not have any success on this problem either; in any case, he did not report any new result on the hydrogen problem in his subsequent paper. Other reasons must be sought why both Heisenberg and Kronig did not continue their common project. Heisenberg, as he remarked later, just shifted his interest to another, more fundamental aspect of the same problem.[237] He did not really intend to give up the calculation of the intensities of hydrogen lines, but expected to return to it after he had clarified the fundamental mathematical problem. Kronig, on the other had, turned his attention again to the intensities of the complex multiplets. One may ask, naturally, why Kronig—then in Copenhagen —did not join forces with Hendrik Kramers, the best available expert on the earlier approach to the problem of the hydrogen spectrum? The reason probably was the following. Kramers, like Bohr, did not believe that the sharpening of the correspondence principle, which had been so successful in dispersion theory, could be extended similarly to problems—such as the Stark effect—in which electromagnetic fields of very small frequencies were involved. Bohr expressed

[236] See, for example, the paper of Sommerfeld and Heisenberg on the intensities of the multiplet components; there the authors noted explicitly the restriction of their treatment to give only the relative intensities within a multiplet because they wanted to avoid any use of Bessel functions (Sommerfeld and Heisenberg, 1922b, p. 140).

[237] Heisenberg's recollection was confirmed by his Göttingen colleague Pascual Jordan, who wrote: 'Then Heisenberg tried to guess the exact formula for the intensity laws of the Balmer series, based on correspondence considerations; that [he thought] should be done in a way similar to the case of multiplets and Zeeman effects. The problem, however, turned out to be too difficult, and he then preferred to think in a more fundamental way about the question of an exactly formulated quantum mechanics' (Jordan to van der Waerden, 1 December 1961).

this feeling very clearly in his address to the sixth Scandinavian Mathematical Congress. 'It is impossible,' he stated there, 'on the basis of the scattering activity of illuminated atoms demanded by the dispersion theory to construct an asymptotic connection between the reaction of an atom in alternating fields of smaller and smaller frequency and the reaction in constant fields as calculated from quantization rules of the theory of periodicity systems' (Bohr, 1925b, p. 851). Kramers, at that time, still seemed to be preoccupied with the old dispersion problems (see Kramers, 1925b). Other research work, performed in Copenhagen during spring and summer 1925, was also devoted to pursuing the consequences of dispersion theory, as for instance the formulation of the quantum-theoretical sum-rule in dispersion by Werner Kuhn (1925a). Thus the Copenhagen attitude in mid-1925 was one of conservative innovation by means of a step-by-step experimental verification of quantum formulae that had been arrived at by systematic guessing.

In Copenhagen, Bohr was still too deeply committed to understanding the difficulties arising from the failure of the Bohr–Kramers–Slater radiation theory. This commitment prevented him from engaging himself in other problems, such as the question of calculating quantum-theoretical intensities of the hydrogen lines. Moreover, the atmosphere, in which one could hope to achieve a definite push forward in atomic theory by discovering a new mathematical tool for treating the problems of quantum theory, did not then prevail in Copenhagen; there the emphasis was on 'physical' as against 'mathematical' reasoning. Heisenberg was fortunate to be left alone in Göttingen with his thoughts, which soon turned to the mathematical structure of quantum-theoretical Fourier series.

IV.2 The Quantum-Theoretical Fourier Series and an Unexpected Difficulty

At the end of his paper on complex spectra Heisenberg had stated that the simplicity of the quantum-theoretical laws derived from the observations was not sufficiently represented in the classical models describing the atomic systems under consideration (Heisenberg, 1925b, p. 860). While studying the problem of the intensities of hydrogen lines in May 1925, and employing the well-known mechanical model of a one-electron atom, he was again reminded of that insufficiency. Heisenberg felt intuitively that something in his procedure was not correct, or at least too complicated to allow one to guess the right results. However, in the classical calculation of the hydrogen line intensities the following feature showed up: the intermediate steps of the calculation were very complicated, but finally the expressions led to a simple result. Heisenberg expected that somehow the same must happen in the quantum-theoretical treatment, and this expectation was responsible for the procedure he adopted in the hydrogen problem. There, after some preliminary attempts, he did not seek to translate the classical formula for the intensities into the corresponding quantum-theoretical one—he had followed this procedure with Kramers in the dispersion formulae—

but rather he turned to solving the equation of motion of the electron in the hydrogen atom by a consistent quantum-theoretical method. The task was just to find the quantum-theoretical counterpart of the classical equation of motion, especially to figure out what was meant by 'motion' in a consistent quantum theory.

Another reason for formulating the quantum-theoretical problem on a more fundamental level, and to begin by investigating the kinematical aspects, was the following. Heisenberg realized that, even if he had succeeded in obtaining the intensities of the Balmer lines by intelligent guessing, he would have solved the problem of calculating the quantum-theoretical intensities in an approximation only. In classical theory this approximation meant the restriction to dipole radiation. On the other hand, Heisenberg knew from his own work that quadrupole radiation also played a role in quantum theory. For example, in discussing the so-called combination frequencies occurring in complex spectra, he had found that they were associated with changes of two units in the azimuthal quantum number of an electron, and such transitions corresponded in classical theory to quadrupole transitions (Heisenberg, 1925b, p. 857). Given the motion of an accelerated electron, it was straightforward in classical electrodynamics to compute the intensity of quadrupole and higher-order radiation. While the amplitude of the electric field vector of dipole radiation was proportional to the acceleration of the electron, the amplitude of the quadrupole radiation was proportional to the (scalar) product of velocity and acceleration, and in the amplitude of higher-order radiation there appeared more composite products of velocity and acceleration. Thus, after thinking for some time about higher-order corrections, Heisenberg concluded:

> In order to calculate the radiation [emitted by electrons in atoms] one has not only to calculate the dipole radiation, etc. These parts of the interaction between the electrons in the atoms and the radiation do not depend linearly on the coordinates, but on the square or higher powers of the coordinates and their derivatives. Now I asked myself: "Even if I had solved my whole problem of guessing the hydrogen spectrum intensities, how would I calculate [the intensities of] the quadrupole radiation instead of the dipole radiation?" In this way I came to the idea that one should ask whether, in quantum theory, one can in general multiply the quantities X and Y. (Heisenberg, Conversations, p. 261; also AHQP Interview)

Here the X and Y were supposed to represent, for example, the velocity and acceleration of an electron in quantum theory. Now, in dispersion theory one had already considered special cases of such quantities and their products; hence one could hope to learn from there about the more general cases.

Thus, Heisenberg, after his failure to guess the intensities of the spectral lines of hydrogen, entered into a general programme of working out quantum-theoretical relations with the help of two fundamental assumptions. First, he demanded that for any classical description of an atomic system there exists an analogous quantum-theoretical description, in which the classical variables are replaced by quantum-theoretical variables and the algebraic operations (like summation and multiplication) are taken over. Second, for obtaining the quantum-theoretical solution of an atomic problem, one must start from something like an equation of motion. Moreover, in quantum theory, in carrying out operations such as the solution of the equation of motion, one had, of course, to observe the correspondence principle. Thus, the quantum-theoretical procedure was supposed to follow the classical procedure as closely as possible. Since all hitherto successful attempts to reformulate classical equations into quantum theory had made use of the Fourier expansions of the classical coordinates describing the multiply periodic motion of electrons in atoms, Heisenberg was convinced that in quantum theory those quantities, which corresponded to the classical Fourier expansions, would play a crucial role. But what did these quantities look like? The method for obtaining an answer to this question seemed to be clear: one had to go back to the earlier treatment of atomic systems, especially in dispersion theory, and find out what one could learn from it.

Already in the reformulation of his theory of atomic models, completed in November 1917, Niels Bohr had emphasized the central role of the Fourier decomposition of the coordinates representing the dynamical variables of atomic systems (Bohr, 1918a). Ascribing f degrees of freedom to an atomic system, one may express the position coordinate q_k $(k = 1, \ldots, f)$, in terms of the f-fold Fourier sum,

$$q_k = \sum_{\tau_1} \ldots \sum_{\tau_f} C^{(k)}_{\tau_1, \ldots, \tau_f} \exp\{2\pi i[(\tau\nu)t + (\tau\delta)]\}, \tag{60}$$

with

$$(\tau\nu) = \tau_1\nu_1 + \ldots + \tau_f\nu_f, \tag{60a}$$

and

$$(\tau\delta) = \tau_1\delta_1 + \ldots + \tau_f\delta_f, \tag{60b}$$

where ν_i are the frequencies, and δ_i the phases $(i = 1, \ldots, f)$. The τ-sums extend to integral values of $\tau_1, \ldots, \tau_f$. In classical mechanics, q_k, the coordinates $(k = 1, \ldots, f)$, obey a dynamical equation of motion, with the help of which the Fourier coefficients $C_{\tau_1, \ldots, \tau_f}$, the frequencies ν_i, and the phases δ_i may be determined. For instance, in the case of the one-dimensional harmonic oscillator this equation is

$$\frac{d^2q}{dt^2} + (2\pi\nu)^2 q = 0, \tag{61}$$

and it is satisfied by the single Fourier term

$$q = C\exp[2\pi i(\nu t + \delta)], \tag{62}$$

with an arbitrary amplitude C and an arbitrary phase δ.

The introduction of h, Planck's quantum of action, in the description of nature had changed the classical situation in which ν_{cl}, the frequency of the mechanical motion, was considered to be identical with the frequency of the emitted radiation. Whereas in Planck's harmonic oscillator (or resonator) this equality still held accidentally, it became a crucial assumption of Bohr's theory that in atomic systems ν_{qu}, the frequency of radiation of quantum frequency, *did not coincide* with ν_{cl}, the frequency of the mechanical motion of the electron in its orbit (Bohr, 1913b). In accordance with Bohr's *analogy argument*, expounded first in 1913 (Bohr, 1913b, p. 13), only in the limit of high quantum numbers should these two frequencies become equal to each other. However, it had been gradually realized that the quantum frequency emitted during the transition from one stationary state to another assumed a certain average value between the mechanical frequencies of the electron in the two states under consideration (e.g., Kramers, 1919, p. 327). Yet, no prescription was available for how to compute the quantum frequencies of an atom from the mechanical model. The state of affairs in mid-1925 was still characterized by the assumption expounded in the paper of Bohr, Kramers and Slater in early 1924, namely, that *the atoms behave as if they were surrounded at all times by fields of virtual oscillators having the frequencies*, ν_{qu}, *emitted by the atoms* (Bohr, Kramers and Slater, 1924). Starting from this idea—which was due to Slater (1924a)—Kramers had devised an explicit formula for the dispersion of radiation by atoms, which seemed to be confirmed by the known data (Kramers, 1924a). Then Born had shown that Kramers' result could be derived by using a quantum-theoretical perturbation scheme analogous to a classical one for multiply periodic systems (Born, 1924b). The central point in Born's scheme consisted in a replacement of classical quantities; e.g., for $|C_\tau|^2/(\tau\nu)$ (where $|C_\tau|^2$ is the absolute square of the Fourier amplitudes of Eq. (60), and $\tau\nu$ a given higher harmonic frequency) he substituted the quantum-theoretical expression $\Gamma(n+\tau,n)/\nu(n+\tau,n)$, where $\Gamma(n+\tau,n)$ is related to the Einstein absorption coefficient and $\nu(n+\tau,n)$ is the absorption frequency, and he replaced the differential quotients of such expressions by the corresponding difference quotients (see Eq. (51)).

The idea of the field of virtual oscillators attached to each atom, together with Bohr's reformulation prescription, had contributed importantly to the successful work on atomic theory, carried out both in Copenhagen and Göttingen in winter 1924–1925 and spring 1925. In this context the Copenhagen theoreticians had been especially busy in constructing mechanical models to describe adequately the phenomena of atomic physics, such as resonance fluorescence or the Zeeman effects of complex spectra. Yet another aspect of these endeavours, though less obvious in publications, had emerged slowly: the hope that some of the mathematical formulae obtained would represent the physical situation much better than any mechanical model that could be constructed. 'At any rate,' Heisenberg remarked later, 'I found in the formulae, which were the result of my collaboration with Kramers [on dispersion theory], a mathematics which in a certain sense worked automatically independently of all physical models. This mathematical scheme had for me a magical attraction, and I was fascinated by the thought that perhaps here could be seen the first threads of an enormous net of deep-set relations' (Heisenberg in Rozental, 1967, p. 98). The mathematical scheme which Heisenberg had in mind had very much to do with those quantities, which in atomic theory corresponded to the Fourier series representing the variables of classical multiply periodic systems. By late fall 1924 Heisenberg had become convinced that the quantum-theoretical analogues of the Fourier series should have a fundamental importance for the atomic theory of the future. In spring 1925 he had returned to Göttingen from Copenhagen with this conviction further strengthened and had started the investigation of the hydrogen atom by studying the well-known classical Fourier series, occurring in the description of the motion of that system, and by imitating the reformulation procedure which Kramers and he had employed in dispersion theory (Kramers and Heisenberg, 1925).

There were especially two points which Heisenberg could learn from his previous engagement in dispersion theory. These were: First, what the quantity corresponding to a classical Fourier series describing a variable of a multiply periodic system would look like? Second, how would one have to operate with these new quantities? For example, Kramers and Heisenberg had introduced the Fourier series representing $P_0(t)$, the electric moment of the unperturbed atom (see Eq. (53a)). In quantum theory they had substituted for it the expression $P_0(n,t)$, where

$$P_0(n,t) = \sum_{\tau} A(n, n-\tau) \exp\left[2\pi i \nu(n, n-\tau)t\right], \tag{63}$$

and $P_0(t)$ was supposed to represent the electric moment of the atom in the stationary state denoted by the quantum number (or numbers) n. Clearly, the quantities $A(n, n-\tau)$ replace the corresponding classical Fourier amplitudes, and $\nu(n, n-\tau)$, with $\tau = \pm 1, \pm 2, \ldots$, replaces the classical harmonic frequencies, $\pm \tau\nu$; for positive integer τ, $\nu(n, n-\tau)$ represent the emission frequencies, and for negative integer τ the absorption frequencies. Since the τ-sum (or τ-sums, if the quantum system has several degrees of freedom) on the right-hand side of Eq. (63) includes both positive and negative integral values (unless the system is in its ground state, then only negative values are adopted), it contains, apart from the constant term (i.e., $A(n,n)$ because the associated frequency $\nu(n,n)$ is zero) emission ($\tau > 0$) and absorption ($\tau < 0$) terms.[238] The quantum-theoretical reformulation of the classical Fourier series for $P_0(t)$, i.e., Eq. (63), followed immediately from the assumption of the virtual oscillators having the observed absorption and emission frequencies of the atom under consideration. In quantum theory the $A(n, n-\tau)$ were called the transition amplitudes and the $\nu(n, n-\tau)$ the associated transition frequencies.

[238] Although Eq. (63) does not occur explicitly in the paper of Kramers and Heisenberg (1925), it was known to them, as they used it to derive, e.g., their equation (27) for the scattering moment of the atom responsible for coherent scattering (Kramers and Heisenberg, 1925, p. 692). We should also note that in the Kramers–Heisenberg paper the quantum-theoretical amplitudes corresponding to the classical C_τ are not written as $A(n, n-\tau)$ but as A_a $(= A(n, n-\tau)$ for $\tau = -1, -2, \ldots)$, or absorption amplitudes, and A_e $(= A(n, n-\tau)$ for $\tau = 1, 2, \ldots)$ or emission amplitudes. The notation with two integers was introduced, however, earlier in the paper of Born (1924b, p. 389, Eq. (32)).

To reach his final goal of obtaining the intensities of the hydrogen lines, or of even more complicated models of radiation, Heisenberg had to perform certain operations with series like the one representing $P_0(n, t)$. He especially wanted to know what corresponded to the sum and the product of two classical Fourier series. While the addition of two expansions of the type given by the right-hand side of Eq. (63) seemed obvious, this was not so with the multiplication. Hence Heisenberg was primarily interested in the question: 'Can one, in general, multiply two quantum amplitudes X and Y, which, in classical physics, correspond to Fourier series? If X and Y are known as representing such patterns of quantum-theoretical amplitudes [i.e., sums like the ones on the right-hand side of Eq. (63)], can I then guess the pattern of X times Y? Actually that had already been done in the paper on dispersion theory' (Heisenberg, Conversations, p. 261; also AHQP Interview).

In the paper on dispersion theory Kramers and Heisenberg had started from the classical expression for the induced electric moment, Eq. (53b). This expression contained a quantity, which could be considered as representing the product of two Fourier series, that is,[239]

$$\left[\sum_{\tau} C_{\tau} \exp(2\pi i \omega t)\right]\left[\sum_{\tau'} C_{\tau'} \exp(2\pi i \omega' t)\right] = \sum_{\tau}\sum_{\tau'} C_{\tau} C_{\tau'} \exp\left[2\pi i(\omega + \omega')t\right]. \quad (64a)$$

Hence the quantum-theoretical analogue should represent a special example of the desired product, which we shall represent by XX, where X is the quantum-theoretical expansion corresponding to the classical Fourier series with amplitudes C_{τ}. In the process of rewriting the classical formula for the induced electric (or scattering) moment, Kramers and Heisenberg had obtained for the product XX the expression

$$XX = \sum_{Q}\sum_{R} A_p A_q \exp\left[2\pi i(\nu_p + \nu_q)t\right]. \quad (64b)$$

(See Eq. (56).[240]) The right-hand side of Eq. (64b) consists of a double sum, which is physically interpreted in the following way. The scattering atom, originally in the state P, passes, under the influence of the incident radiation, into stationary states denoted by Q. If one takes a given state Q, the transition from P occurs in two steps: first, the atom goes over from P into an intermediate state R, with transition amplitude A_p and transition frequency ν_p; second, the atom goes over from the state R to the state Q, the transition amplitude being A_q and the transition frequency ν_q. If, instead of the Kramers–Heisenberg amplitudes and frequencies, one introduces notations by which the analogy to the classical

[239]On the right-hand side of Eq. (53b) these products appear as multiplied with factors and differentiated with respect to the action variables. We have neglected these extra terms and operations at first in order to simplify the discussion; we shall return to them in the following.

[240]From the quantum-theoretical formula for the scattering moment of the atom, $P_1^{\mathrm{qu}}(t)$, Eqs. (56) and (56a), we take out those terms which correspond to the rewriting of the classical product Fourier series, Eq. (64a).

Fourier series becomes more transparent—i.e., one replaces A_p by $A(n,n-\tau)$ and ν_p by $\nu(n,n-\tau)$, also A_q by $A(n-\tau,n-\tau-\tau')$, where, obviously, the quantum numbers of the states R and Q are $n-\tau$ and $n-\tau-\tau'$, respectively—the expression XX assumes the form

$$XX=\sum_\tau\sum_{\tau'}A(n,n-\tau)A(n-\tau,n-\tau-\tau')\exp\left[2\pi i\nu(n,n-\tau-\tau')t\right], \quad (65)$$

with

$$\nu(n,n-\tau-\tau')=\nu(n,n-\tau)+\nu(n-\tau,n-\tau-\tau'). \quad (65a)$$

This result gives the expression for the simplest product of two quantum-theoretical variables, or 'patterns of quantum-theoretical amplitudes,' namely, of X with itself ($X=\sum_\tau A(n,n-\tau)\exp[2\pi i\nu(n,n-\tau)t]$). However, Kramers and Heisenberg's paper on dispersion theory also provided an answer in the more general case of multiplying X and Y, two arbitrary quantum-theoretical variables of the above type, that is,

$$XY=\sum_\tau\sum_{\tau'}A(n,n-\tau)B(n-\tau,n-\tau-\tau')\exp\left[2\pi i\nu(n,n-\tau-\tau')\right], \quad (66)$$

where the $A(n,n-\tau)$ denote the amplitudes of X and the $B(n-\tau,n-\tau-\tau')$ denote the amplitudes of Y.[241]

In a letter to Kronig, dated 5 June 1925, Heisenberg stated that the multiplication rule, expressed in Eqs. (65) and (66), followed 'inevitably' ('*zwangsläufig*'), if one had to reformulate the multiplication law of classical Fourier series. The question how 'inevitable' the quantum-theoretical multiplication rule was may be regarded as being serious. In dispersion theory Kramers and Heisenberg had worked with the quantum-theoretical transition amplitudes, $A(n,n-\tau)\times\exp[2\pi i\nu(n,n-\tau)t]$, and their complex conjugates, $\bar{A}(n,n-\tau)\exp[-2\pi i\cdot\nu(n,n-\tau)t]$; these corresponded to the classical amplitudes C_τ and $C_{-\tau}$, consistent with the translation code for the frequencies

$$(\tau\nu)_n=-(-\tau\nu)_n \qquad \text{in classical theory} \quad (67a)$$

and

$$\nu(n,n-\tau)=-\nu(n-\tau,n) \qquad \text{in quantum theory.} \quad (67b)$$

The suffix n in the classical formula (which is identical with the n in the quantum formula) denotes the stationary state of the multiply periodic system emitting or

[241] The occurrence of this generalized product in the Kramers–Heisenberg dispersion theory is obvious. In Eqs. (56) and (56a) one has simply to consider the series consisting of terms, $A_p\cdot A_q/(\nu_q+\nu)\cdot\exp[2\pi i(\nu_p+\nu_q)t]$, which may be interpreted as the product of two patterns, X and Y, with $X=\sum_\tau A(n,n-\tau)\exp[2\pi i\nu(n,n-\tau)t]$ and $Y=\sum_{\tau'}B(n-\tau,n-\tau-\tau')\exp[2\pi i\nu(n-\tau,n-\tau-\tau')t]$, with $B(n-\tau,n-\tau-\tau')=A(n-\tau,n-\tau-\tau')/[\nu(n-\tau,n-\tau-\tau')+\nu]$.

absorbing radiation. Relation (67a) follows from the physical interpretation of the classical Fourier terms: in case of absorption the frequency appearing in the periodic exponential bears the negative sign of the frequency appearing in the periodic exponential describing the emission of radiation of the same frequency. Since the position variables of multiply periodic functions are supposed to be real, we have *classically* the relation between the Fourier coefficients,

$$(C_\tau)_n = (\bar{C}_{-\tau})_n. \tag{68a}$$

The corresponding *quantum-theoretical* translation of this equation had been found in dispersion theory to be

$$A(n, n-\tau) = \bar{A}(n-\tau, n). \tag{68b}$$

(This result may be derived, for example, from Eq. (56a) by putting the state P equal to Q; hence $\nu_q = -\nu_p$ and Eq. (68b) follows immediately if one denotes A_q by $\bar{A}_p$ in agreement with the classical notation.) In order to represent a *real* quantum-theoretical property of a multiply periodic system, it is not enough to replace the right-hand side of Eq. (68b) by $\bar{A}(n, n+\tau)$, primarily because in quantum theory the frequencies, $\nu(n, n-\tau)$ and $\nu(n+\tau, n)$, are not identical as in classical theory. Thus, from the physical point of view, the prescription for translating Eq. (68a) into Eq. (68b) becomes inevitable.

So far, only the quantum-theoretical reformulation of a single amplitude was given, involving the transition from a state denoted by the quantum number (or quantum numbers, if the system possesses several degrees of freedom) n to a state denoted by $n-\tau$, and vice versa. However, in dispersion theory more complicated expressions had arisen when the electric moment of the atom was considered. For instance, Born had suggested the following replacement:

$$\tau \frac{\partial}{\partial J}(C_\tau C_{-\tau})_n \rightarrow \frac{1}{h}\left[\Gamma(n+\tau, n) - \Gamma(n, n-\tau)\right], \tag{69}$$

where, on the left-hand side, $\partial/\partial J$ denotes the sum of the partial derivatives (i.e., $\tau\partial/\partial J = \tau_1\partial/\partial J_1 + \tau_2\partial/\partial J_2 + \dots$) with respect to the action variables of the multiply periodic classical system (Born, 1924b, p. 389). $\Gamma(n+\tau, n)$ and $\Gamma(n, n-\tau)$ denote the absolute squares of quantum-theoretical absorption and emission amplitudes, respectively. Again the prescription (69) may not appear, at first sight, as being necessary or inevitable since in its stead one could think of

$$\tau \frac{\partial}{\partial J}(C_\tau C_{-\tau})_n \rightarrow \frac{1}{h}\left[\Gamma(n, n-\tau) - \Gamma(n-\tau, n-\tau-\tau)\right]. \tag{70}$$

The physical reason for preferring the prescription (69) over (70) is obvious, since the first term should refer to the absorption and the second to emission of radiation from a system in the state n; this interpretation is guaranteed by the reformulation (69) but not (70). Moreover, the classical quantity,

$\tau\partial(C_\tau C_{-\tau})_n/\partial J$, remains invariant under the change of sign of τ, and the analogous quantum formula should have the same behaviour; only the prescription (69) gives it, not (70).

While these experiences from dispersion theory helped Heisenberg in deriving his multiplication rule, expressed by Eqs. (65) and (66), the most important guiding principle was much older. Indeed, Eq. (65) or (66) could only be obtained by assuming the frequency composition rule contained in Eq. (65a). This quantum-theoretical relation is different from the corresponding classical one, that is,

$$(\tau\nu)_n + (\tau'\nu)_n = ((\tau + \tau')\nu)_n, \tag{71}$$

because in the latter *both* terms on the left-hand side refer to harmonic frequencies connected with the stationary state of the system denoted by the subscript n. On the other hand, the composition rule, Eq. (65a), was one of the oldest results of modern spectroscopy, expounded as early as 1908 by Walther Ritz and later called Ritz' combination principle (Ritz, 1908b). The fundamental importance of the combination principle in atomic theory had already been recognized by Bohr in his first paper on the atomic constitution (Bohr, 1913b). Sommerfeld, who more than anyone else sought to establish the validity of this principle, thought that it had 'maintained itself in the whole region of spectroscopy from infra-red to X-ray spectra as an *exact* physical law with the degree of accuracy that characterizes spectroscopic measurements' (Sommerfeld, 1924d, p. 88). The Ritz combination principle, as interpreted by Bohr, had also been the guideline in the attempts, again mainly by Sommerfeld and his disciples, to explain the band spectra of molecules and the anomalous Zeeman effects. Heisenberg, at the crucial juncture in his work when he had discarded most of the premises of the Bohr–Sommerfeld theory, did not hesitate in appealing to the combination principle. It gave the necessary help in determining the rule for the quantum-theoretical reformulation of the multiplication of two classical Fourier series, a rule which could even be extended to products of more than two factors. Hence the multiplication rule, expressed by Eqs. (65) and (66), followed 'inevitably' from the Ritz combination principle and the procedure leading to the Kramers–Heisenberg formulae in dispersion theory, while the combination principle received strong support from dispersion theory, too.[242]

In writing down the quantum-theoretical product, Heisenberg immediately noticed an important fact: X and Y, the quantum-theoretical expressions representing the variables of a multiply periodic system, had to be taken for *different stationary states* of the atom. This fact followed directly from the physical interpretation of the Kramers–Heisenberg dispersion formulae, Eqs. (56) and (56a). There the products of two patterns of quantum-theoretical amplitudes were connected with the transition in two steps, from a state P of the atom to a state R and then from the state R to the state Q; for example, on the right-hand

[242]If the phase factors in the products, Eqs. (66), were not the correct ones, the unphysical terms obtained in the Kramers–Heisenberg formula would not drop out, as discussed in Section III.5.

side of Eq. (64b), the amplitudes A_p represent the 'pattern' of transition amplitudes from the state P, and the amplitudes A_q represent the pattern of transition amplitudes from the state R. Thus, the situation with the multiplication of two variables in quantum theory deviated strongly from the one in classical theory, where the variables, which are multiplied, refer always to the same state of the system.[243] And this difference between the multiplication of classical and quantum-theoretical variables led to an important consequence. If one tried to multiply two different patterns of quantum-theoretical amplitudes according to Eq. (66)—i.e., calculated the product XY—the result did not, as a rule, agree with the product YX, in which the factors were commuted. The reason was that the product amplitude multiplying a given exponential function, $\exp[2\pi i\nu(n, n-\tau-\tau')t]$, was composed of sums of different factors, $A(n,n-\tau)B(n-\tau, n-\tau-\tau')$ and $B(n,n-\tau)A(n-\tau,n-\tau-\tau')$, in the two cases.[244] When Heisenberg first hit upon this difficulty in connection with the problem of the intensities of hydrogen lines, he hoped to be able to overcome it, perhaps with the help of an argument based on dispersion theory.[245] However, after investigating the general problem of multiplication in quantum theory, he found that—for the moment at least—the difficulty could not be removed, and he concluded: 'There was a very disagreeable feature about it, that YX was not equal to XY. I noticed that, and I was very dissatisfied with the situation' (Heisenberg, Conversations, p. 262).

Heisenberg's discovery of the noncommutation of the product of quantum-theoretical amplitudes would turn out to be the crucial step in the mathematical transition from classical to quantum mechanics, and one might wonder why it had not been noticed earlier, especially in connection with dispersion theory. Indeed, the easiest way of recognizing noncommutation is to take the limit of very high incident frequency ν (much larger than all absorption and emission frequencies of the atom) of Kramers' dispersion formula describing $P_{\mathrm{qu}}(\nu, t)$, the induced electric moment for coherent scattering, that is,[246]

$$P_{\mathrm{qu}}(\nu,t) = -\frac{E}{4h\nu^2}\cos(2\pi\nu t)\cdot\sum_{\tau}\Big[|A(n+\tau,n)|^2\nu(n+\tau,n) - |A(n,n-\tau)|^2\nu(n,n-\tau)\Big]. \tag{72}$$

[243] The same difference does not show up in the other operation of adding two quantum-theoretical analogues of classical Fourier series. There one evidently adds the amplitudes belonging to the same stationary state.

[244] In classical theory the factors in the products of two Fourier series commute, of course, since both factors in each product term, say $(A_\tau B_{\tau'})$, belong to the same stationary state of the system, i.e., $(A_\tau B_{\tau'}) = (A_\tau B_{\tau'})_n$.

[245] See Heisenberg's letter to Bohr, dated 16 May 1925, to which we have referred above (in Section IV.1), in particular the remark that 'the conditions so far did not suffice completely to determine the intensities uniquely' and Heisenberg's assertion that he still wanted 'to try to get on [with the problem].'

[246] Equation (72) is derived from Eq. (55) by assuming the vector of the electric field of the incident radiation to be parallel to the vectors of the transition amplitudes A^a and A^e, and by writing the quantum-theoretical amplitudes as $A^a = A(n, n+\tau)$ and $A^e = A(n, n-\tau)$—with positive τ—and the absorption and emission frequencies, respectively, as $\nu_a = \nu(n+\tau, n)$ and $\nu_e = \nu(n, n-\tau)$.

The square bracket in Eq. (72) may be considered in terms of the product of two quantities, X and Y, if one defines their amplitudes by the equations

$$X(n+\tau,n) = A(n+\tau,n), \qquad X(n,n-\tau) = A(n,n-\tau),$$

$$Y(n,n+\tau) = \bar{A}(n+\tau,n)\nu(n+\tau,n), \tag{73}$$

$$Y(n-\tau,n) = \bar{A}(n,n-\tau)\nu(n,n-\tau).$$

Now in the dispersion formula the amplitude $A(n,n-\tau)$ denoted the transition amplitude which turns the state, designated by the quantum numbers n, into the state designated by the quantum numbers $n-\tau$, while the complex conjugate amplitude $\bar{A}(n,n-\tau)$ is connected with the return from the state $(n-\tau)$ to the state n. The interpretation of the first term in the square bracket is then clear: the amplitude $\bar{A}(n+\tau,n)$ changes the state n (whose electric moment is considered) into the state $n+\tau$, while the amplitude $A(n+\tau,n)$ gives back the state n. Hence the order of factors X and Y in the sum on the right-hand side of Eq. (72) becomes

$$\sum_{\tau}\left[Y(n,n+\tau)X(n+\tau,n) - X(n,n-\tau)Y(n-\tau,n)\right]. \tag{74}$$

The summation over τ in Eq. (74) includes positive and negative integers; hence expression (74) is identical to the difference of products $XY - YX$ in the stationary state of the atom denoted by the quantum numbers n. And since this expression, according to Kramers' dispersion formula, should be proportional to the electric moment of the atom, induced by the incident light of large frequency, this difference cannot be zero. Although a demonstration of noncommutation of the product of quantum-theoretical variables was possible as early as 1924, and Kramers, Born, or Kramers and Heisenberg could have come across it on the basis of the simplest dispersion formula, all these authors missed it. The main reason was that it did not occur to anybody at that time to write it all down in terms of what corresponded to the classical Fourier series. That was done only by Heisenberg in spring 1925.

Interestingly enough, even Heisenberg in May 1925 did not seem to have noticed the occurrence of noncommutation in Kramers' dispersion formula, at least in the beginning; he rather hoped to avoid the problem of noncommutation altogether. As he recalled later on about the moment when he discovered noncommutation: 'Then I felt, "After all, can I not try a problem in which I have to deal with only simple multiplications—not so complicated as the Kepler motion?" And so I came to the anharmonic oscillator—the anharmonic oscillator being a system where, if you do the perturbation theory, you just have a few powers of X, and that is, of course, very simple' (Heisenberg, Conversations, p. 262). Heisenberg was so involved in his programme of trying to treat a quantum system, in which he could work out the results consistently and uniquely, that he did not occupy himself with exploring the possible consequences of the property of noncommutation. He rather shifted the difficulty into the background. He felt that his basic programme had to be given absolute priority and thought: 'Now I must try to see whether I can get a complete quantum mechanics of the anharmonic oscillator, just by using that kind of multiplication which was taken from the dispersion paper. And that I did' (Heisenberg, Conversations, p. 262).

IV.3 The Anharmonic Oscillator and a Letter to Kronig

In May 1925 Heisenberg proposed a new approach to the problems of atomic theory. For this purpose he first introduced for each system a quantum-theoretically adapted kinematic description, which was obtained by a proper reformulation of the Fourier series representing the position coordinates of the corresponding classical multiply periodic system as patterns of quantum-theoretical amplitudes. Second, he established relations in order to determine the quantum-theoretical amplitudes and associated phases. Since in classical theory the parameters of the Fourier series in question could be obtained by using the equation of motion for the position variables (together with the available initial or boundary conditions), the question arose whether there also existed in quantum theory equations corresponding to the classical equations of motion. Heisenberg assumed that the answer to this question was affirmative. Many years later he tried to recall the steps and thoughts leading to this assumption, and said:

> I could not say that this idea [i.e., the idea of employing an equation of motion in quantum theory] occurred to me in a single instance which I would remember. It came gradually as something that was almost obvious. In that sense one would say, "What does $X(t)$ [i.e., the quantity corresponding to the position coordinate of a multiply periodic system] mean?" Now $X(t)$ in an atom does not really mean an orbit because such an orbit has the wrong frequency. So what does it mean? Well, it means certainly some kind of radiation, so all of these frequencies which are possibly emitted [and their amplitudes] do somehow represent this $X(t)$. "Somehow represent" was, of course, a very vague term, but it wasn't more than that. Still I felt, "Why shouldn't this representation of $X(t)$ also obey the laws of motion?" I saw that there was really no reason why it shouldn't. At least one should try [to investigate that point]. (Heisenberg, Conversations, p. 263; also AHQP Interview)

Although this development of Heisenberg's thoughts appears to be natural and logical, in fact this reasoning was quite audacious in spring 1925. After the theoreticians had discovered the breakdown of classical mechanics in atomic theory, they had abandoned the use of equations of motion altogether. They confined themselves to taking up mathematical relations from the classical theory of multiply periodic systems—relations, which had been obtained after going through the details of integration of the equations of motion—and translating them by more or less systematic guessing into the corresponding quantum-theoretical formulae. Heisenberg did not leave this successful avenue of obtaining quantum-theoretical results voluntarily; he was forced to do so because the standard methods failed to provide an answer to the problem of the intensities of hydrogen lines. It was this failure which brought him to reconsider the kinematical description of atoms. This step was audacious by itself, for people had avoided talking about the kinematics of electron motions in atoms for quite a while. Only one thing was known for certain: the electron orbits in atoms, assumed by the Bohr–Sommerfeld theory, should not be taken literally. Heisenberg was fully aware of this, but he hoped to avoid the earlier difficulties in his new approach. He found that his views were supported by the observation, made during the previous year, especially in Copenhagen and Göttingen, that in some cases mechanical models seemed to account in a formal way—i.e., after a quantum-theoretical translation—for the properties of atomic systems. The next step then consisted in the following assumption: that it was possible to account for the kinematical situation in a given atom by formally rewriting the classical Fourier series, which described the position coordinates of the corresponding multiply periodic classical system. Heisenberg thus arrived at the conclusion that the patterns of quantum-theoretical amplitudes would represent the position coordinates in quantum theory; with this conclusion, the following procedure was almost determined. In particular, one had to search for the equation, which, in quantum theory, would assume the place of the classical equation of motion. Heisenberg did not have the slightest doubt that the quantum equation should be obtained from a reformulation of the classical equation by taking into account the rules that he had worked out before. Of course, he had to demonstrate how the quantum equation would be used. Hence he thought: 'Can't I find a single example in which I could see how things work out?' (Heisenberg, Conversations, p. 263). And he decided to treat the anharmonic oscillator in one space dimension, the simplest interesting example he could imagine.

The anharmonic oscillator had already served as a useful model for specific problems of quantum theory in the places where Heisenberg had grown up scientifically. For instance, Adolf Kratzer, a student of Sommerfeld's at Munich, had employed the anharmonic oscillator model to explain the systematic deviations of certain molecular band spectra—i.e., the shift of the frequencies of the band heads—from the prediction of Henri Deslandres' formula (Kratzer, 1920a). Kratzer continued to work on this problem after he went from Munich to Göttingen in fall 1920. In Göttingen, there already existed some tradition of employing the anharmonic oscillator in quantum theory. Peter Debye, Sommerfeld's first research pupil and collaborator, had, in his Wolfskehl lectures of 1913, dealt with the quantization of anharmonic motions occurring in the thermal vibrations of crystals close to their melting points (Debye, 1914). A year later Sergei Boguslawski, a former student of Born's, employed the anharmonic oscillator model (Boguslawski, 1914a, b, c), to explain the observations by Walter Ackermann, a student of Woldemar Voigt's, on the pyroelectricity of certain compounds (Ackermann, 1915). In 1921 Max Born, together with his collaborator E. Brody, discussed systematically the problem of anharmonic oscillations in crystal lattices by first developing a rigorous approach based on the perturbation theory of celestial mechanics (Born and Brody, 1921b).

Heisenberg, who might have become acquainted with the anharmonic oscillator in Sommerfeld's courses, but certainly knew about Kratzer's application of it to explain the anomalies in band spectra, did not refer to it in his publications earlier than 1925. However, he had used the anharmonic oscillator in a calculation, which he reported in a letter to Pauli, written in fall 1922. The occasion for this letter had been to answer some of Pauli's questions concerning a paper by Sommerfeld and Heisenberg (1922a); this paper dealt with the relation between the 'classical duration of radiation,' as observed in the experiments of Wilhelm Wien (1919, 1921) and the 'instability' of stationary states of atomic systems. In particular, Sommerfeld and Heisenberg had concluded that $\Delta\nu_{\text{qu}}$, the quantum-theoretical linewidth of the spectral lines of hydrogen, was proportional to the difference of two energies, U_1 and U_2, which represented the radiation losses in the two stationary states involved in the transition.[247] In the limit of high quantum numbers, the inverse of $\Delta\nu_{\text{qu}}$ passed over into the classical decay time ('*Abklingzeit*'). Pauli had not believed the result of Sommerfeld and Heisenberg and had written to Heisenberg about it in September 1922. Heisenberg had replied to those doubts by stating: 'I am convinced that the formula, $U_1/h - U_2/h = \Delta\nu_{\text{qu}}$, always yields, in the limiting case of large quantum numbers, the classical damping. I have not thought about a general proof, but I shall briefly treat the case of the anharmonic oscillator, from which one may probably obtain the general case' (Heisenberg to Pauli, 29 September 1922).

To demonstrate his point, Heisenberg chose a one-dimensional anharmonic oscillator, whose position coordinate x satisfied the equation of motion

$$\ddot{x} + \omega_0^2 x + \lambda x^2 = 0, \tag{75}$$

with $\ddot{x}$ denoting the second time derivative of x, ω_0^2 the force constant, and λ a small parameter characterizing the anharmonicity. He then integrated this equation and obtained the following solution:

$$\begin{aligned} x = a\exp(2\pi i\nu t) &+ \frac{1}{3}\left(\frac{\lambda}{4\pi^2\nu^2}\right)a^2\exp[2\pi i(2\nu)t] \\ &+ \frac{1}{12}\left(\frac{\lambda}{4\pi^2\nu^2}\right)^2 a^3\exp[2\pi i(3\nu)t] + \dots, \end{aligned} \tag{76}$$

where a is an arbitrary amplitude and ν ($\approx \omega_0/2\pi$) a frequency. If one adds a damping term (i.e., a term proportional to the time derivative of x, whose

[247] We have discussed the paper of Sommerfeld and Heisenberg referred to in Section I.6. The last statement may be derived from Eq. (12).

strength is determined by a parameter γ, the inverse of a damping time) in the equation of motion, its solution is again given formally by the right-hand side of Eq. (76); however, one has to replace the amplitude a by the product $a \cdot \exp(-\gamma t)$. From this result Heisenberg concluded: 'Indeed, the damping of the τ-th harmonic becomes τ-times as large as that of the fundamental frequency. As one can see, this follows in the above calculation from a kind of dimensional equation; I believe that the general proof can be carried through in this way, though I cannot say for sure. The calculation is already rather complicated in the case of the ellipse' (Heisenberg to Pauli, 29 September 1922). In any case, the same result as the one derived in the case of the anharmonic oscillator also followed from the Sommerfeld–Heisenberg equation for the quantum-theoretical linewidths. Heisenberg had thus succeeded in giving it some credibility, although he had not been able to extend his calculations to the hydrogen atom in the Bohr–Sommerfeld theory.

This old problem of the quantum-theoretical linewidths had certain aspects in common with the problem of transition amplitudes, which Heisenberg investigated in May 1925. As in 1922, he now began by considering the hydrogen atom and finally ended up by replacing the one-electron problem by the anharmonic oscillator because he could not perform the calculations he had in mind on the hydrogen atom. In fact, the reappearance of the anharmonic oscillator at this time could be regarded as a feature characteristic of Heisenberg's manner of attacking difficult problems. On both occasions—in 1922 and 1925—Heisenberg did not use the anharmonic oscillator to imitate realistically a phenomenon of nature, as Debye, Kratzer, or Born and Brody had intended earlier. He rather took it as the simplest mechanical model in which he could demonstrate the validity of new physical ideas and test the applicability of mathematical methods. He hardly imagined that the anharmonic oscillator would provide an approximate representation of the radiation from a Bohr orbit in the hydrogen atom. What attracted him, however, were the few properties which the system exhibited in spite of its simplicity. In particular, it gave rise to more than a single frequency; already the classical system allowed for the occurrence of a full spectrum of harmonics, ν, 2ν, 3ν, . . . , which could be regarded as having a faint similarity to the discrete spectra of atoms. It seemed, therefore, worthwhile to test the new ideas concerning quantum-theoretical amplitudes and their multiplication in the case of the anharmonic oscillator, and to see whether the results obtained in this way made sense. The reappearance of the anharmonic oscillator in Heisenberg's endeavours following the unsuccessful attempt at guessing the intensities of hydrogen lines was thus a logical step.[248]

From these considerations it becomes quite evident in which aspects of the anharmonic oscillator problem Heisenberg was primarily interested. He did not

[248] Heisenberg would use the anharmonic oscillator quite often in his later work, as the simplest example of a nonlinear system, and try out with it new ideas and methods. For example, in 1953, he tested an approximation method (developed for treating many-body problems, the so-called Tamm–Dancoff method) first with the anharmonic oscillator before applying it to nonlinear quantum field theory (Heisenberg, 1953).

want to take the anharmonic oscillator as a kind of substitute ('*Ersatz*') model for a realistic atom. It would be wrong to look at Heisenberg's procedure in late spring 1925 as a continuation of the programme of constructing specific models to describe complicated atomic properties, which he had carried out with reasonable success in Copenhagen during the previous months. He had left behind this stage of the development and had become allergic to any kind of '*Ersatz*' model or system. Thus, he wrote to Kronig about the latter's notion of 'substitute radiator' ('*Ersatzstrahler*') the following critical words: 'The phrase "*Ersatz* [substitute] radiator" has assumed in literature (Landé) the meaning: "Unclean application of the correspondence principle, which one cannot understand". I request you urgently to eliminate this phrase, which evokes in me the memories of wartime *Ersatz*-marmalade, etc.' (Heisenberg to Kronig, 20 May 1925). In the use of the anharmonic oscillator model Heisenberg did not see such a 'wartime substitute' for the real atomic systems. He was convinced that it provided the unique case of being the first example, for which a consistent quantum theory could be given and in which the differences from the classical theory could be easily demonstrated. Thus far, the differences between quantum and classical theory had only been recognized when one compared the semiempirical formulae for multiplet intensities with the corresponding formulae derived from the classical models of multi-electron atoms. In Copenhagen, Kronig was working on the latter question; and Heisenberg, who regularly exchanged letters with him on this subject, expected that he might also be interested in the more general aspects of the intensity problems. Hence Kronig had the opportunity of witnessing the earliest steps in the development of a consistent quantum theory.

On a postcard, dated 3 June 1925, Heisenberg announced to Kronig the first indication of any progress in the problem of getting closer to the solution of the intensity problem. He remarked: 'I shall write to you soon in greater detail about intensity questions; I believe I have made a little progress in principle and would like to discuss it with you' (Heisenberg to Kronig, 3 June 1925). The promised letter arrived a couple of days later, and in it Heisenberg presented a complete report on his recent endeavours. He began by saying:

> Now I want to tell you a bit about my own intensity considerations, and I would like your sharpest possible ("*möglichst scharfe*") criticism. The basic idea is this: In classical theory the knowledge of the Fourier series for the motion [of a system] is sufficient to calculate *everything* [i.e., all the properties of the system]; not only, say, the dipole moment (and the [associated] emission), but also the quadrupole moment, higher moments, etc. (Heisenberg to Kronig, 5 June 1925)

These statements just referred to the conclusions of his previous considerations on the general problem of the intensities of radiation emitted by atoms. But then he proceeded to treat two examples in which he employed an anharmonic oscillator system. In the first example Heisenberg treated a problem related to the intensity of the spectral lines of a many-electron atom, in which he expected particular interest and criticism from Kronig; in the second he tried to demon-

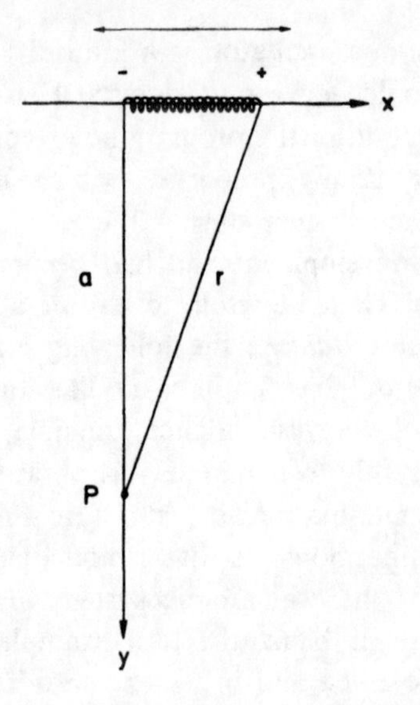

Figure 1. Example 1 from Heisenberg's Letter to Kronig, 5 June 1925.

strate how one could obtain the solution of a problem in quantum theory by using his new ideas and methods.[249]

In order to make the treatment of the anharmonic oscillator similar to the well-known procedure applied previously for calculating the intensity ratios of complex line multiplet components, Heisenberg considered the following problem. A negative charge ($-e$) is situated at a point P, and another negative charge is placed at a distance a from it (along the y-axis). In the direction perpendicular to the line joining the two negative charges, a positive charge ($+e$) performs anharmonic oscillations in the x-direction. Heisenberg illustrated the system as in Fig. 1. The position coordinate, x, of the positive charge is described in classical theory by the Fourier series

$$x = a_0 + a_1 \cos(2\pi\nu t) + a_2 \cos\left[2\pi(2\nu)t\right] + \dots \tag{77}$$

The oscillator system, consisting of the two charges on the x-axis, exerts a periodic force K on the charge at the point P, given by

$$K = -\frac{e^2}{a^2} + \frac{e^2}{a^2 + x^2} = \frac{e^2}{a^2}\left(-1 + \frac{1}{1 + x^2/a^2}\right). \tag{78}$$

By expanding the right-hand side of Eq. (78) in powers of the small parameter x^2/a^2, and inserting for x the Fourier series, Eq. (77), Heisenberg obtained a Fourier series representing K, i.e.,

$$K = \frac{e^2}{a^2}\left\{-1 + b_0 + b_1 \cos(2\pi\nu t) + b_2 \cos\left[2\pi(2\nu)t\right] + \dots\right\}, \tag{79}$$

[249] Part of Heisenberg's letter to Kronig, dealing with the anharmonic oscillator calculations, is reproduced in Kronig's contribution to the Pauli memorial volume (Kronig, 1960, pp. 23–25).

where the first two coefficients are given by

$$-1 + b_0 = \frac{a_0^2 + \frac{1}{2}a_1^2 + \dots}{a^2} + \dots, \tag{79a}$$

and

$$b_1 = -\frac{2(a_0 a_1 + \frac{1}{2}a_1 a_2 + \dots)}{a^2} + \dots. \tag{79b}$$

(The dots indicate further terms like those arising from the power x^4/a^4, etc.) Thus, the Fourier coefficients of the force may be expressed with the help of the Fourier coefficients of the position coordinate of the anharmonic oscillator.

At first sight one might wonder how Heisenberg came to this particular example, and what did it have to do with the problem of radiation in atoms. With a little fantasy one might take the model of Fig. 1 as a simplified representation of the atomic core–series electron model of doublet atoms, whose consequences Heisenberg had pursued since 1921. For this purpose one has just to identify the oscillator system on the x-axis with the core of an atom (consisting of the nucleus and the inner electrons), and the negative charge at point P with the series electron. Thus, K would represent the force of the core on the series electron and give rise to the interaction term, which, in turn, is responsible for the complex structure of the radiation emitted by the atom.

After establishing the classical expression for the force K, Heisenberg assumed that the relation, Eq. (79), could be taken over into quantum theory by applying the translation prescriptions which he had developed earlier. Hence he reported to Kronig: 'One would therefore try to re-interpret Eq. [(79)] in quantum theory, and indeed a re-interpretation follows inevitably' (Heisenberg to Kronig, 5 June 1925). With these words, then, he proposed to take over the classical relation between K and the position coordinate x, as expressed by Eq. (79), but to rewrite the Fourier amplitudes b_n in agreement with the quantum-theoretical multiplication rule. Thus, e.g., for $b_1(n, n-1)$ he found the relation[250]

$$\begin{aligned} b_1(n, n-1) = -\frac{1}{a^2}\Big[& a_0(n)a_1(n, n-1) + a_1(n, n-1)a_0(n-1) \\ & + \frac{1}{2}a_1(n, n+1)a_2(n+1, n-1) \\ & + \frac{1}{2}a_2(n, n-2)a_1(n-2, n-1) + \dots \Big]. \end{aligned} \tag{80}$$

Equation (80) for the transition coefficients $b_1(n, n-1)$, leading from the state denoted by the quantum number n to the state denoted by $n-1$, shows

[250] Heisenberg wrote the right-hand side of Eq. (80) a little differently without, however, changing its content. He also forgot a factor $\frac{1}{2}$ in front of the second and third terms. One should also note that a corresponding equation exists for b_0, which is not as interesting as the one for b_1. Both b_0 and a_0 depend only on the quantum number n of the one state under consideration.

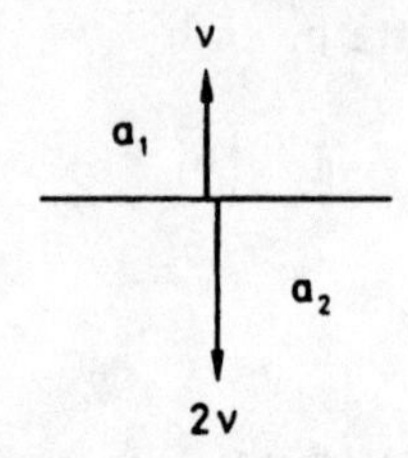

Figure 2. The Classical Combination of Two Amplitudes, a_1 and a_2, with Frequencies ν and 2ν to Form a Product Amplitude with Frequency ν.

characteristic differences as compared to the classical Eq. (79), which may be explained by the difference in the physical situation in quantum and classical theory, respectively. In classical theory the radiation is emitted from a charged accelerated particle moving in a given orbit, described by the label n (i.e., the value of the action variable up to a constant). If one considers that from this level the first harmonic (frequency $\nu \approx \omega_0/2\pi$) and the second harmonic (frequency 2ν) are emitted, there is only one combination of these two harmonics having the frequency ν; hence in the expression for b_1 there will be only one product term involving the Fourier amplitudes a_1 and a_2. The classical situation is indicated in Fig. 2. In quantum theory the situation is basically different, because the differences between the energy terms and, therefore, the transition frequencies between neighbouring states, are not the same. The term corresponding to the product $a_1 a_2$, which contributes to the quantum frequency $\omega(n, n-1)$, can be realized in two ways, (i) and (ii), as indicated in Fig. 3. In both cases the transition in the product term takes place from the initial state n to the final state $n-1$. Due to the different levels involved in the quantum case—e.g., in product

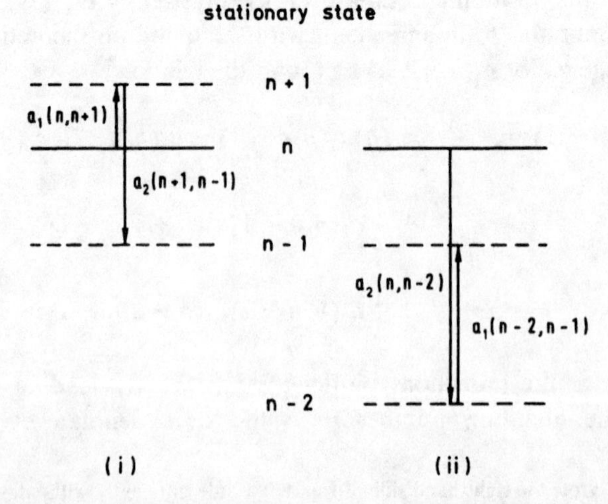

Figure 3. The Quantum-Theoretical Combination of Two Amplitudes to Form a Product Amplitude with Frequency $\nu(n, n-1)$.

(ii) the state with quantum number $n - 2$ comes in, while in product (i) the state with quantum number $(n + 1)$ is reached—the products (i) and (ii) must be counted as separate contributions to $b_1(n, n-1)$.

Two principal assumptions entered into the above considerations, as Heisenberg stressed in his letter to Kronig. First, there was the assumption that 'also in the quantum theory everything is determined by the knowledge of the transition probabilities or the corresponding amplitudes.' Second, one works with the prescription that 'the arguments of the quantum-theoretical amplitudes must be so chosen as to correspond to the connection with the frequencies' (Heisenberg to Kronig, 5 June 1925). Heisenberg had already arrived at both of these assumptions before treating the example of the anharmonic oscillator; the results from the anharmonic oscillator system only served to illustrate his general conclusion: 'If one now admits that such a calculation and quantum-theoretical conversion really makes sense, then one also already has the quantum-theoretical intensity laws' (Heisenberg to Kronig, 5 June 1925). Indeed, the structure of Eq. (80) was similar to the one required to yield the intensity rules for components of complex multiplets, such as those derived by Kronig through clever guessing in agreement with the correspondence principle. Hence, the investigation performed so far convinced Heisenberg that he was on the right track towards solving his problem of calculating the intensity of radiation of atomic systems. One task remained, namely, to determine for a given mechanical system the quantum-theoretical transition amplitudes and transition frequencies, the quantities appearing in the quantum-theoretical series corresponding to the classical Fourier series that describe the position coordinates of multiply periodic systems. This task involved the systematic solution of the quantum-theoretical equation of motion of the system.

With the difficulty arising from the noncommutativity of quantum-theoretical products, which we have discussed above, Heisenberg had to restrict himself for the moment to treating the anharmonic oscillator in one space dimension. Therefore, in his letter to Kronig, he suggested that he intended to try first the integration of the equation of motion, Eq. (75), the same equation which he had already used in 1922. In classical theory Heisenberg expected the position coordinate x to be described by the Fourier series,

$$x = \lambda a_0 + a_1 \cos(2\pi\nu t) + \lambda a_2 \cos\left[2\pi(2\nu)t\right] + \lambda^2 a_3 \cos\left[2\pi(3\nu)t\right] + \dots . \quad (81)$$

That is, the general term on the right-hand side has the structure, $\lambda^{\tau-1} a_\tau \times \cos[2\pi(\tau\nu)t]$, with λ being the (small) parameter of anharmonicity.[251] Now Heisenberg substituted the Fourier series for x, Eq. (81), into Eq. (75) and solved

[251] The anharmonic oscillator, given by Eq. (75), had occurred in the literature before Heisenberg made use of it. For example, it was applied by S. Boguslawski (1914a, b, c) and by Born and Brody (1921b). Born also discussed the details of the perturbation treatment, which he had developed with Brody, in his encyclopedia article on crystal lattices (Born, 1923b, pp. 667–674) and in his lectures on atomic mechanics (Born, 1925, §12). Heisenberg, who had helped Born with the lectures (which were delivered in the winter semester, 1923–1924), was certainly familiar with the results.

the resulting equation containing an infinite number of Fourier amplitudes in successive approximations; i.e., he considered the separate equations for terms multiplied by a given power of λ. Thus, he obtained an infinite number of equations, each involving only a finite number of terms. The first three equations, which are associated with the powers λ^0, λ^1, and λ^2, respectively, could be written as[252]

$$4\pi^2\nu^2 = \omega_0^2, \tag{82a}$$

$$a_2(4 \cdot 4\pi^2\nu^2 - \omega_0^2) = \tfrac{1}{2}a_1^2, \tag{82b}$$

$$a_3(9 \cdot 4\pi^2\nu^2 - \omega_0^2) = a_1 a_2. \tag{82c}$$

Heisenberg translated these equations into the following quantum-theoretical equations:

$$4\pi^2\nu^2(n, n-1) = \omega_0^2, \tag{83a}$$

$$a_2(n, n-2)3\omega_0^2 = \tfrac{1}{2}a_1(n, n-1)a_1(n-1, n-2), \tag{83b}$$

$$a_3(n, n-3)8\omega_0^2 = \tfrac{1}{2}\big[a_1(n, n-1)a_2(n-1, n-3) + a_2(n, n-2)a_1(n-2, n-3)\big]. \tag{83c}$$

The first of these equations states that the transition frequency, $\nu(n, n-1)$, is given by the frequency of the harmonic term in Eq. (75), $\omega_0/2\pi$, up to corrections of the order λ, λ^2, etc. The next two equations give the relations between the quantum-theoretical transition amplitudes. (In Eqs. (83b) and (83c) the squares of the frequencies $\nu^2(n, n-2)$ and $\nu^2(n, n-3)$ were taken in the first approximation, i.e., to order λ^0, as $4\omega_0^2/(4\pi^2)$ and $9\omega_0^2/(4\pi^2)$, respectively.) The equations involving higher amplitudes, e.g., $a_4(n, n-4)$, $a_5(n, n-5)$, etc., may be derived by retaining terms of the order of λ^3, λ^4, etc. All these relations exhibit a structure similar to that of Eq. (82c) in classical theory, or Eq. (83c) in quantum theory, except that the right-hand sides contain many more product terms.

In order to solve the quantum equations for the coefficients $a_\tau(n, n-\tau)$, $\tau = 1, 2, \ldots$, Heisenberg referred to the procedure in classical theory. There, a_1 had been found to be given by the equation

$$a_1 = k(1)\sqrt{n} + \ldots, \qquad \text{with } k(1) = \sqrt{\frac{h}{\pi\omega_0 m}}. \tag{84}$$

[252]There are two further relations between the amplitudes a_0, a_1 and a_2; one (in the approximation to order λ) relates a_0 and a_1, the other (in the approximation to order λ^2) relates a_0, a_1 and a_2. All these have to be satisfied as well. Heisenberg, in his letter to Kronig, dated 5 June 1925, did not refer to them because they had no impact on his conclusions. In this letter Heisenberg also gave the right-hand side of Eq. (82c) the wrong (negative) sign. This error had no consequences, especially since it was corrected in the quantum-theoretical reformulation.

In Eq. (84), n is the label (i.e., the quantum number in the Bohr–Sommerfeld theory of atomic systems) determining the state of the oscillator which emits the radiation; h and m in the expression for the factor $k(1)$ denote, respectively, Planck's constant and the oscillating mass. The Fourier coefficient a_1 possesses, besides the $\sqrt{n}$-term, additional terms having the orders of magnitude λ^2, λ^3, etc., which are indicated in Eq. (84) by the dots.[253] Now, by examining Eqs. (82b) and (82c), and the subsequent ones for higher amplitudes, one sees immediately that in classical theory the general amplitude a_τ assumes the form

$$a_\tau = k(\tau)\sqrt{n^\tau} + \ldots, \tag{85a}$$

$k(\tau)$ being a numerical factor; e.g., $k(2)$ is $k^2(1)/6\omega_0^2$. In quantum theory the same arguments led to the equation

$$a_\tau(n, n-\tau) = k(\tau)\sqrt{n(n+1)\ldots(n-\tau+1)} + \ldots . \tag{85b}$$

The proportionality factor, $k(\tau)$, which depends on the transition (i.e., the number τ) should be identical in both theories.[254]

Expressions of the type of Eq. (85b), which replace the analogous classical ones, Eq. (85a), had occurred before in atomic physics, in particular in the analysis of empirical intensity formulae. For this reason Heisenberg had begun his enterprise of obtaining the intensities of hydrogen lines exactly by trying to guess a replacement of that sort, as we have discussed in Section IV.1. The important progress that followed from the study of the anharmonic oscillator was that it strongly suggested a possibility of *proving* the earlier guesses by means of a systematic and straightforward procedure in quantum theory; i.e., by integrating the quantum-theoretical equation of motion for the systems under investigation. Evidently, Heisenberg was happy about such prospects, and he commented about Eq. (85b) in his letter to Kronig as follows:

> Now this formula is indeed the one we have discussed earlier (compare it with the discussions with Pauli [in Copenhagen]), and I can well imagine that with it one has really a general law for calculating the intensities in hand. From the equations of motion there follow simple relations between the a_τ, which determine the a_τ (up to f independent constants for f degrees of freedom). After a quantum-theoretical re-interpretation, these relations can be taken over directly into quantum theory; thereby one obtains the intensities (again up to f independent constants). The determination of the constants is a chapter by itself, and I shall not write anything about it today. But one can demonstrate, for instance, that your intensity formulae for multiplets and Zeeman effects seem to follow also from the scheme just sketched. (Heisenberg to Kronig, 5 June 1925)

[253] See, for instance, the expression for x in Born, 1925, p. 79, Eq. (11).

[254] Although the quantum-theoretical expressions for $a_\tau(n, n-\tau)$ involve more terms (because each classical product is replaced in general by a sum of quantum-theoretical products), the factors of individual terms add up to yield the classical factor $k(\tau)$ in all cases.

Altogether the initial results from calculations with the anharmonic oscillator made Heisenberg very optimistic about achieving his goal of obtaining a consistent method for computing the intensities of radiation from atoms. The procedure which he had chosen—namely, the introduction of a quantum-theoretical equation of motion as a first step, and its evaluation by using his quantum-theoretical product rule—seemed to be correct. Thus, he remarked to Kronig: 'What I like in this scheme is that one can really reduce all interactions between atoms and the external world (apart from the problems of degeneracy) to transition probabilities' (Heisenberg to Kronig, 5 June 1925). However, he was aware that he had not yet proceeded very far with his enterprise, as he had not even been able to arrive at the complete solution of the anharmonic oscillator problem. 'For the moment I am not content with the mathematical aspect (up to now I do not see any possibility of calculating the intensities in a simple manner) and the determination of the f constants,' he confessed in his letter to Kronig. Not only did there remain mathematical problems, the physical interpretation of the scheme had also to be explored. Thus, Heisenberg concluded: 'The physical interpretation of the above-mentioned scheme also yields again very strange points of view' (Heisenberg to Kronig, 5 June 1925). He was very curious to hear Kronig's opinion about his scheme, with which he planned to go ahead as fast as he could.

IV.4 The Determination of the Constants and a Severe Attack of Hay Fever

For integrating a classical equation of motion of a given mechanical system and obtaining a well-defined solution, one needs to know a certain number of integration constants. In astronomical problems, for example, these integration constants are given by the orbital parameters of the planets. Frequently, the constants fall into two classes: one class of parameters determines momentum variables, the other the initial values of the position variables. For multiply periodic systems, in particular, one may take the action variables (which are time independent) and the values of the canonically conjugate angles at a given instant, say, at an instant $t = 0$, as integration constants of the equations of motion. A very simple periodic system is the harmonic oscillator in one dimension, whose equation of motion is represented by Eq. (75) with λ, the anharmonicity parameter, being put equal to zero. In the solution, i.e., the equation for x,

$$x = a\cos(\omega_0 t + \delta), \tag{86}$$

there appear two constants of integration, the amplitude a and the phase δ. Only after the values of both are known does the solution, Eq. (86), become completely determined. The amplitude a can be replaced by the phase integral J, defined as the integral $\oint p\,dx$ (with $p = m\dot{x}$, the momentum of the oscillating mass m) over a full period of motion, according to the equation

$$J = \int_0^{2\pi/\omega_0} m\dot{x}^2\, dt = \pi m a^2 \omega_0. \tag{87}$$

J, the phase integral given by Eq. (87), represents the action variable of the one-dimensional harmonic oscillator. In the case of a one-dimensional anharmonic oscillator, the action variable is expressed as a sum by the equation

$$J = \frac{\pi m}{2} \sum_{\tau=-\infty}^{+\infty} \lambda^{2|\tau|-2}\left(\tau^2 a_\tau^2 \omega_0\right), \tag{88}$$

i.e., the factor $a^2\omega_0$ must be replaced by a sum of terms, each involving one Fourier amplitude ($a_{-\tau} = a_\tau$).

In his letter to Ralph Kronig, dated 5 June 1925, Heisenberg had mentioned that he did not know how to obtain the constants of motion in the treatment of the quantum-theoretical anharmonic oscillator. Of course, he did not refer by this remark to the value of the phase at the instant of time $t = 0$, but rather to the value of what corresponded to the action variable J. In classical theory the value of this variable can be calculated with the help of Eq. (88). Bohr and Sommerfeld had postulated in their theory of atomic structure that the phase integral assumed values that were integral multiples of Planck's constant h; they had thus fixed the values of J in quantum theory. The analysis of data, derived from atomic systems since the early 1920s, had shown, however, that the Bohr–Sommerfeld postulate was not always valid. As Heisenberg recalled: 'One had seen that $\oint p\, dq$ was sometimes an integral, and sometimes a half-integral multiple of h' (Heisenberg, Conversations, p. 266). Heisenberg felt that this situation in quantum theory prevented one from learning anything from the method of determining constants in classical theory. But then he got a helpful idea: 'I felt that perhaps only the difference of the integral, $\oint dpq$, between one quantum state and the next quantum state is an important quantity' (Heisenberg, Conversations, p. 266). Indeed, taking the difference, $J(n) - J(n-1)$, would remove the indeterminancy of the integration constant $J(n)$ in the Bohr–Sommerfeld theory.[255] However, it did not yet help in the question of obtaining the integration constant of the anharmonic oscillator. Now, Heisenberg realized that he could go one step further and, instead of the difference of phase integrals, he could write a differential quotient—i.e., differentiate both sides of Eq. (88) with respect to the action variable J. This procedure yielded the equation

$$1 = \frac{\pi m}{2} \sum_{\tau=-\infty}^{+\infty} \lambda^{2|\tau|-2}\tau \frac{\partial}{\partial J}\left(a_\tau^2 \tau \omega_0\right), \tag{89}$$

[255] The indeterminacy of the phase integral of the harmonic oscillator was first noticed by Max Planck, who then proposed to remove it by his second quantum hypothesis; he suggested that J should be given by the half-integral multiples of Planck's constant, $(n + \frac{1}{2})h$, n being an integer (Planck, 1911a). Bohr and Sommerfeld did not admit half-integral quantum numbers in their theory of atomic structure; however, in the problems of analyzing band spectra (Kratzer, 1922) and the anomalous Zeeman effects (Heisenberg, 1922a), they sneaked in.

provided the corresponding exchange of the summation and integration could be performed, which Heisenberg assumed tacitly. Equation (89) possesses the structure of those classical equations, which one had been able to reformulate successfully according to Max Born's discretization rule, Eq. (51). The application of Born's rule yielded immediately the quantum-theoretical equation,

$$1 = \frac{\pi m}{h} \sum_{\tau=1}^{\infty} \lambda^{2\tau-2}\left[|a(n+\tau,n)|^2 2\pi\nu(n+\tau,n) - |a(n,n-\tau)|^2 2\pi\nu(n,n-\tau) \right]. \tag{90}$$

Commenting later upon the steps leading from the recognition of the importance of differences of action variables associated with neighbouring states to Eq. (90), Heisenberg remarked: 'I saw that if I wrote down this and tried to translate it according to the scheme of dispersion theory, then I got the Thomas–Kuhn sum rule. And that is the point. Then I thought, "That is apparently the way how it is done"' (Heisenberg, Conversations, p. 266).

In calling his quantum-theoretical Eq. (90) the Thomas–Kuhn sum-rule, Heisenberg referred to results published independently by Willy Thomas, a student of Fritz Reiche's from Breslau (Thomas, 1925a), and Werner Kuhn from Copenhagen (Kuhn, 1925a). Thomas stated in his paper, which would appear in the issue of *Naturwissenschaften*, dated 10 July 1925, his sum-rule as the equation

$$\sum f_a - \sum f_e = \text{degree of periodicity}. \tag{91}$$

That is, he claimed that the degree of periodicity of a multiply periodic, nondegenerate system, containing one electron, is given by a difference of sums, whose terms are the oscillator strengths, f_a and f_e, of all transitions occurring in the system. The f's (subscripts a and e stand for absorption and emission, respectively) are products of the Einstein absorption or emission coefficients and τ, the decay times ('*Abklingzeiten*') of classically oscillating electrons with the corresponding frequencies (i.e., $\tau_{\nu_a} = 3c^3 m_e / 8\pi^2 e^2 \nu_a$, where c is the velocity of light, and m_e and $-e$ are the mass and charge of the electron). The left-hand side of Thomas' formula can be identified, up to a factor, with $|P|$, the amplitude of the induced electric moment (P) of the atomic system responsible for coherent scattering of incident light having very high frequency ν ($\gg \nu_a, \nu_e$), i.e.,[256]

$$\sum f_a - \sum f_e = \left(- \frac{4\pi^2 m \nu^2}{E e^2} \right) |P|, \tag{92}$$

and $P = |P| \cos(2\pi\nu t)$. In this limit the binding of the electron in the system becomes negligible and does not play a role. As a result, $|P|$ on the right-hand side of Eq. (92) should be replaced by the classical expression, which had been

[256] Equation (92) is identical with a formulation which Kramers and Heisenberg gave to Kramers' dispersion formula for coherent scattering (Kramers and Heisenberg, 1925, p. 693, Eq. (29)).

obtained a long time ago by Joseph John Thomson (1903, p. 263). In this way one finds that Thomas' difference of sums of oscillator strengths is unity, provided only one electron is contained in the system. Kuhn, considering the hydrogen atom, also arrived at the same result in his note (Kuhn, 1925a, p. 409, Eq. (2)). The connection between the Thomas–Kuhn sum-rule and Heisenberg's quantum-theoretical reformulation of $\oint p\,dq$-rule, Eq. (90), is now easily established. The left-hand side of Eq. (91) can be written as

$$\sum f_a - \sum f_e = \sum a_{\nu_a}\tau_{\nu_a} - \sum a_{\nu_e}\tau_{\nu_e}$$

$$= \frac{\pi m}{h}\sum_{\tau=1}^{\infty}\lambda^{2\tau-2}\Big[|a(n+\tau,n)|^2 2\pi\nu(n+\tau,n) \tag{93}$$

$$- |a(n,n-\tau)|^2 2\pi\nu(n,n-\tau)\Big],$$

where a_{ν_a} and a_{ν_e} are the Einstein absorption and emission coefficients (e.g., $a_{\nu_a} = ((2\pi\nu_a)^4 e^2/3c^3)A_a\bar{A}_a$; the absorption amplitude, $A_a = a(n+\tau,n)$, the emission amplitude, $A_e = a(n,n-\tau)$, $\tau = 1,2,3,\ldots,$ and the associated frequencies are $\nu_a = \nu(n+\tau,n)$ and $\nu_e = \nu(n,n-\tau)$). In the case of the anharmonic oscillator in one dimension, the right-hand side of Eq. (93) becomes identical to the right-hand side of Eq. (90). Hence, according to Thomas' considerations (Thomas, 1925a), Heisenberg's quantum-theoretical equation demands simply that the anharmonic oscillator possesses just one periodic degree of freedom, in obvious agreement with the facts.

Heisenberg did not know about Thomas' result before it was published later in July 1925. He had also not seen Kuhn's paper on the sum-rule (Kuhn, 1925a). The latter was received by *Zeitschrift für Physik* on 14 May 1925; that is, it was submitted by the author more than one month after Heisenberg had left Copenhagen.[257] However, Heisenberg knew Werner Kuhn from his stay in Copenhagen and was familiar with the kind of work the Swiss scientist did there. Kuhn had graduated in summer 1923 with a doctoral thesis on the photomechanical dissociation of ammoniac under the supervision of Victor Henri in the Chemistry Department of the University of Zurich. Then he spent nearly the entire period between spring 1924 and spring 1926 in Copenhagen.[258] He had

[257] In none of the existing letters, which Heisenberg wrote in May and June 1925 to Kronig, Bohr or Pauli, did he mention Kuhn's sum-rule.

[258] Werner Kuhn was born on 6 February 1899 in Maur, Canton Zurich. He studied chemistry at the *Eidgenössische Technische Hochschule* (E.T.H.) in Zurich (1917–1921); however, he took particular interest at the same time in the lectures of Mieczyslaw Wolfke on statistical mechanics and of Hermann Weyl on mathematics. On receiving his *Diplom* in chemistry, he became assistant to Victor Henri at the Physico-Chemical Institute of the University of Zurich (1921–1924). After completing his doctorate in 1924 he inquired of Niels Bohr whether he might continue his studies under him. Bohr agreed and proposed that Kuhn should come to Copenhagen after his (Bohr's) return from his visit to the United States, which took place in fall and winter 1923 (Bohr to Kuhn, 11 August 1923). Kuhn's stay in Copenhagen was supported by a grant from the International Education Board.

gone to Bohr's Institute with definite ideas in mind, especially to work on a problem of statistical mechanics of his own. 'Besides, I am also interested in general questions,' he had written to Bohr, 'and if the opportunity arises I would also like to tackle a small experimental problem, if possible in a field in which I have not worked so far' (Kuhn to Bohr, 25 February 1924). We do not know whether Kuhn succeeded in solving his statistical mechanical problem in Copenhagen, but he certainly undertook an experimental investigation in a field that was new to him; namely, the magneto-rotation of thallium vapour. However, he could not complete the experimental project within a year, though, as Bohr wrote to Victor Henri on 23 June 1925, he had shown 'unusual skill and perseverance in obtaining suitable conditions for the measurements.' Kuhn's stay in Copenhagen was extended until early March 1926, and then he used the results of his experimental investigations on the anomalous dispersion of thallium and cadium vapours for his *Habilitation* thesis at the University of Zurich (Kuhn, 1926b).[259] Besides preparing the experimental setup for his measurements, Kuhn became acquainted with the general theoretical problems being discussed at Bohr's Institute. 'Dr. Kuhn has given much thought to the general problem of the quantum-theoretical interpretation of the dispersion phenomena,' Bohr reported to Henri, 'and has made an interesting contribution to this subject which he has described in a note to appear soon' (Bohr to Henri, 23 June 1925). The contribution to which Bohr referred was, of course, the sum-rule. Kuhn expected that by taking p_i, the oscillator strengths of individual absorption lines of an atom in the ground state, from dispersion measurements and substituting them in the left-hand side of Eq. (92), one would obtain the number of radiating electrons in the atom. Thus, for example, he found in the case of helium a result which was 'not in contradiction with the expected value 2 of the p-sum' (Kuhn, 1925a, p. 412).

In 1925 Werner Kuhn was mainly concerned with exploring the consequences of the sum-rule for one of the fundamental questions of atomic structure, i.e., to determine the number of electrons occupying different stationary states. The answer to this question with the help of dispersion measurements was of special interest to Niels Bohr and Hendrik Kramers, who thought that dispersion theory (and the sum-rule derived from it) was the most reliable part of quantum theory

[259] After submitting his thesis for the *Habilitation*, Werner Kuhn delivered his *Probevorlesung* (trial or inaugural lecture) in early December 1925 to become a *Privatdozent* at the University of Zurich. As a result of a visit to George de Hevesy in Freiburg, he turned his attention to the problem of emission and absorption of radiation from nuclei. In two papers, submitted in spring 1927, he discussed the polarizability of nuclei, thus transferring the quantum-theoretical dispersion treatment from atoms to nuclei, and came to the conclusion that the γ-radiation cannot arise from electrons bound in nuclei, but from hydrogen or helium nuclei, which might occur as parts of heavy nuclei (Kuhn, 1927a, b). Kuhn's papers must be regarded as an early contribution to the solution of the problem of nuclear structure. Kuhn left Zurich in 1928 to become assistant and *Privatdozent* at the University of Heidelberg. From there he moved in 1930 to the Technical University of Karlsruhe and was promoted in 1931 to *extraordinary* (or associate) professor. In 1936 he became a full professor at the University of Kiel, and in 1939 he obtained a similar position at the University of Basle. He died in Basle on 27 August 1963.

at that time. That a sum-rule of the kind of Eq. (91) was valid, the Copenhagen physicists knew already much earlier. As Heisenberg remarked: 'Possibly the sum rule belongs among the results, which first emerged in the many Copenhagen discussions and which were then brought into a definite form only much later' (Heisenberg to van der Waerden, 2 July 1965). Heisenberg did not recall who in Copenhagen first proposed the sum-rule, but Kuhn certainly participated in the discussions there in fall and winter 1924 and gave the sum-rule its final form.[260] He showed that the right-hand side of Eq. (92) assumed, in analogy with Thomson's classical theory of the dispersion of X-rays, the integral values 1, 2, etc., according to the number of radiating electrons in an atom. It is very likely that he found this result only in spring 1925, probably even after Heisenberg had left Copenhagen, because otherwise he would have submitted his note on the sum-rule for publication earlier. On the other hand, Heisenberg did not need Kuhn's final formulation of the sum-rule in order to establish his quantum condition, Eq. (90). As we have pointed out, he obtained this quantum condition from the quantum-theoretical re-interpretation of the Bohr–Sommerfeld quantization rule for phase integrals of multiply periodic systems. That method of derivation was entirely Heisenberg's, for Kuhn did not mention it at all in his note on the total intensity of absorption lines (Kuhn, 1925a).[261] However, *after* he had found Eq. (90) for the anharmonic oscillator, Heisenberg remembered with satisfaction the earlier discussions in Copenhagen on the sum-rules for the oscillator strengths. He realized that the very existence of such a sum-rule—independent of what value the right-hand side of the sum-rule assumed—supported his new quantum-theoretical version of the old Bohr–Sommerfeld quantum condition. And he desperately needed the new quantum condition in order to proceed with his programme of obtaining a complete solution of the equation of motion of the anharmonic oscillator in quantum theory.

Since Eq. (90) contained just the difference of the squares of amplitudes, Heisenberg had not yet reached his goal. Even in the case of the harmonic oscillator, where the only nonzero amplitudes were $a(n+1,n)$ and $a(n,n-1)$, he would not be able to determine the squares of the amplitudes uniquely, and an additional term, a constant independent of the quantum number n, would remain undetermined. And the same situation would hold for the anharmonic oscillator. However, this shortcoming did not come as a surprise to Heisenberg; it already existed in the classical equation (89), from which he had obtained the quantum condition, Eq. (90). That is, he had assumed that already in classical theory the Fourier amplitudes, $a_\tau(n)$, were determined only up to a constant; e.g.,

[260] Heisenberg mentioned Kuhn's name in a letter to Pauli, dated 8 October 1924. He wrote: 'Bohr and Kramers send their warm regards, as does Dr. Kuhn, who was just here.' (It is not clear whether Heisenberg meant that Kuhn had just been in Copenhagen or in his office. In any case, he had had the opportunity of meeting and having discussions with him in fall 1924.)

[261] Note also Heisenberg's later remark: 'I certainly obtained the Thomas–Kuhn sum-rule at that time from the quantum conditions by formal re-interpretation' (Heisenberg to van der Waerden, 2 July 1965).

in the classical harmonic oscillator a_1^2 was given by the equation

$$|a_1(n)|^2 = \frac{(n + \text{const.})h}{\pi m \omega_0}. \qquad (94)$$

The unknown constant in Eq. (94) expresses the fact that J, the action variable of the system, is not necessarily given by nh, the integral multiples of Planck's constant; in certain empirical situations, described by the Bohr–Sommerfeld theory, other values also seemed to represent the data. Heisenberg regarded it as one of the serious disadvantages of the Bohr–Sommerfeld theory that it did not provide a rule for fixing the constant uniquely for each mechanical system; one of the requirements, which a consistent quantum theory had to satisfy, was to remove that ambiguity. But as hard as he tried in May and early June 1925, he did not find any possibility of obtaining the desired additional rule. Hence the ultimate success in calculating the intensities of the anharmonic oscillator seemed to be remote, as he confessed in his letter to Kronig, dated 5 June 1925. And yet, with the sum-rule like Eq. (90) in hand, the determination of the constant was quite easy, and a brief reflection would provide the key to the solution. At that moment, however, Heisenberg could not really keep his head clear: this was late spring in Göttingen, with fresh grass and flowering bushes, and Heisenberg was interrupted in his work by a severe attack of hay fever. Since he could hardly do anything, he had to ask his director, Max Born, for a leave of about two weeks, which he decided to spend on the rocky island of Helgoland to effect a cure.

On 7 June 1925 Heisenberg took the night train from Göttingen to Cuxhaven, where he had to catch the ferryboat for Helgoland in the morning.[262] On arrival at Cuxhaven, 'I was extremely tired and my face was swollen. I went to get breakfast in a small inn and the landlady said, "You must have had a pretty bad night. Somebody must have beaten you." She thought I had had a fight with somebody. I told her that I was ill and that I had to take the boat, but she was still worried about me"' (Heisenberg, Conversations, p. 264). A few hours later he reached Helgoland.

Helgoland, a rocky island in the North Sea, consists of a mass of red sandstone, rising abruptly to an elevation of about 160 feet, and there is nearly no vegetation on it.[263] Heisenberg rented a room on the second floor of a house

[262]The date of Heisenberg's departure is confirmed by a letter, which he wrote to Niels Bohr in Copenhagen from Helgoland on 8 June 1925.

[263]Helgoland, or Heligoland, part of the North Frisian Islands, lies about 55 miles northwest of Bremerhaven. It has an area of about 380 acres and a permanent population of several hundred inhabitants. On the lower section of the island lies a fishing village, while the upper section serves as a summer resort for tourists.

The history of Helgoland is also interesting. From 1402 to 1714 it formed a part of Schleswig-Holstein, then became Danish until it was seized by the English fleet in 1807. It was formally ceded to Great Britain in 1814. Britain gave it to Germany in exchange for Zanzibar and some territory in Africa (1890). Helgoland was an important base for the German Navy. In accordance with the Treaty of Versailles the military and naval fortifications were demolished in 1920–1922. Under the

situated high above the southern edge of the island, which offered him a 'glorious view over the village, and the dunes and the sea beyond' (Heisenberg, 1971, p. 60). 'As I sat on my balcony,' he recalled more than forty years later, 'I had ample opportunity to reflect on Bohr's remark that part of infinity seems to lie within the grasp of those who look across the sea' (Heisenberg, 1971, p. 60). He began to take walks to the upper end of the island and swam daily in the sea. Soon he felt much better, and he began to divide his time into three parts. The first he still used for walking and swimming; the second he spent in reading Goethe's *West-Östlicher Divan*; and the third he devoted to work on physics (Heisenberg, Conversations). Having nothing else to distract him, he could reflect with great concentration on the problems and difficulties which had been occupying him until a few days earlier in Göttingen.

At his refuge in Helgoland, Heisenberg made quick progress within a few days. He first went over the mathematical formulation of the entire problem of the anharmonic oscillator and tried to jettison all the unnecessary ballast that had sneaked into it. Almost immediately he found the solution to the problem of determining the unknown constant in the expressions for the squares of transition amplitudes. He described the situation in retrospect as follows:

> I turned my attention to the "lowest state" of the atomic system. From the Thomas–Kuhn sum rule, which just gives the difference between the phase-space integrals, it was clear that one additive constant was missing. I noticed that this constant could be obtained from the assumption of the "lowest state." There must be a state in the atomic system defined by the fact that the amplitude of the transition to the next lower state must be zero, because the latter does not exist. That, of course, fixed the constant and then everything was clear. (Heisenberg, Conversations, p. 267; also AHQP Interview)

In considering the lowest state of a system Heisenberg could actually get some hint from the existing dispersion theory. For that purpose he had just to go back from Kramers' dispersion formula (Kramers, 1924a) to Ladenburg's (1921a). Ladenburg had taken into account only the absorption of radiation by atoms, and Kramers had remarked: 'In the case where the dispersing atoms are present in the normal states and only positive oscillators come into play, his [Ladenburg's] formula is thus equivalent to ours' (Kramers, 1924a, p. 674). The reason for this was that the normal state in the atom was the lowest state, and no radiation could be emitted from it. As a consequence, the difference of sums in Kramers' formula, Eq. (50), reduced to the single sum in Ladenburg's formula, Eq. (49). The same reduction, Heisenberg now argued, would happen in the case of the anharmonic oscillator. In particular, the right-hand side of the quantum-

---

Nazi régime Helgoland again became a military stronghold and was a target for heavy Allied bombing towards the end of World War II. From 1947 to 1 March 1952, when it was handed back to Germany, the island was used as a bombing range by the Royal Air Force. Then it was restored as a tourist and fishing centre. (See *Encyclopaedia Britannica*, Vol. 11, 1967.)

theoretical equation (90) would have to be replaced by a single sum involving only absorption amplitudes. Denoting the lowest state of the anharmonic oscillator by the quantum number n_0, he thus obtained the equation

$$1 = \frac{\pi m}{h} \sum_{\tau=1}^{\infty} \lambda^{2\tau-2} |a(n_0+\tau, n_0)|^2 2\pi\nu(n_0+\tau, n_0). \tag{95}$$

Hence in the approximation, in which terms of order λ^2 and higher powers of the anharmonicity parameter are neglected, the square of the absorption amplitude associated with the transition from the ground state to the first excited state is given by the equation

$$|a(n_0+1, n_0)|^2 = \frac{h}{2\pi^2 m\nu(n_0+1, n_0)}, \tag{96}$$

in which no unknown constant has now remained. By identifying n_0, the quantum number of the lowest state, with zero, and substituting ω_0 for $2\pi \times \nu(n_0+1, n_0)$, according to Eq. (83a), one may rewrite Eq. (96) as

$$|a(1,0)|^2 = \frac{h}{\pi m\omega_0}. \tag{96a}$$

Heisenberg proceeded further and calculated also the square of the amplitude $a(2,1)$ from Eq. (90) in the same approximation; thus he obtained, successively, equations for $a(3,2)$, $a(4,3)$, etc. The general equation reads

$$|a(n, n-1)|^2 = \frac{nh}{\pi m\omega_0} + \ldots, \tag{97}$$

where the dots indicate corrections of the order λ^2 (which imply corrections of the order λ to the amplitudes). It might be noted that this result, Eq. (97), seems to correspond to taking the constant in the classical expression for $a(n)$, Eq. (94), as zero; however, this analogy should not be taken too literally.[264]

By introducing the lowest state into the anharmonic oscillator system, a natural and obvious physical assumption, Heisenberg had succeeded in solving his problem. But his success represented only the beginning of the long way he still had to go. One question, naturally, was whether one could also treat, according to the same principles, other systems in quantum theory that represent atoms more realistically than the anharmonic oscillator. In Helgoland, however, Heisenberg did not concern himself much with this question; he had to discuss

[264] In particular, it does not imply that the action variable J assumes values that are only integral multiples of Planck's constant. The reason is, of course, that Eq. (97) passes into the classical equation (94) only in the limit of high quantum numbers n; in that case one may neglect the constant in Eq. (94), which has the order of magnitude of unity.

another problem of principle, still connected with the anharmonic oscillator. He described it as follows: 'Then I noticed that there was no guarantee that the new mathematical scheme could be put into operation without contradictions. In particular, it was completely uncertain whether the principle of conservation of energy would still apply, and I knew only too well that my scheme stood or fell by that principle' (Heisenberg, 1971, p. 61). This problem of proving energy conservation in quantum theory, especially in the example of the anharmonic oscillator, became Heisenberg's most urgent goal while he recovered from hay fever in Helgoland.

IV.5 Energy Conservation and Stationary States

In all the efforts to work out a theoretical description of atoms and their properties, mechanical concepts had played an important role. In 1913 Niels Bohr had built his model of the hydrogen atom entirely on classical mechanics; e.g., he had calculated the energy with the help of the well-known mechanical equations. Some properties of electron orbits, such as the angular momentum, had then been given discrete values, obeying certain quantum conditions. But these conditions did not restrict the use of classical mechanics; they just determined free parameters in the theory. The early successes of the Bohr–Sommerfeld theory of atomic structure had thus nourished the expectation that a consistent atomic mechanics could be established on the basis of classical mechanics and suitable additional prescriptions derived from quantum theory, the latter fixing the initial or boundary values of mechanical variables. However, this dream had been destroyed in the early 1920s, mainly because of two reasons. First, the application of the theory to calculate the properties of many-electron atoms had yielded results in irreparable disagreement with observation—as, for example, in the case of the energy states of the helium atom or the complex structure of spectral lines. Second, the analysis of the old semiclassical quantum theory had revealed internal contradictions between the fundamental principles on which the theory was based. Thus, Bohr, in particular, had emphasized again and again that the prescription for deriving the frequencies of atomic spectra from differences of discrete energy states was at variance with the dynamical explanation of the origin of classical radiation. These contradictions had become even more serious after the discovery of the Compton effect, which called attention to the importance of the nonclassical aspects of radiation. One may summarize the situation by saying that the first dozen years of research following the discovery of Bohr's model of the hydrogen atom had succeeded eventually in eroding completely its physical foundation.

The quantum physicists, had, of course, tried to meet the challenges posed to atomic theory by the empirical observations. They had mainly followed one of the two attitudes. The first was: one effectively gave up the mechanical description of atoms; e.g., one did not take the orbits of electrons seriously anymore and avoided talking about mechanical frequencies; only those mechanical concepts were (formally) kept that were associated with observed quantities (phenomenologically described by quantum numbers). The other alternative led in the opposite direction: one considered the mechanical relations as the foundation of the description of atomic phenomena, though interpreting them in a formal way; one could then write down equations between 'reformulated' quantities, which could be compared to the empirical data. Both procedures had, until summer 1925, allowed some progress to be made in limited areas of atomic theory, such as the organization of multiplet spectra or dispersion theory, but only at the expense of giving up any deeper understanding of the principles underlying the whole theory. The final blow had been dealt in April 1925 in an attempt to explain at least the phenomena arising from the interaction of radiation and atoms on the basis of a general philosophy; at that date the result of the experiment of Walther Bothe and Hans Geiger had disproved one of the crucial assumptions of the Bohr–Kramers–Slater theory of radiation. This result re-established the validity of an old mechanical principle—the conservation of energy; but at the same time it seemed to remove the foundation of the formal relations that had been obtained so far. Hence the situation in atomic theory in June 1925 was characterized by two facts: a comeback of certain mechanical concepts and principles, which one had already more or less given up; and an increasing strangeness involved in the mechanical and physical description of atomic phenomena.

Heisenberg was completely aware of the unsatisfactory situation in the existing theoretical description of atomic systems. His strong optimism, and his naïve and pragmatic belief in being able to overcome the hardest problems, prevented him from becoming depressed or overwhelmed by the difficulties facing him. Thus, he continued to pursue his programme of calculating the intensity of atomic radiation, apparently undisturbed by the fact that Bohr's philosophy, which had originally guided that programme, had broken down. He hoped that possibly the re-interpretation of mechanical equations and relations, which he had attempted since May in Göttingen, had not been affected by the failure of the Bohr–Kramers–Slater theory. In this respect, Heisenberg could refer to an earlier observation of Max Born's, namely, that one should be able to derive Kramers' dispersion formula 'independently of the critically important and still debatable conceptual framework of that theory [i.e., the Bohr–Kramers–Slater theory], such as the statistical interpretation of energy and momentum transfer' (Born, 1924b, p. 386). Heisenberg regarded his re-interpretation as a systematic continuation and generalization of Born's and his own attempts of 1924 to create a discrete mechanics or quantum mechanics; hence, what was valid for the earlier scheme should be valid *a fortiori* for the new re-interpretation scheme. At the same time, however, he knew perfectly that the quantities entering into the new relations—such as the patterns of transition amplitudes—satisfied relations that looked strange and presented difficulties for any physical interpretation. Could one demand at all that some quantity existed in the re-interpreted mechanics that would represent the physical energy and even be conserved in time? Only then, however, would the attempted quantum mechanics be a consistent theory, as had become obvious since the Bothe–Geiger experiment of April 1925.

In order to answer the question of whether energy was conserved in his quantum-theoretical scheme, Heisenberg had first to turn to another question. What would the properties characterizing a (stationary) state of an atomic system, e.g., the angular momentum of a series electron (in an orbit) or the energy, look like? Surely they were not represented by the patterns of quantum-theoretical amplitudes because these patterns exhibited explicitly not only the properties of the state under consideration but also the properties of those states that could be reached through transitions. For example, the quantum-theoretical series corresponding to the Fourier series, which represents the position coordinate of an electron in a given orbit, contains transition amplitudes from the given stationary state (associated with the given orbit) to another stationary state of the atom—i.e., it involves orbits other than the given one. In spite of this apparent confusion brought into the new formalism by the pattern-of-amplitudes representation of mechanical variables, Heisenberg quickly realized how to represent properties associated with a single stationary state of an atom in his theory. He started from the fact that these properties could, in classical theory, always be written in terms of products having at least two Fourier series as factors. For example, the energy of the one-dimensional harmonic oscillator, W, was expressed normally as a function of x, the position coordinate, and $\dot{x}$, its time derivative, with the help of the equation

$$W = \tfrac{1}{2}m\left[\dot{x}^2 + \omega_0^2 x\right]. \tag{98}$$

In classical theory x was given by the single harmonic term, $a(n)\cos(\omega_0 t)$, with $a(n)$ as the amplitude of the state denoted by the action variable $J = nh$. Hence the energy of the oscillator eventually became

$$W = \tfrac{1}{2}m\omega_0^2\left[|a(n)|^2\sin^2(\omega_0 t) + |a(n)|^2\cos^2(\omega_0 t)\right] = \tfrac{1}{2}m\omega_0^2|a(n)|^2, \tag{99}$$

i.e., it depended only on the amplitude $a(n)$. In order to rewrite the right-hand side of Eq. (98) according to his reformulation prescription, Heisenberg had to take squares of the patterns of quantum-theoretical amplitudes. These squares had, in principle, the same structure as the original patterns because they were obtained according to the multiplication rule, Eq. (66). How could one be certain that the resulting expression, say for the harmonic oscillator, was time independent? Heisenberg had to demand that requirement, however, not only because he believed in the conservation of energy in atomic processes; he had also to demand it because, without it, he was not able to describe stationary states in quantum theory. A stationary state of the harmonic oscillator was given by the values which the reformulated energy expression (98) would assume; and these values had to be independent of time, as long as one had not included a damping term in the equation of motion. Moreover, the same conclusion had to hold in the case of the anharmonic oscillator. It was this aspect of energy conservation, the very existence of stationary states, which played a crucial role in Heisenberg's thinking in June 1925.

It should be noted in passing that Heisenberg was not allowed to derive the energy conservation, say of the anharmonic oscillator given by the equation of motion (75), in the same way as one does in classical mechanics. There one multiplies the equation of motion with $m\dot{x}$, and writes the left-hand side as a total derivative with respect to time, to obtain the equation

$$\frac{d}{dt}\left(\frac{1}{2}m\dot{x}^2 + \frac{1}{2}m\omega_0^2x^2 + \frac{1}{3}mx^3\right) = 0. \tag{100}$$

Evidently, the expression in the parentheses represents the classcial energy of the anharmonic oscillator; hence Eq. (100) automatically establishes energy conservation.[265] In Heisenberg's quantum-theoretical scheme, Eq. (100) cannot be obtained by means of the above steps from Eq. (75), the reason being that the operation with quantum-theoretical products involves the product rule, Eq. (66). Still, Heisenberg set himself the goal of proving the time independence of the re-interpreted classical energy expression of the anharmonic oscillator. Of course, he had no criterion by which to decide that this would be the correct energy expression in quantum theory, and it was a miracle that he got anywhere. Indeed, to an outsider he would seem to be moving along like a sleepwalker on rooftops: on the one hand, taking a few careful steps, such as the quantum-theoretical re-interpretation of the Fourier series; on the other, jumping across the abysses of deep problems, such as what the exact energy expressions should be, with closed eyes. Only the ultimate success could, and would, justify his steps.

Heisenberg did not give himself much time to ponder; he was just interested in carrying out his plan to prove the time independence of the energy expression for the anharmonic oscillator as quickly as possible. He was aware of the fact that he had to go through successive approximations and that he might be able to demonstrate the results only to a limited order of powers in the anharmonicity parameter λ. Therefore, he looked for an example in which the calculations were particularly simple; thus, he turned to a slightly different type of anharmonic oscillator than he had considered earlier, the one satisfying the equation of motion,

$$\ddot{x} + \omega_0^2x + \lambda x^3 = 0. \tag{101}$$

In this special case the Fourier series representing the position coordinate x

[265] A damping term $(-\gamma\dot{x})$ in the equation of motion does not lead to a total time derivative, as in Eq. (100); in this case the energy would not be conserved.

contains no terms with even coefficients, i.e., all a_0, a_2, etc., have to be zero; hence x may be written

$$x = a_1 \cos(\omega t) + \lambda a_3 \cos(3\omega t) + \lambda^2 a_5(5\omega t) + \dots \\ + \lambda^n a_{2n+1} \cos\left[(2n+1)\omega t\right] + \dots . \tag{102}$$

In Eq. (102) we have replaced the term $2\pi\nu$, where ν is the harmonic frequency, by ω, in order to save on the writing.[266] By retaining terms up to the order λ^2 Heisenberg obtained the following relations between the odd coefficients:

$$\left[(-\omega^2 + \omega_0^2)a_1 \qquad + \tfrac{3}{4}\lambda a_1^3 + \tfrac{3}{4}\lambda^2 a_1^2 a_3 \qquad \right]\cos(\omega t) = 0, \tag{103a}$$

$$\left[(-9\omega^2 + \omega_0^2)a_3 + \tfrac{1}{4}a_1^3 \qquad + \tfrac{3}{2}\lambda a_1^2 a_3 \qquad \right]\cos(3\omega t) = 0, \tag{103b}$$

$$\left[(-25\omega^2 + \omega_0^2)a_5 + \tfrac{3}{4}a_1^2 a_3 + \tfrac{3}{4}\lambda a_1 a_3^2 + \tfrac{3}{4}\lambda^2 a_1^2 a_7\right]\cos(5\omega t) = 0, \tag{103c}$$

$$\vdots$$

He re-interpreted these relations quantum-theoretically by writing the quantum-theoretical amplitudes $a(n, n-1)$, $a(n, n-3)$ and $a(n, n-5)$ instead of the

[266] The same notation was used by Heisenberg, both in his letters to Kronig and later to Pauli, as well as in his published paper (Heisenberg, 1925c).

We do not know exactly when Heisenberg began to use the anharmonic oscillator given by the classical equation (101), rather than the one given by Eq. (75), because the anharmonic oscillator of Eq. (101) was never treated in detail in the available letters of Heisenberg. However, we may conclude why Heisenberg employed Eq. (101) to prove energy conservation from several indications. First, he tried in Helgoland to simplify his mathematical treatment of the anharmonic oscillator problem, and the nonexistence of the even terms, a_0, a_2, etc., in the Fourier series for x represented a great mathematical simplification. Second, in three letters written shortly after his return from Helgoland (Heisenberg to Pauli, 21 June, 24 June, and 4 July 1925), Heisenberg mentioned an energy expression for the harmonic oscillator that belonged to the oscillator obeying Eq. (101). Third, although Heisenberg employed both types of anharmonic oscillator in his paper, he presented there the energy expression only for the case of Eq. (101) (Heisenberg, 1925c, p. 890, Eq. (27)).

The only statement which might contradict the assumption that Heisenberg immediately employed the oscillator of Eq. (101) for the purpose of demonstrating energy conservation in Helgoland may be found in a much later letter of Heisenberg. He remarked there: 'I investigated at that time in Helgoland first the matrix element of the energy, leading to the transitions $n \to \pm 2$, and found that it vanishes as it should' (Heisenberg to van der Waerden, 18 January 1966). The matrix element in question, however, arises only when treating the anharmonic oscillator with the equation of motion (75). Thus, one cannot entirely exclude the possibility that Heisenberg at least started the energy calculations with his standard anharmonic oscillator, Eq. (75), rather than with the oscillator of Eq. (101). However, if one assumes that the first energy calculation in quantum mechanics was performed with the anharmonic oscillator having the equation of motion, Eq. (75), then one wonders about two things. First, why did Heisenberg not report the result of this calculation in his paper? Second, why did Max Born and Pascual Jordan concern themselves many weeks later with the integration of the equation of motion (75) by using matrix mechanics (see Born and Jordan, 1925b, Chapter III)? These authors gave a result (Born and Jordan, 1925b, p. 883, Eq. (88)), which Heisenberg never mentioned before.

classical Fourier amplitudes a_1, a_3 and a_5, and the quantum-theoretical frequencies $\omega(n,n-1)$ $(=2\pi\nu(n,n-1))$, $\omega(n,n-3)$ and $\omega(n,n-5)$ instead of the classical frequencies ω, 3ω, and 5ω. For the products of classical coefficients he used the reformulations:

$$\begin{aligned}
a_1^3 \text{ in Eq. (103a)} &\to \tfrac{1}{3}\big[a(n,n-1)a(n-1,n)a(n,n-1) \\
&\quad + a(n,n+1)a(n+1,n)a(n,n-1) \\
&\quad + a(n,n-1)a(n-1,n-2)a(n-2,n-1)\big], \\
a_1^2a_3 \text{ in Eq. (103a)} &\to \tfrac{1}{3}\big[a(n,n+1)a(n+1,n+2)a(n+2,n-1) \\
&\quad + a(n,n+1)a(n+1,n-2)a(n-2,n-1) \\
&\quad + a(n,n-3)a(n-3,n-2)a(n-2,n-1)\big], \\
a_1^3 \text{ in Eq. (103b)} &\to a(n,n-1)a(n-1,n-2)a(n-2,n-3), \\
a_1^2a_3 \text{ in Eq. (103b)} &\to \tfrac{1}{3}\big[a(n,n+1)a(n+1,n)a(n,n-3) \\
&\quad + a(n,n+1)a(n+1,n-2)a(n-2,n-3) \\
&\quad + a(n,n-3)a(n-3,n-2)a(n-2,n-3)\big], \qquad \text{etc.}
\end{aligned} \tag{104}$$

This procedure takes into account the noncommutativity of the quantum-theoretical reformulation of the Fourier series; hence, e.g., the classical product a_1^3 splits into a sum of three products, in which the same amplitudes referring to different stationary states are commuted.

The next task was to calculate the Fourier coefficients and the corresponding transition amplitudes. The classical equations (103a, b, c) could be satisfied by an *Ansatz*, Eq. (85a), apart from corrections to the order λ and its higher powers.[267] In the quantum-theoretical case, however, both the transition amplitudes, $a(n,n-1), a(n,n-3), \ldots,$ and the associated frequencies, $\omega(n,n-1)$, $\omega(n,n-3), \ldots,$ had to be obtained. While the amplitudes were given by expressions of the form given in Eq. (85b), the frequencies of the anharmonic oscillator became

$$\omega(n,n-\tau) = \tau\omega_0 + \text{corrections}(\lambda, \lambda^2, \ldots), \tag{105}$$

[267] The structure of Eqs. (103) is the same as that of Eqs. (83) for the anharmonic oscillator having the equation of motion (75).

where the dots denote corrections of higher order, i.e., λ^3, λ^4, etc. By taking successive and systematic approximations, Heisenberg arrived at the following results:[268]

$$\omega(n,n-1) = \omega_0 + \lambda \frac{3nh}{8\pi\omega_0^2 m} - \lambda^2 \frac{3h^2(17n^2+7)}{256\pi^2\omega_0^5 m^2} + \dots, \tag{106a}$$

$$a(n,n-1) = \sqrt{\frac{hn}{\pi\omega_0 m}}\left(1 - \lambda \frac{3nh}{16\pi\omega_0^3 m} + \dots\right), \tag{106b}$$

$$a(n,n-3) = \frac{1}{32}\sqrt{\frac{h^3 n(n-1)(n-2)}{\pi^3\omega_0^7 m^3}}\left(1 - \lambda \frac{39(n-1)h}{32\pi\omega_0^3 m} + \dots\right), \tag{106c}$$

where the dots again denote corrections of order higher than λ^2.

After determining the Fourier coefficients and the corresponding amplitudes and frequencies, Heisenberg proceeded to study the expression for the energy of his anharmonic oscillator,

$$W = \tfrac{1}{2}m\left(\dot{x}^2 + \omega_0^2 x^2 + \tfrac{1}{2}\lambda x^4\right). \tag{107}$$

Substituting in it for x the Fourier series, Eq. (102), he obtained an equation in which the energy was expressed in terms of Fourier amplitudes and frequencies, that is,

$$\begin{aligned} W = {} & \tfrac{1}{2}m\left[\tfrac{1}{2}a_1^2(\omega^2+\omega_0^2) + \tfrac{3}{16}\lambda a_1^4 + \dots\right] \\ & + \tfrac{1}{2}m\left[\tfrac{1}{2}a_1^2(-\omega^2+\omega_0^2) + \lambda a_1 a_3(3\omega^2+\omega_0^2) + \tfrac{1}{4}\lambda a_1^4 + \dots\right]\cos(2\omega t) \\ & + \tfrac{1}{2}m\left[\lambda a_1 a_3(-3\omega^2+\omega_0^2) + \left(\tfrac{1}{16}\lambda\right)a_1^4 + \dots\right]\cos(4\omega t) + \dots . \end{aligned} \tag{108}$$

In Eq. (108) the dots indicate terms of order higher than λ, i.e., terms of order λ^2, λ^3, etc. There was, of course, no doubt about energy conservation in classical theory. It could be easily demonstrated by the above-mentioned procedure of multiplying Eq. (101) by $m\dot{x}$ and expressing the left-hand side of the resulting equation as a derivative of an expression—which is again the energy of the oscillator—with respect to time. Another way of proving energy conservation is to use Eq. (108) and to show that the factors multiplying the periodic functions, $\cos(2\omega t)$, $\cos(4\omega t)$, etc., are zero. This procedure naturally requires a great amount of computational work. Moreover, one will not be able to carry out the

[268] Equations (106a)–(106c) can be found in Heisenberg's published paper (1925c, p. 890, Eqs. (24)–(26)).

complete proof because one has to confine oneself to finite orders in the anharmonicity parameter λ. In the approximation (to order λ), which is written down in Eq. (108), the factors multiplying $\cos(2\omega t)$ and $\cos(4\omega t)$ indeed turn out to be zero after a short calculation, if the expressions for a_1, a_3 and ω of the classical anharmonic oscillator are inserted.[269]

In quantum theory Heisenberg had no choice but to prove energy conservation in a laborious way. He was thus forced to start from the formal expression, Eq. (107), for the energy of the anharmonic oscillator and to substitute for x the quantum-theoretical series corresponding to the Fourier series, Eq. (102). In this way he obtained an expression for what he thought was the quantum-theoretical energy in terms of transition amplitudes and transition frequencies. The same expression would also follow from a careful quantum-theoretical re-interpretation of the right-hand side of Eq. (108). Heisenberg's task was now to check whether the factors multiplying the periodic functions, $\cos[\omega, (n, n-\tau)t]$, vanished or not. For that purpose he had to calculate, for example, the factor multiplying $\cos[\omega(n, n-2)t]$. This factor, including terms of the order λ, turned out to be

$$\begin{aligned}
\frac{1}{2} m \Big\{ & \frac{1}{2}\left[-\omega(n,n-1)\omega(n-1,n-2)+\omega_0^2\right]a(n,n-1)a(n-1,n-2) \\
& +\frac{\lambda}{2}\left[-\omega(n,n+1)\omega(n+1,n-2)+\omega_0^2\right]a(n,n+1)a(n+1,n-2) \\
& +\frac{\lambda}{2}\left[-\omega(n,n-3)\omega(n-3,n-2)+\omega_0^2\right]a(n,n-3)a(n-3,n-2) \\
& +\frac{\lambda}{16}\big[a(n,n-1)a(n-1,n-2)a(n-2,n-3)a(n-3,n-2) \\
& \qquad +a(n,n-1)a(n-1,n-2)a(n-2,n-1)a(n-1,n-2) \\
& \qquad +a(n,n-1)a(n-1,n)a(n,n-1)a(n-1,n-2) \\
& \qquad +a(n,n+1)a(n+1,n)a(n,n-1)a(n-1,n-2)\big]\Big\}.
\end{aligned} \tag{109}$$

By substituting for the amplitudes and frequencies the expressions from Eqs. (106), Heisenberg found that the entire factor (109) was indeed zero in the approximation to the order λ. This result might not appear to be very surprising because the quantum-theoretical amplitudes and frequencies entering here were very similar to the classical ones if one neglected corrections of the orders λ^2, λ^3,

[269]The expressions for a_1, a_3 and ω correspond to the quantum-theoretical expressions of Eq. (106); for example, $a_1(n)$ is identical with $a(n, n-1)$ in the approximation including only terms of first order in λ, and the same holds for the frequency ω.

etc. In particular, the transition frequency $\omega(n, n-1)$ was then identical with ω, while $\omega(n, n-3)$ and $\omega(n+1, n-2)$ were approximated by $3\omega_0$, like the third harmonic in the classical case. Only the dependence of the amplitudes on the quantum number n deviated from the classical situation. Therefore, Heisenberg had to check carefully the contributions of the order λ to the factor (109). He noticed, in particular, that besides the contribution from terms, which are already multiplied with λ in the factor (109), another term proportional to λ had to be taken into account; it arose from the first-order corrections to the quantum frequencies $\omega(n, n-1)$ and $\omega(n-1, n-2)$. He found this contribution to be

$$\frac{3}{16}\lambda(2n-1)\left(\frac{h}{\pi\omega_0 m}\right)^2. \tag{110}$$

It exactly cancelled the other terms of order λ, which multiplied the periodic function $\cos[\omega(n, n-2)t]$. Similarly, he proved that the factor multiplying the periodic function $\cos[\omega(n, n-4)t]$ became zero in the approximation to order λ. Energy conservation was thus proved to be valid for the quantum-theoretical anharmonic oscillator in the first approximation in the anharmonicity parameter λ.[269a]

Having noticed from the evaluation of the first terms that energy conservation seemed to be upheld in his scheme, Heisenberg began to work even harder than before. He recalled later: 'I got into a state of great excitement because it worked out so nicely. So I continued to work all night and I made numerous slips in the calculation.' Around three o'clock in the morning Heisenberg realized that, 'The energy principle had held for all the terms, and I could no longer doubt the mathematical consistency and coherence of the kind of quantum mechanics to which my calculations pointed' (Heisenberg, 1971, p. 61). He had completed the proof of energy conservation for terms up to the second order in the anharmonicity parameter λ.

One can grasp Heisenberg's excitement when one goes through the details of his energy calculation. It called for the full exercise of physical intuition in writing down the mathematical equations and for strenuous care in evaluating all the terms that occurred. Heisenberg had only a vague idea of the scheme which, he believed, would express the formal physical content of quantum theory. It was an intellectual adventure to seek to employ this scheme, and the fact that he was able to establish the conservation of energy without insurmountable difficulties was for him 'a gift from heaven' (Heisenberg, Conversations, p. 267). The result seemed too beautiful to be true. As he recalled:

> At first, I was deeply alarmed. I had the feeling that, through the surface of atomic phenomena, I was looking at a strangely beautiful interior, and felt almost giddy at

[269a] In the published paper, the energy was given by including terms of order λ^2 (Heisenberg, 1925c, p. 890). Since the first approximation was comparatively easy to calculate, it is evident that Heisenberg went to the second order in Helgoland.

the thought that I now had to probe this wealth of mathematical structures nature had so generously spread out before me. I was far too excited to sleep, and so, as the new day dawned, I made for the southern tip of the island, where I had been longing to climb a rock jutting out into the sea. I now did so without much trouble, and waited for the sun to rise. (Heisenberg, 1971, p. 61)

Chapter V
The Discovery of Quantum Mechanics

In summer 1925 Heisenberg's earlier tentative attempts led him to formulate the fundamental ideas of the new quantum mechanics. Paul Dirac summarized this important progress in a lecture in the following words:

> The great advance was made by Heisenberg in 1925. He made a very bold step. He had the idea that physical theory should concentrate on quantities which are closely related to observed quantities. Now, the things you observe are only very remotely connected with the Bohr orbits. So Heisenberg said that the Bohr orbits are not very important. The things that are observed, or which are connected closely with the observed quantities, are all associated with two Bohr orbits and not with just one Bohr orbit: *two* instead of *one*. Now, what is the effect of that?
>
> Suppose we consider all the quantities of a certain kind associated with two orbits, and we want to write them down. The natural way of writing down a set of quantities, each associated with two elements, is in a form like this:
>
> $$\begin{pmatrix} \times & \times & \times & \times & \cdot & \cdot & \cdot \\ \times & \times & \times & \times & \cdot & \cdot & \cdot \\ \times & \times & \times & \times & \cdot & \cdot & \cdot \\ \times & \times & \times & \times & \cdot & \cdot & \cdot \\ \cdot & \cdot & \cdot & \cdot & \cdot & \cdot & \cdot \end{pmatrix},$$
>
> an array of quantities, like this, which one sets up in terms of rows and columns. One has the rows connected with one of the states, the columns connected with the other. Mathematicians call a set of quantities like this a matrix.
>
> Now, Heisenberg assumed that one should deal with such a set of quantities and consider the whole set together as corresponding to one of the dynamical variables of the Newtonian theory. These dynamical variables are, of course, the coordinates of the particles, or the velocities, or momenta. Each of these quantities is to be replaced by a matrix, according to Heisenberg. The underlying idea that led Heisenberg to think of this was that one should construct the theory in terms of observable quantities and that the observable quantities are these matrix elements, each associated with two orbits. (Dirac, 1978, pp. 4–5)

What Dirac described so clearly, Heisenberg had to find by trial and error more than fifty years earlier, i.e., in the second half of June and beginning of July 1925. At that time he occupied himself with incorporating his results from the anharmonic oscillator calculation into a general and systematic scheme. He hoped that this scheme would represent a step forward in the direction of the future quantum mechanics, but he found the path to be very cumbersome. Indeed, he did not obtain many further results beyond the ones he had obtained already before going to Helgoland or during his stay there. Nevertheless he soon became quite optimistic, for he became aware of a philosophical principle which seemed to underlie all successful quantum formulae discovered in Copenhagen and Göttingen since 1924: this was the prescription to use only observable quantities in the description of atomic phenomena. The new scheme, involving patterns of quantum-theoretical amplitudes and the product rule, the reformulated equation of motion and the reformulated quantum condition, also satisfied that principle. Hence it became a good candidate for a consistent atomic mechanics. While his paper containing the new results was in print, Heisenberg propagated the new theory on his travels to Leyden and Cambridge.

V.1 Pauli's Attitude Towards Attempts to Formulate a Quantum Mechanics

After a stay of less than two weeks in Helgoland, where he recovered from hay fever, Heisenberg returned to Göttingen. On his way back he visited Wolfgang Pauli in Hamburg and had a long conversation with him covering many topics of common interest in quantum theory. A few days later Heisenberg wrote to Pauli at length.

> Again many thanks [he said] for your kind reception and hospitality in Hamburg. I completely agree with your letter as far as Part I is concerned; it is very nice that you have considered the problem so thoroughly, and I have no doubt that everything will turn out to be empirically correct. I am just surprised that you wonder about the "failure of mechanics." If anything like mechanics were true, one would never understand the existence of atoms. Evidently, there exists another, a "quantum mechanics," and one must only wonder about the fact that the hydrogen atom agrees, as far as the energy constant is concerned, with something computed classically. With respect to Part II, dealing with the Hanle effect, I have arrived at just the opposite conclusion than yours; hence I shall write about it in some more detail. (Heisenberg to Pauli, 21 June 1925)[270]

At that time Pauli was particularly interested in investigating the influence of external electric and magnetic fields on the scattering of light by atoms from the point of view of the Kramers–Heisenberg dispersion theory. Naturally, Heisenberg was an ideal partner to talk to about this subject; hence Pauli already had mentioned the problem in the discussions during Heisenberg's visit, and he continued to deal with it in letters. Part I of his letter, to which Heisenberg referred above, dealt with some considerations concerning the influence of crossed electric and magnetic fields on the scattering of light by the hydrogen

[270] From Heisenberg's letter it is clear that he had returned to Göttingen several days before 21 June and had in the meanwhile received a letter from Pauli. As with most of the letters of Pauli to Heisenberg of this period, it was lost. In this letter Pauli had reported certain conclusions on physical topics, to which he had arrived after the meeting with Heisenberg in Hamburg.

atom.[271] Part II, to which Heisenberg replied in detail on 21 June 1925, concerned the influence of an external electric field on the polarization of the scattered light. Pauli had concluded that a depolarization should occur in case that the direction of the electric field was neither parallel nor perpendicular to the electric vector of the incident radiation. However, Heisenberg, on the basis of a different argument, claimed the absence of any depolarizing effect. The two friends carried on the discussion on this problem in later letters without reaching a definite agreement (Heisenberg to Pauli, 24 June 1925; 4 July 1925).[272]

While the above problem had emerged from Pauli's work, and it was Pauli who profited from Heisenberg's answers, Heisenberg also received Pauli's help in his own investigations. In Hamburg he had already mentioned his new ideas to Pauli about how to obtain quantum-theoretical intensities and, especially, his solution of the anharmonic oscillator problem. He had then expected sharp and sarcastic opposition from Pauli, but to his great surprise the reaction had been quite different. 'When I reported my results to Wolfgang Pauli, generally my severest critic, he warmly encouraged me to continue along the path that I had taken,' recalled Heisenberg later (Heisenberg, 1971, p. 62).[273] One can hardly imagine how much Pauli's support meant to Heisenberg at this moment, when he was groping along a completely unexplored path to a consistent quantum theory. He decided to keep Pauli informed about every detail, no matter how small, of his progress. Thus he wrote:

> In my attempts to fabricate a quantum mechanics there is only slow progress; but I do not care any more about how far I move away from the theory of conditionally periodic systems. For instance, what I said the other day remains valid, i.e., that already the energy of the oscillator must be $(n + \frac{1}{2})h\nu$, just as that of the rotator must be $(m + \frac{1}{2})^2 h^2/4\pi^2 A$. I do not believe it all that this contradicts any experience (n and m *must* be integral numbers, otherwise everything here becomes meaningless; one has no longer any choice, other than the cooking recipes of the erstwhile theory); in the former case one cannot check the result, while in the latter case the bands of the hydrogen halogenides (where the angular momentum of the electrons is certainly zero) rather seem to favour it. Moreover, the new "formalism" is not so terribly formal at all; e.g., one can derive from it certain results concerning the kinematics in quantum theory' (Heisenberg to Pauli, 21 June 1925).

[271] The consideration of the influence of crossed electric and magnetic fields on the hydrogen atom had been crucial in showing the failure of the Bohr–Sommerfeld theory of atomic structure. In part I of his letter, Pauli had apparently suggested a possibility of treating atomic systems with the help of dispersion-theoretic arguments. This possibility implied, then, that the Bohr–Sommerfeld model of the hydrogen atom was wrong.

[272] Pauli repeated his arguments in his *Handbuch* article on quantum theory (Pauli, 1926b, pp. 107–108). There he also referred (p. 108, footnote (1)) to an experiment of Wilhelm Hanle's in Göttingen, who had obtained a depolarization by external electric fields but interpreted his results in a way different from Pauli's (Hanle, 1926).

[273] The fact that Heisenberg mentioned his most recent results on the anharmonic oscillator problem during his visit to Hamburg in June 1925 was confirmed by Rudolph Minkowski, then Otto Stern's collaborator. He remembered both Heisenberg's visit and that 'Heisenberg had the basic elements of his new theory' (Minkowski, AHQP Interview, p. 8).

And during the following three short weeks, while he worked impatiently on a systematic formulation of his quantum-theoretical re-interpretation of kinematical and dynamical relations, Heisenberg again reported to Pauli on 24 June, 29 June and 4 July, until he finally completed the manuscript of his paper on 9 July and sent it to Pauli. Thus, Pauli became the closest witness of the endeavours to 'fabricate quantum mechanics,' and his comments on a few points of the formulation contributed invaluably, at least in Heisenberg's mind, to the birth of the new theory.

It is easy to understand why Heisenberg felt so happy with Pauli's positive and helpful attitude towards his programme of summer 1925. One must wonder, however, about the reasons which caused this attitude. An examination of the statements, which Pauli made in the period between spring 1924 and spring 1925, does not reveal any change in his critical views on Heisenberg's theoretical work; he remained as critical of it as he had been in previous years. Certainly, Pauli had been very pleased that Heisenberg had been able to collaborate with Niels Bohr for some time. As a result of this collaboration Heisenberg had even developed a 'philosophical' attitude towards the treatment of problems of atomic theory. But, at the same time, he had become more of a partisan of Bohr's views than perhaps Pauli thought was desirable. In particular, there was the question of the Bohr–Kramers–Slater radiation theory, which stood in the foreground of the discussions about atomic physics in Copenhagen since spring 1924 and remained the physical foundation of the dispersion-theoretic approach. With regard to the Copenhagen radiation theory—although he had briefly succumbed to Bohr's persuasion against his 'scientific conscience' during his visit to Copenhagen in April 1924 (Pauli to Bohr, 2 October 1924)—Pauli had soon become convinced that it had not brought forth 'any progress worth mentioning' (Pauli to Sommerfeld, November 1924). Heisenberg, on the other hand, had developed a feeling of optimism about the progress achieved in atomic theory in 1924, without taking too seriously the assumption of the statistical conservation of energy.[274] Thus he had defended, against Pauli's onslaughts, the correspondence principle as a *strict* guideline for further progress in atomic theory, the same principle that seemed to have led inevitably to the Bohr–Kramers–Slater theory. The reason for Heisenberg's strong stand, at least in part, was the success he had begun to score by the method of sharpening the correspondence principle. Surely, in order to achieve successful results, he had felt obliged to invent some queer substitute ('*Ersatz*') mechanical models to describe the atomic systems.

At the same time, when Heisenberg proposed his queer models, Pauli had come to the conclusion that one must give up the construction of mechanical

[274] As we have mentioned earlier, one of the reasons for Heisenberg's cautious attitude towards the assumption of statistical energy and momentum conservation in the Bohr–Kramers–Slater theory was the meeting with Albert Einstein in Göttingen in June 1924. Einstein had raised serious objections, which one could not easily push aside. It was probably Heisenberg's good fortune that he did not become as consistent and 'philosophical' in his thinking as Niels Bohr. This saved him the trouble of being too depressed by the outcome of the Bothe–Geiger experiment, which destroyed the foundation of the Bohr–Kramers–Slater theory.

models altogether.[275] Consistent with this conclusion he had criticized the detailed models, which Heisenberg had constructed in early 1925 to describe the intensities of multiplets (Heisenberg, 1925b); they were, said Pauli, 'all rather equally wrong' ('*alle ziemlich gleich unrichtig*,' Pauli to Heisenberg, 28 February 1925). And he used to call the quantum-theoretical formalism following from these models simply '*Schimmel*' (mould; '*schimmlig*' = mouldy, musty), a phrase, which he and Heisenberg had earlier attached jokingly to the formal theoretical schemes favoured by Max Born.[276] Pauli was afraid that one might go too far into the details of these '*Ersatz*' models, and then *mistake them* (if they gave a correct description of the phenomena) *as the final answer to the problems of atomic theory*. He remembered only too well that, just one year previously, Heisenberg seemed to be convinced that he had obtained the correct theory of the anomalous Zeeman effects by applying the difference calculus to the old core model of many-electron atoms (Heisenberg, 1924e), *without invoking any new physical idea*. Pauli expected real progress to come only by discovering the central relationship between all existing outstanding problems of atomic theory, especially by bringing together the difficulties of the theory of the anomalous Zeeman effects with those of the dispersion theory of degenerate systems (Pauli to Heisenberg, 28 February 1925).[277] He did not see any salvation in extending the existing models or building new fancier ones. 'I really consider the mere extension of formal zoology to more and more complicated cases to be a fruitless procedure,' he wrote to Kronig on 21 May 1925. Pauli had felt for a long time that Heisenberg had to be persuaded of this point of view.

In spite of these grave disagreements between their respective attitudes towards the problems of atomic theory, Pauli and Heisenberg had developed over the years some common ideas. One was connected with the fundamental question of emission of radiation by atoms. 'In spite of the Compton effect,' Heisenberg had written to Pauli in December 1923, 'it appears increasingly more reasonable to say that the entire Balmer series is emitted (simultaneously?) by a single electron and atom' (Heisenberg to Pauli, 7 December 1923). And Pauli had emphasized the same point soon afterwards in a letter to Bohr. He wrote:

> The most important question seems to me to be the following: *To what extent is one allowed at all to speak about definite orbits of the electrons in the stationary states*? In

[275] Pauli expressed this attitude most candidly in his letter to Sommerfeld, dated 6 December 1924, where he praised his former professor for having abandoned in the fourth edition of *Atombau und Spektrallinien* the model-dependent description of many-electron atoms, claiming that such a description could not account for the 'simplicity and beauty of the quantum world' ('*Einfachheit und Schönheit der Quantenwelt*').

[276] The phrase '*Schimmel*' had been used, for instance, by Heisenberg as follows: 'Born has thought about the following: If one considers Kramers' dispersion formula in terms of the usual perturbation *mould* ("*Störungsschimmel*"), . . .' (our italics, Heisenberg to Pauli, 8 June 1924).

[277] To deal with degenerate systems, as, for example, in the problem of resonance fluorescence, Bohr (1924) and Heisenberg (1925a) had to introduce rather artificial assumptions, which their systems had to satisfy (see Section III.4).

> my opinion, Heisenberg has expressed just the right viewpoint in this regard: he doubts that one can speak about the possibility of definite orbits. Kramers has never admitted to me earlier that such doubts were reasonable. Nevertheless I must insist on them, for the matter appears too important to me. (Pauli to Bohr, 21 February 1924)

Later in 1924, motivated by his study of the relativistic doublet spectra (Pauli, 1925a) and the new organization of electron states in many-electron atoms (Pauli, 1925b), Pauli had concluded that the idea of electrons moving in atoms had to be given up completely. Thus, he had written to Bohr:

> In my view, the relativistic doublet formula seems to show without doubt that not only the dynamical concept of force [because it did not explain the non-mechanical stress responsible for the doublet splitting of atomic spectra], but also the kinematic concept of motion of the classical theory will have to undergo profound modifications . . . Since this concept of motion also lies at the basis of the correspondence principle, the efforts of the theoreticians must above all be devoted to its clarification. (Pauli to Bohr, 12 December 1924)

A little later he had come back to the same point; he expressed his expectation that the most serious difficulties of atomic theory would not be removed by the application of the correspondence principle—as Bohr had had in mind—but 'from a physical analysis of the concepts of motion and force in the sense of quantum theory, in which one should let oneself be guided by empirical criteria' (Pauli to Bohr, 31 December 1925).

Heisenberg had never really denied the need for a revision of the (classical) kinematical and dynamical concepts in quantum theory. He had also arrived, at least in February 1925, at the conclusion that the queer mechanical models, or '*Ersatz*' models, by which he had described the empirical observations on complex spectra, made such a revision necessary. He had even adopted another opinion held by Pauli, for he remarked: 'By the way, I now believe that every single atom possesses a definite angular momentum, namely—in contrast to your view?—Landé's (i.e., J and not Sommerfeld's j)' (Heisenberg to Pauli, 26 February 1925). That is, he agreed with the assumption that every atom in a given stationary state possessed a definite value of its total angular momentum. That indeed meant a conversion, for previously Heisenberg had achieved the successful description of complex spectra (e.g., in Landé and Heisenberg, 1924) only by assuming that the angular momenta of the core and the atom were double-valued. Heisenberg's conversion had made Pauli rather happy. He informed Heisenberg that he would not quarrel much as to which description (whether Landé's or Sommerfeld's angular momentum) was to be preferred since this appeared to be a matter of taste. He continued: 'I have definite reasons to believe in a uniquely defined value of the [angular] momentum of an atom in a stationary state, provided this momentum is considered as a dynamical (not as a kinematic) quantity (in the same way as there also exists a well-defined energy value of the atom, i.e., the term value). [I shall give you the] details in person. I am glad that you now also share this view' (Pauli to Heisenberg, 28 February 1925). During their following meeting in Copenhagen, in March and April 1925, the two friends had ample occasion to discuss the questions of fundamental concepts in atomic theory in detail and to get closer to each other's point of view.

The agreement about completely abolishing electron orbits in atoms and replacing the classical kinematic concepts by adequate quantum-theoretical ones encouraged Pauli to look with sympathy at Heisenberg's way of calculating the intensities in summer 1925. To begin with, Pauli had himself been interested for some time in the general problem of how to calculate the intensities of spectral lines in quantum theory.[278] Just recently, during his spring visit to Copenhagen, he had taken the opportunity of discussing the intensity problem with Heisenberg and Kronig, and had continued to think about it on his return to Hamburg. In May 1925 he informed Kronig: 'I am interested . . . in the general formal problem of determining the transition probabilities, especially in reformulating and extending Born's formalism [for that purpose], about which we spoke in Copenhagen. When I come to this formalism in my article for the Springer *Handbuch* (which makes progress meanwhile), I shall again think about it' (Pauli to Kronig, 21 May 1925). Thus, Pauli, at the same time as Heisenberg and Kronig, thought about the consistent quantum-theoretical calculation of the intensities of atomic spectra.[279] Now Heisenberg's new formalism for the quantum-theoretical amplitudes seemed to be much more revolutionary than Born's discretization formalism of the previous year (Born, 1924b). Moreover, it contained a proposal about how to reformulate classical kinematic and dynamical relations, i.e., about how to introduce new quantum-theoretical concepts for the old classical concepts of force and motion, which had been found to be inadequate for the description of atoms. In this situation Pauli decided to halt his own endeavours and to encourage Heisenberg in his programme. And he sharpened his tools to perform what he considered his duty; namely, to offer as clear criticism as possible against all the details of Heisenberg's procedure and results in order to prevent him from obtaining merely a 'formal' solution of the underlying profound quantum-mechanical problem.

[278] For example, in summer 1923 Pauli had begun to study the intensities of combination lines, observed in the mercury spectrum when an external electric field is present (Hansen, Takamine and Werner, 1923). 'As far as my theoretical considerations on the observations of Hansen, etc., are concerned, they deal primarily with the intensities of the new (p, p)-combinations and their complex-structure components, which arise in the presence of a field,' he had written to Landé in a letter, dated 23 September 1923. He had taken up this work again in spring 1925, during his Copenhagen visit, and completed the manuscript of a paper (Pauli, 1925d).

[279] Pauli, in his letter dated 21 May 1925, quoted above, had also asked Kronig whether he had obtained any progress with the same problem. However, neither Kronig nor Pauli really proceeded with the calculation of the intensities of lines emitted by atoms in summer 1925. When Pauli completed his *Handbuch* article (Pauli, 1926b), he was so convinced about Heisenberg's success with the new quantum-mechanical scheme that he did not include anything about his idea of extending Born's formalism (i.e., the scheme contained in Born, 1924b) to calculate intensities.

Pauli had often criticized Heisenberg's earlier theories of atomic structure as being too formal. Though he himself possessed great skill in handling mathematical formalism, he had—after completing his doctoral thesis (Pauli, 1922) and the work with Born on perturbation theory (Born and Pauli, 1922)—protected himself against an excessively brilliant use of mathematical machinery. He believed that the latter often served to hide the essential physical difficulties. Hence, in much of his own work for several years following his thesis, he had practised an almost Bohr-like abstinence of mathematical formalism. Pauli had come to the conclusion that in atomic theory, in particular, one had to distrust formalism where the physical conceptions were still unclear.[280] Therefore, he was extremely concerned about the possibility that Heisenberg, in one of his usual outbursts of optimism, would consider a particular scheme—which merely seemed to be formally consistent—as *the* solution of the quantum problem. Heisenberg, chiefly through Pauli's goading, was aware of this danger. In summer 1925 he tried to convince Pauli that indeed he was not going to settle for a cheap solution. Thus he wrote: 'Moreover, the new "formalism" is not so terribly formal at all; e.g., one can derive from it certain results concerning the kinematics in quantum theory' (Heisenberg to Pauli, 21 June 1925).

If one reads the letters, which Heisenberg wrote to Pauli in June and July 1925, one gets the impression that Pauli did not criticize him very often at that time. There were only a few specific points in Heisenberg's results against which he raised objections. One such point was the appearance of half-integral quantum numbers in Heisenberg's expression for the energy of the anharmonic oscillator. Heisenberg had reported it to be

$$W_{\mathrm{qu}} = (n + \tfrac{1}{2})h\nu + \beta\left(n^2 + n + \tfrac{1}{2}\right) + \dots, \qquad (n = 0, 1, 2, \text{etc.}), \qquad (111)$$

where $\nu = \omega_0/2\pi$, β is proportional to the anharmonicity parameter λ, and the dots indicate corrections of order λ^2, λ^3, etc. (Heisenberg to Pauli, 24 June 1925).[281] At first Pauli thought that on the basis of the 'vapour-pressure argument' a different second term, namely, βn^2, followed on the right-hand side of Eq. (111).[282] He had in mind Otto Stern's old formula for the chemical constant,

[280]This attitude was particularly reflected in Pauli's review article on quantum theory for the *Handbuch* (Pauli, 1926b), which represented a masterly treatment of the groping approach to the problems of atomic systems, cautiously emphasizing the physical principles and putting the formalism into the background. His other paper on atomic theory, also written in 1925, dealing with the intensities of combination lines, showed the same style (Pauli, 1925d).

[281]In a later letter, Heisenberg gave a different expression for the quantum-theoretical energy of the anharmonic oscillator, in which the second term on the right-hand side of Eq. (111) was replaced by $\beta(n^2 + n + \frac{1}{4})$ (Heisenberg to Pauli, 4 July 1925). The energy, given by Eq. (111), is the correct one if the anharmonic oscillator satisfies the equation of motion (101). The anharmonic oscillator, which satisfies Eq. (75), leads to a different second term in the energy formula, namely, $\beta(n^2 + n + 17/30)$. (See Born and Jordan, 1925b, p. 883, Eq. (88).)

[282]In the margin of Heisenberg's letter Pauli noted: '*Anharmonischer Oszillator rein empirisch* $(n + \frac{1}{2})\,h\nu + \beta n^2$, *nicht* $\beta(n + \frac{1}{2})^2$. *Dampfdruck Argument*.' ('Anharmonic oscillator, purely empirically $(n + \frac{1}{2})h\nu + \beta n^2$, not $\beta(n + \frac{1}{2})^2$. Vapour-pressure argument.')

which had been derived by calculating the vapour pressure existing above the surface of a solid (Stern, 1913, 1919). Stern had obtained agreement with the experimental data by considering harmonic forces between the atoms and incorporating a zero-point energy of magnitude $\frac{3}{2}h\nu$ for each atom, which can vibrate in space with frequency ν. However, the anharmonic forces may also play a role in the thermal vibrations of atoms in a solid. Hence, in the energy, contributions like the second term in Eq. (111) have to be taken into account, and these terms would give rise to a different zero-point energy, $(\frac{1}{2}h\nu + \frac{1}{2}\beta)$, than the one yielding Stern's formula. For many years the latter formula had been 'considered as one of the best founded results of quantum theory' (Born, 1923b, pp. 705–706); therefore, Pauli could indeed argue that Stern's vapour pressure considerations might speak against the validity of Eq. (111).

At the same time, Pauli was himself quite familiar with the theory of the anharmonic oscillator. For instance, he had been interested in the absorption of infrared radiation of characteristic frequencies by crystals (Pauli, 1925c); in the solution of this problem the anharmonic part of the interatomic forces could not be neglected.[283] As for the argument concerning the vapour pressure, Pauli was able to discuss it in detail with Stern himself, who was then professor of experimental physics in Hamburg. From such conversations he probably learned that no decision was possible empirically either in favour or against Heisenberg's expression for the energy of the anharmonic oscillator.

But Pauli cited still another argument against the energy formula, Eq. (111), namely, the information derived from the band spectra of molecules.[284] Pauli had some experience of his own in this regard, as two years previously he had written a paper, jointly with Hendrik Kramers, dealing with the rotational part of band spectra (Kramers and Pauli, 1923). The authors had intended to cast doubt on a previous analysis of certain band spectra by Adolf Kratzer, who had employed half-integral quantum numbers in the expression for the rotational energy (Kratzer, 1922). Pauli and Kramers, who subscribed to the deeply held Copenhagen view against the occurrence of half-integral quantum numbers (in the quantum conditions and the energy terms), had shown that a closer inspection of many band spectra did not support Kratzer's replacement of integral by half-integral quantum numbers, and that the situation was more complex than Kratzer believed insofar as numbers different from integral or half-integral might occur in a phenomenological description. However, in 1925 Pauli did not address the question of the rotational energy of molecules—although Heisenberg, in his

[283] The solution of this problem, which represented a part of the problem of heat conduction in crystals, had been mentioned by Born in his encyclopedia article on '*Atomtheorie des festen Zustandes*' ('Atomic Theory of the Solid State') as an important goal in the kinetic theory of solids. (See Born, 1923b, p. 709 at the end of Section 35, which also deals with Stern's method of determining the chemical constants of monatomic gases.)

[284] That Pauli used this argument follows from a sentence in a postcard of Heisenberg to Pauli. There he wrote: 'You once wrote about arguments against the quantization, $W = (n + \frac{1}{2})h\nu + \beta(n + \frac{1}{2})^2$, for band spectra; how does it stand now?' (Heisenberg to Pauli, 4 July 1925).

letter to Pauli, dated 21 June 1925, had given an expression for this energy as well—but thought of another part of the energy arising from the oscillations of the atoms within the molecules. These oscillations also showed up in the structure of band spectra, especially in the so-called rotation–vibration bands. These spectra, which had been observed in the region of visible and near-infrared radiation, consisted of a repetition of structures, i.e., sequences of equally distant band lines led by a band-head. The corresponding energy terms of molecules had been described by two quantum numbers, a rotational quantum number (m) determining the position of the band lines in the sequence, and an oscillation quantum number (n) determining the position of the band-head. From an analysis of the infrared band spectra of cyanide, Kratzer had concluded that even the small anharmonic part of the forces between the atoms in these molecules played a role; that is, he explained with its help the observed deviations from the equal spacing of band-heads (Kratzer, 1920a). Thus, in principle, the rotation–vibration band spectra of molecules could be used to test Heisenberg's oscillator energy formula (111). As a matter of fact, a paper of Kratzer and his student Elisabeth Sudholt had just appeared in an issue of *Zeitschrift für Physik* in late June, in which the authors had analyzed carefully the resonance spectra of iodine vapour excited by the green mercury line (Kratzer and Sudholt, 1925). They had succeeded in fitting the data beautifully by assuming integral values for the oscillation quantum number, in contrast to Heisenberg's quantum-theoretical result.[285] Interestingly enough, neither Kratzer and Sudholt, nor Pauli and Heisenberg, referred in this context to a slightly earlier work, namely, Robert Sanderson Mulliken's detailed investigation of the band spectra of boron oxide, from which the preference for half-integral values of the oscillation quantum number followed (Mulliken, 1924a, b; 1925a, b). This work must have been noticed by the experts on band spectra, among them Wilhelm Lenz, the professor of theoretical physics in Hamburg.[286]

In his criticism of Heisenberg's attempts to formulate the quantum-theoretical relations, Pauli's deepest concern was not that he considered this or that equation as wrong or doubtful, but that Heisenberg should not squander his evidently great physical intuition on a too narrow or even conceptually wrong formal scheme. Heisenberg himself was aware of this danger, and that was also the reason why he invited Pauli's criticism. He also did not try to hide any unclear point; on the contrary, he drew attention to all the difficulties of the entire scheme. Thus, for example, he wrote: 'The strongest objection appears to me to be that the energy, written as a function of q and $\dot{q}$ [i.e., the position coordinate of the oscillator and its time derivative], need not turn out to be a constant.

[285] A year later Francis Wheeler Loomis reanalyzed basically the same data and found excellent agreement with half-integral values for the oscillation quantum number (Loomis, 1927).

[286] We do not know to what extent Pauli followed the details of the literature on band spectra. But we do know that he appreciated discussions on physical questions with Lenz, whom he knew since his student days in Munich. However, in June 1925 Lenz was sick and stayed at a sanatorium in Heilbronn, Württemberg. (See Pauli to Sommerfeld, 22 June 1925.)

Ultimately this is caused by the fact that the product of two Fourier series is not yet uniquely defined' (Heisenberg to Pauli, 24 June 1925).

It is at first difficult to understand Heisenberg's last comment about the nonuniqueness of the product of two Fourier series. After all, in classical theory the product of two classical Fourier series is, of course, unique; and if one reformulates the product by applying the product rule, Eq. (66), the quantum-theoretical product is also unique. However, one has to remind oneself that in classical theory the product of two quantities, X and Y, does not depend on the order in which it is taken, i.e.,

$$XY = YX. \tag{112}$$

In Heisenberg's new scheme, Eq. (112) was not always true. Hence, if one started from a product XY in classical theory, one did not know *a priori* how to translate it into quantum theory; either as

$$(XY)_{\rm cl} \to (XY)_{\rm qu} \tag{113a}$$

or as

$$(XY)_{\rm cl} \to \tfrac{1}{2}(XY + YX)_{\rm qu}, \tag{113b}$$

or still in another way (such that the correspondence principle was satisfied).[287] In that sense the product of two Fourier series was indeed *not* defined uniquely, and Heisenberg had good reason for his worry. And yet, he had successfully avoided such ambiguities in his treatment of the anharmonic oscillator. A few days after he had expressed the above doubt to Pauli, he wrote to him again: 'As far as the energy calculation in my quantum-theoretical formalism is concerned, it is in my opinion as inevitably determined as the multiplication of the Fourier series. That is, *if* one believes in the quantum-theoretical reformulation of this multiplication and, *further*, in the form of the energy, $E = p^2/2m + \omega^2 q^2$, then it must be admitted that $W = (n + \frac{1}{2})h\nu$' (Heisenberg to Pauli, 29 June 1925). That is, the energy of the harmonic oscillator was uniquely determined, since in the formal expression for the energy (E denoting the reformulated Hamiltonian) the terms, namely, $p^2/2m$ and $\omega^2 q^2/2m$ (with p the momentum and q the position variable), were uniquely defined. The same was true, of course, of the anharmonic oscillator, whose energy was formally given by Eq. (107). Since Heisenberg had proved energy conservation—at least in the approximation in which he had been able to carry out the calculations—with this uniquely defined energy expression for the anharmonic oscillator, he became confident again that it might also work in other cases. Thus, he wrote to Pauli: 'In the meantime I have made

[287] For example, the quantum-theoretical expression for $(XY)_{\rm cl}$ could be $(aXY + bYX)_{\rm qu}$, where a and b are numbers whose sum adds up to 1. Of course, the correspondence principle would be satisfied in all such cases.

some, but not much, progress, and I am convinced in my heart that this quantum mechanics is already the correct theory; hence Kramers accuses me of being [too] optimistic' (Heisenberg to Pauli, 29 June 1925).

The 'fabrication of quantum mechanics' turned out to be a slow and cumbersome process, and Heisenberg frequently jumped between depressing doubts and joyful optimism. However, between 21 June, when he first wrote to Pauli after returning to Göttingen, and 29 June, when he informed him about his 'heartfelt conviction,' Heisenberg clarified his ideas about the new quantum-theoretical formalism considerably. During this period he had some discussions with Hendrik Kramers, who spent a week in Göttingen at that time.[288] However, in spite of the fact that Kramers, Bohr's close associate since 1916, knew the formalism of atomic theory extremely well, he could not discuss with Heisenberg in the same manner as Pauli did. The main reason was that he had occupied himself too much with translating the physical ideas and insights of Bohr into mathematical language. In 1925 Kramers would just not deviate from the canonical path of Copenhagen physics: applying the correspondence principle together with the dispersion-theoretic methods to describe an increasing number of atomic phenomena. The pursuit of a new theory, radically different from the previous atomic theory, was not in his line. How could he encourage or object to Heisenberg's attempt 'to fabricate a new quantum mechanics' on the basis of strange mathematical rules, such as the multiplication rule for patterns of quantum-theoretical amplitudes? All he could do was to urge Heisenberg not to build extravagant hopes regarding the final solution of the quantum problem.[289]

Pauli—though, like Kramers, experienced in the methods of Bohr–Sommerfeld atomic theory—exerted a far greater influence on Heisenberg. Moreover, he was psychologically in a different situation than Kramers. For years he had become increasingly convinced of the untenability of all known methods to solve the principal difficulties of the existing theory. He had regarded the earlier attempts, like the discretization procedure of Born and Heisenberg, as too tame and conservative for curing the ills. And he had argued that the Bohr–Kramers–Slater theory of radiation led into the wrong direction. However, Heisenberg seemed to follow the right path with his radical method of reformulating the kinematical and mechanical relations of classical theory. Hence Pauli considered it his duty to encourage him in trying out unconventional methods. And Heisenberg took his advice seriously, for he wrote: 'If you believe that I have read your letter only with a scornful laughter, then you are mistaken. On the contrary, since Helgoland my opinion about mechanics has become more radical from day to day, and I am firmly convinced that Bohr's theory of the hydrogen

[288] Kramers' visit to Göttingen is confirmed by a letter of Born. He remarked: 'Kramers was here for eight days' (Born to Einstein, 15 July 1925).

[289] It should be noted that Kramers, because of his visit to Göttingen in summer 1925, was able to take the first news about Heisenberg's successful treatment of the anharmonic oscillator including energy conservation to Copenhagen, thereby informing Bohr and Kronig about the progress of Heisenberg's work.

atom in its present form is not better than "Landé's theory of the Zeeman effect"' (Heisenberg to Pauli, 9 July 1925).

Having denounced all previous descriptions of atomic structure, Heisenberg had to reconstruct atomic theory from scratch. But what should the new quantum mechanics look like? The only positive hints so far were provided by the quantum-theoretical reformulation of classical equations, which had led to the solution of the anharmonic oscillator problem. The question arose whether one could also apply the same formal method to other mechanical problems—to the rotating electron, for example. At least on first inspection, the answer seemed to be affirmative. Still, even many successful applications of the quantum-theoretical reformulation scheme would not answer the questions concerning the fundamental principles underlying it. What principle made it necessary to use exactly the method which Heisenberg had proposed? Heisenberg answered this question by referring to an important philosophical guiding principle in quantum theory, namely, to use only observable quantities in the description of atomic phenomena.

V.2 A Guiding Philosophical Principle

Three decades after the formulation of quantum mechanics Max Born, recalling the decisive steps which had led to it, remarked:

> In Göttingen we also took part in efforts to distil the unknown mechanics of the atom from the experimental results. The logical difficulty became ever sharper. Investigations into the scattering and dispersion of light showed that Einstein's conception of transition probability as a measure of the strength of an oscillation did not meet the case, and the idea of an *amplitude* of oscillation associated with each transition was indispensable. In this connection, work by Ladenburg [1921a], Kramers [1924a, b], Heisenberg [Kramers and Heisenberg, 1925], Jordan and me [Born, 1924b; Born and Jordan, 1925a] should be mentioned. The art of guessing correct formulae, which deviate from the classical formulae, yet contain them as a limiting case according to the correspondence principle, was brought to a high degree of perfection. A paper of mine, which introduced, for the first time I think, the expression *quantum mechanics* in its title, contains a rather involved formula (still valid today) for the reciprocal disturbance of atomic systems [Born, 1924b, Eq. (33)].
>
> Heisenberg, who at that time was my assistant, brought this period to a sudden end. He cut the Gordian knot by means of a philosophical principle and replaced guesswork by a mathematical rule. The principle states that concepts and representations that do not correspond to physically observable facts are not to be used in theoretical description. Einstein used the same principle when, in setting up his theory of relativity, he eliminated the concepts of absolute velocity of a body and of absolute simultaneity of events at different places. Heisenberg banished the picture of electron orbits with definite radii and periods of rotation because these quantities are not observable. (Born, 1955b; English translation, pp. 258–259)

Born wanted to draw attention to the importance of the idea of employing only observable quantities in physical theory. This idea is often considered as the philosophical basis of Heisenberg's new theory, and Heisenberg is held responsible for having introduced it as a guiding principle in quantum theory. However, this idea had been appealed to earlier occasionally even in writings, including Born's, on atomic physics. It seems impossible to find out who actually first enunciated it or formulated it as a principle. Heisenberg never claimed any priority, but rather believed that,

> The idea of having a new [quantum] theory in terms of observables did indeed originate in Göttingen and was closely connected with the interest in relativity theory that existed there. Hermann Minkowski had been very interested in special relativity. When people spoke about it they always said that, "There was this very famous point of Einstein that one should only speak about those things which one can observe, that actually the time entering in the Lorentz transformation was the *real* time." In some way that was an essential turn which Einstein had given to Lorentz' idea. Lorentz had the right formulas, but he thought that the time which entered into the transformation equations was the "apparent" time. Einstein said, however, that there is not one "apparent" and another "real" time; there is just one "real" time, and that is what Lorentz called "apparent" time. So this turning around of the physical picture by saying that the real things are those which you can observe, and everything else has no meaning, was very much in the minds of people at Göttingen. (Heisenberg, Conversations, p. 174; also AHQP Interview)

With these remarks Heisenberg referred to a crucial step in the development of special relativity theory, a step which had been especially noticed by the Göttingen mathematican Hermann Minkowski. Minkowski discussed in his famous address on space and time ('*Raum und Zeit*'), given on 21 September 1908 to the 80th Assembly of the Association of German Scientists and Physicians (*Versammlung der Gesellschaft Deutscher Naturforscher und Ärzte*) at Cologne. He considered the case of two electrons, one at rest and the other in uniform motion, to which Lorentz had attributed the times t and t', respectively. Then he said:

> Lorentz called the t' combination of x and t the *local time* of the electron in uniform motion, and applied a physical construction of this concept, for the better understanding of the [Lorentz–Fitzgerald] contraction [of rigid bodies]. But the credit of first recognizing clearly that the time of the one electron is just as good as that of the other, that is to say, that t and t' are to be treated identically, belongs to A. Einstein. Thus time, as a concept unequivocally determined by phenomena, was first deposed from its high seat. (Minkowski, 1909, p. 107; English translation in *Lorentz*, *Einstein*, *Minkowski and Weyl*, 1923, Dover reprint, pp. 82–83)

The recognition of time as a physical concept which cannot be *uniquely* determined by the observed phenomena indeed provided Einstein one of the keys to special relativity and a consistent electrodynamics of moving bodies. He had realized that the old definition of time and of absolute motion caused the principal difficulties in the previous theory. Einstein then built the new theory entirely by applying two fundamental postulates, namely, the principle of relativity of inertial frames of reference and the requirement of a constant velocity of light *in vacuo*, to the description of natural phenomena.

The arguments, which had led Einstein to abolish the concept of a preferred time or motion in physical theories, had already been considered by Ernst Mach in his historico-critical analysis of mechanics (Mach, 1883). In this book, entitled '*Die Mechanik in ihrer Entwicklung*,' Mach had criticized, in particular, Newton's concepts of time and motion. Thus he claimed that Newton, with his concepts of 'absolute time' and 'uniform motion without reference to any external object,' had violated his own principle in natural philosophy of sticking merely to facts. Mach had argued: 'The question whether a motion *per se* can be uniform does *not make any sense at all*. Just as little can we speak about an "absolute" time (independently of each change). This absolute time cannot be measured with the help of any motion, hence it does not have any practical or scientific value; nobody has the right to say that he knows anything about it, for it is just a useless "metaphysical" concept' (Mach, 1883, 7th ed. Chapter 2, Section 6, p. 217). Similarly, about 'absolute space' and 'absolute motion,' he had concluded:

> About absolute space and absolute motion nobody can say anything; they are merely objects of thought, which cannot be exhibited in practice. All of our basic assumptions about mechanics are, as we have demonstrated in detail, experiences about *relative* positions and motions of bodies. They could not rightly be taken over without examination into those regions, in which they are considered to be valid. Nobody has the right to extend these basic assumptions beyond the limits of experience. This extension is even meaningless, since nobody knows how to apply it. (Mach, 1883, 7th ed., Chapter 2, Section 6, pp. 222–223)

Mach, in his scientific and popular writings, had put the emphasis on directly observed facts like sensations. Science, he taught, consists in establishing relations between the observed facts. He had argued that, in principle, one could even restrict oneself to pure experiences or facts, but that it was an act of 'economy of thought' to express many experiences in a few mathematical or physical equations. Einstein had read some of Mach's books, especially the *Mechanik*, before he worked on the electrodynamics of moving bodies. Fifty years later he described how Mach had shaken his faith in the validity of Newtonian mechanics: 'This book [i.e., the *Mechanik*] exercised a profound influence on me in this regard while I was a student. I [now] see Mach's greatness in his incorruptible scepticism and independence; in my younger years, however, Mach's epistemological position also influenced me' (Einstein, 1949, p. 21). Because of this influence one can readily understand why the ideas and principles, earlier expressed by Mach, had returned as guiding principles in Einstein's reformulation of Newtonian mechanics and Lorentz' electrodynamics.[290] And through Einstein they even influenced Minkowski, when he reformulated the entire space–time description in physics by introducing the four-dimensional space–time continuum and declared: 'Henceforth space by itself, and time by itself, are doomed to fade away into mere shadows, and only a kind of union of the two will preserve an independent reality' (Minkowski, 1909, p. 104; English translation in *Lorentz, Einstein, Minkowski and Weyl*, 1923, Dover reprint, p. 75).

Max Born had himself been in Göttingen in 1905 when Minkowski presented his own ideas on the electrodynamics of moving bodies—which he had developed independently of Einstein—first in the Hilbert–Minkowski Seminar. After leaving Göttingen and returning to his hometown of Breslau, Born continued to be interested in this subject, especially when he came to know about Einstein's work on special relativity. By connecting Einstein's ideas and Minkowski's mathematical methods he wrote a paper on the electromagnetic mass of the electron (Born, 1909a). He sent this essay to Minkowski, who invited him to return to Göttingen to collaborate with him. Born devoted most of his time during the two years following Minkowski's death (in January 1909) to working on problems of relativistic electron theory; after 1914, however, he stopped publishing original contributions in this field. It is surprising, however, that Einstein's basic philosophical approach to special relativity theory—i.e., that nonobservable concepts should be abolished in the construction of a physical theory—was not mentioned by Born in his early work in Göttingen.[291]

When in 1914 Born went to Berlin to assist Max Planck in his lecturing duties and, later on, to do military service, he frequently had occasion to meet Albert Einstein, who was then working on the formulation of the general relativity theory.[292] Born was the first to write a review of Einstein's new theory (Born,

[290] Einstein did not follow Mach's view in all cases. For example, Mach distrusted atomic theory and regarded it as mere speculation. When somebody talked about atoms in his presence, he usually said: '*Ham'S ans g'sehn?*', i.e., 'Did you see one?' (Heller, 1964, p. 24). Einstein might have taken this disbelief in the existence of atoms as a challenge to prove that one *could* observe atoms. In the same year as he applied Mach's criterion of observability to the electrodynamics of moving bodies, he produced the theory of Brownian motion, which could be explained only by assuming the existence of molecules and their thermal motions (Einstein, 1905c). Mach is said to have become converted to atomism several years later, after seeing the flash on the scintillation screen (caused by α-particles) in Vienna's *Institut für Radiumforschung*. Einstein wrote Mach's obituary in *Physikalische Zeitschrift* (Einstein, 1916a).

[291] Interestingly enough, the 'Mach–Einstein principle' of using only observable quantities in the theoretical description was also not emphasized by Hermann Weyl in his book on relativity theory, entitled *Raum-Zeit-Materie (Space, Time, Matter*, Weyl 1918c). Many people, including Heisenberg, learned about Relativity from this book.

[292] Max Born had first met Einstein at the Salzburg meeting of German Scientists and Physicians (*Versammlung der Gesellschaft Deutscher Naturforscher und Ärzte*) in 1909. After Born went to Berlin as *Extraordinarius* for theoretical physics in 1915, he again had occasion to meet Einstein, who was then with the *Kaiser-Wilhelm-Gesellschaft*. A special friendship developed between them; it was helped by the fact that Born's office (at the Artillery Testing Commission) in *Spichernstraße* and Einstein's home in *Haberlandstraße* were not far from each other, and Born frequently visited Einstein.

1916a), in which Einstein felt himself to have been 'fully understood' (Einstein to Born, 27 February 1916). What distinguished general relativity theory from earlier and contemporary attempts to extend Newton's theory of gravitation was that Einstein—as he had done before in special relativity theory—made two empirical observations the basic principles of his theory: in this case, first the equivalence of gravitational and inertial masses, and, second, the equivalence of all space–time coordinate systems (or the nonobservability, in principle, of accelerated motion). With the help of these principles he formulated the beautiful mathematical scheme of general relativity theory which explained one observed effect, the motion of the perihelion of Mercury, and predicted two new effects, the gravitational red shift of spectral lines and the bending of light in (strong) gravitational fields. Born was so overwhelmed by the 'magnificence of the inner structure' ('*Großartigkeit des inneren Aufbaus*,' Born, 1916a, p. 59) of Einstein's theory that he resolved not to work in this field himself (Born, 1968, p. 30).

Yet the great success of general relativity theory, the formulation of which he had witnessed at close quarters in Berlin, and his friendship with Einstein, influenced Born very deeply. Born continued to work on the problems of lattice dynamics, either alone or in collaboration with Alfred Landé, whom he knew from Göttingen (where Landé had been a student from 1910 to 1912 and had spent the year 1913–1914 as Hilbert's assistant) and who served with Born for a time in the Artillery Testing Commission ('*Artillerie-Prüfungskommission*') under the direction of Rudolf Ladenburg during World War I. In 1919 Landé joined Born as a *Privatdozent* at Frankfurt. Several years later he recalled the direction which Born's thinking had taken under the influence of Relativity:

> Already years ago Born had, in discussions, defended the then heretical notion that the mechanical atomic models in space and time do not possess any physical reality; that is, they cannot be confirmed by *any experiment*, because an experiment would give information only about the energies of stationary states (term values), frequencies and intensities, but never about the instantaneous positions and velocities of the electrons. Instantaneous phases would rather be *unobservable in principle*, and therefore all pictures about space–time changes of the instantaneous phases, i.e., all model conceptions would be superfluous or even inconsistent with the observations. (Landé, 1926c, p. 455)

The question of the use of observable quantities in physical theory was uppermost in Born's mind when he wrote to Pauli:

> I have read your paper in the new issue of *Verhandlungen der Deutschen Physikalischen Gesellschaft* [Pauli, 1919c] on Weyl's theory with great interest; to be sure, at the moment I am not quite familiar with the formulae of relativity theory, for I have not thought about them for months, but I have understood the meaning of your considerations. I have especially been interested in your remark at the end, that you regard the application of the continuum theory to the interior of the electron as meaningless, because one is then dealing with things which are unobservable in principle. I have pursued exactly this idea for some time, though up to now without

> positive success: that is, the way out of all quantum difficulties must be sought by starting from entirely fundamental points of view. One is not allowed to carry over the concepts of space and time as a four-dimensional continuum from the macroscopic world of experience into the atomistic world; the latter evidently demands another type of number-manifold to give an adequate picture . . . Though I am not yet old, I am already too old and burdened to arrive at the solution. That is your task; according to what I have heard about you, to solve such problems is your calling. (Born to Pauli, 21 December 1919)

The work of Pauli to which Born referred had dealt with the question of the motion of the perihelion of Mercury and the deflection of light rays in Weyl's theory of gravitation (Weyl, 1918a). At the close of his paper Pauli had made some general comments about Weyl's theory and raised the following objection:

> There is a physical–conceptual objection which should not be forgotten. In Weyl's theory we continuously operate with the field strength in the interior of the electron. For a physicist this [the field strength] is only defined as a force on a test-body, and since there are no smaller test-bodies than the electron itself, the concept of the electric field strength in a mathematical [space-] point seems to be an empty, meaningless fiction. One should stick to introducing only those quantities in physics which are observable in principle. Can it be that we are following a completely false track when using the *continuum theories* for the field in the interior of the electron? (Pauli, 1919c, pp. 749–750)

The fundamental objection against Weyl's programme of applying a unified field theory of gravitation and electricity in order to derive the existence of the electron must indeed be considered as a clever point, and one may wonder how Wolfgang Pauli came to it. The explanation, however, is not so difficult if one recalls the relationship of Pauli's father, the physiologist Wolfgang Joseph Pauli, to Ernst Mach. He had been greatly influenced by Mach's thinking and had persuaded him to become godfather to his son. Thus, the young Pauli had also, in time, become acquainted with Mach's philosophy about the observability of physical concepts[293]; he was impressed enough by it to use it for criticizing the already famous mathematician Hermann Weyl. Weyl took Pauli's argument quite seriously and replied:

> That one is not able to measure the fields in the interior of the electron can only mean, that the differences [of the fields] in the interior of the electron can never cause such changes in the course of the world which grow to an immediately noticeable magnitude. As soon as such effects occur, I can use them to "measure" those inner differences. But why should it not happen? I believe, for instance, that the fact that the electron does not radiate in the stationary Bohr orbits indicates an inner change of the electron caused by the acceleration, which enables it to preserve its energy. For, why should it [the electron] behave like a rigid sphere with rigid charge, even when a non-quasistationary acceleration acts upon it? (Weyl to Pauli, 9 December 1919)

[293] Wolfgang Pauli, Jr. possessed and read several books of Ernst Mach, for example the *Mechanik* (in an edition of 1912) which Mach had inscribed to him personally (now in the Pauli Memorial Room at CERN, Geneva).

With the above statements Weyl sought to emphasize that one should be careful in applying the criterion of measurability to a theory at a very early stage, when one did not yet have a clear idea about what could eventually be observed. Pauli, on the other hand, does not seem to have been convinced by Weyl's answer at all; he considered his own conjecture to be important enough to raise it again in a discussion with Einstein and Weyl following Weyl's talk on '*Elektrizität und Gravitation*' ('Electricity and Gravitation') in the session on relativity theory at the 86th Assembly of the Association of German Scientists and Physicians (*Versammlung der Gesellschaft Deutscher Naturforscher und Ärzte*) in Bad Nauheim in September 1920 (Weyl, 1920). At this meeting, which would later become well known for the fight between the opponents and partisans of relativity theory, Pauli said:

> None of the erstwhile theories, not even the Einstein theory [Einstein, 1919], has up to now succeeded in solving the problem of the elementary electric quanta in a satisfactory manner; thus it is desirable to look for a deeper reason for this failure. I wish to seek this reason in that it is not allowed at all to describe the electric field in the interior of the electron as a continuous space-function. The electric field strength is defined as the force on a charged test-body, and, if there exists no smaller test-body than the electron, the concept of the electric field strength at a certain [space-] point in the interior of the electron, with which all continuum theories operate, seems to be an empty, meaningless fiction, to which nothing real corresponds. Similar statements might be made about the measurement of space, for arbitrarily small rods do not exist. Therefore I would like to ask Professor Einstein, whether he agrees with the point of view that one may expect the solution of the problem of matter only from a modification of our concepts of space (perhaps also of time) and of the electric field in the sense of atomism, or whether he believes that the above-mentioned objections are not valid and thus holds the opinion that one must stick to the foundations of the continuum theories? (Pauli, in the discussion of Weyl, 1920, p. 650)

Einstein, already before the Bad Nauheim meeting, had decided about his position towards Pauli's question. He had, in particular, considered the question whether the existence of quanta and electrons implied that one had to abandon a continuum description of matter and arrived at the conclusion:

> I myself do not believe that the solution to the quanta [with which Einstein meant not only Planck's quantum of action, but also the light-quantum and the discrete structure of matter, in general] has to be found by giving up the continuum. Similarly, it could be assumed that one could arrive at general relativity by giving up the coordinate system. In principle, the continuum could possibly be dispensed with. But how could the relative movement of *n* points be described without the continuum? [He remarked further:] Pauli's objection is directed not only against Weyl's, but also against anyone else's continuum theory. I believe now, as before, that one has to look for redundancy in determination by using differential equations [for fields] so that the *solutions* themselves have the character of a continuum. But how? (Einstein to Born, 27 January 1920)

At the Bad Nauheim meeting, however, Einstein replied more cautiously to Pauli's argument in favour of using only observable quantities as physical concepts. He said:

> With the increasing refinement of the system of scientific concepts, the manner and procedure of associating the concepts with experiences becomes increasingly more complicated. If, in a certain scientific study, one finds that a definite experience cannot be associated any longer with a concept, then one has the choice whether one wants to drop the concept or keep it; in the latter case one is forced to replace the system of associating concepts [with experiences] by a more complicated one. In the case of the concepts of spatial and temporal distances also we have this alternative. In my opinion, the answer can be given only on the basis of what is functional; what it will be, seems to me to be doubtful. (Einstein, in the discussion of Weyl, 1920, p. 651)

However, in spite of this diplomatic answer, Einstein's preference for the continuum theory was obvious. Pauli, on the other hand, saw no reason to change his opinion on the basis of the response to his argument and, about three months later, he repeated in his article on relativity theory in the *Encyklopädie der mathematischen Wissenschaften* his former criticism of Weyl's theory on the basis of his criterion of the observability of physical quantities.[294] He concluded: 'Whatever may be one's attitude in detail towards these arguments, this much seems fairly certain: new elements which are foreign to the continuum concept of the field will have to be added to the basic structure of the theories developed so far, before one can arrive at a satisfactory solution of the problem of matter' (Pauli, 1921b, p. 775; English translation, 1958a, p. 206).

In fall 1920, when Heisenberg joined Sommerfeld's Seminar, the subject of Relativity was very much under discussion: Sommerfeld had just edited, with notes, the third edition of *Das Relativitätsprinzip* (a collection of papers by *Lorentz*, *Einstein*, *Minkowski and Weyl*, 1920), while Pauli was in the process of completing his article on relativity theory for the *Encyklopädie*. The memory of Pauli's encounter with Einstein and Weyl some weeks previously in Bad Nauheim was still fresh; it had been an event worth talking about among the young researchers around Sommerfeld. Pauli's account of this event and the report of the session on relativity theory at Bad Nauheim published in the *Physikalische Zeitschrift* left, no doubt, a lasting impression on Heisenberg's

[294] Pauli finished his article on relativity theory in December 1920 (Pauli, 1921b). In the last chapter, entitled '*Theorien über die Natur der elektrischen Elementarteilchen*' ('Theories on the Nature of Electrical Elementary Particles'), in which he discussed the theories of matter of Gustav Mie, Albert Einstein and Hermann Weyl, he wrote: 'Finally, a conceptual doubt should be mentioned [at this point he referred to his paper (Pauli, 1919a) and his remarks at the Bad Nauheim meeting]. The continuum theories make direct use of the ordinary concept of electric field strength, even for the fields in the interior of the electron. This field strength is however defined as the force acting on a test particle, and since there are no test particles smaller than an electron or a hydrogen nucleus, the field strength at a given point in the interior of such a particle would seem to be unobservable, by definition, and thus be fictitious and without physical meaning' (Pauli, 1921b, p. 775; English translation, 1958a, p. 206).

mind; he would not easily forget Pauli's clever remark about using observable quantities in physics.

Pauli continued to reiterate his point of view on later occasions as well. For instance, writing to thank Eddington, who had sent him a copy of his new book *The Mathematical Theory of Relativity* (Eddington, 1923b), Pauli expressed his opinion of the problem of matter and remarked:

> Furthermore, I adhere to the (of course, unprovable) viewpoint that each physical theory, which claims to provide a sensible answer to these questions, must start with a definition of the field quantities used; the definition should state how these quantities can be measured. It [the theory] must further reveal relations between electromagnetic quantities and those measured by other methods. (The most beautiful success of relativity theory was indeed that it yielded a deep and tight connection between the results of measurements of rods and clocks, the orbits of freely-falling mass points and those of light rays.) These postulates cannot be proved logically or by means of the theory of cognition (*erkenntnistheoretisch*). However, I am convinced that they are correct. (Pauli to Eddington, 20 September 1923)

While Pauli constantly spoke about what one may call 'the Mach principle' of employing only observable quantities in the theoretical description of phenomena, Born had for some time given up his own speculations on this point. On one hand, he believed that Einstein possessed deeper insight if he refused to take the principle too seriously; on the other, Born felt comfortable only after he could formulate an idea mathematically, and neither Pauli nor he had been able to do so with 'the Mach principle.' However, after the failure of the Bohr–Sommerfeld theory of atomic structure to account for the energy terms of such simple systems as the helium atom or the hydrogen molecule-ion, the necessity of revising the hitherto used concepts of atomic theory became obvious. 'It becomes increasingly more probable,' Born wrote in summer 1923, 'that not only new assumptions will be needed in the sense of [new] physical hypotheses, *but that the entire system of concepts in physics will have to be restructured in its foundations*' (Born, 1923a, p. 542). In this revision of concepts the observability postulate occurred again in the discussions of the Göttingen theoreticians. And Pascual Jordan, a participant in these discussions, recalled later:

> It is an idea, of which it is difficult to say exactly who stated it first. Heisenberg stressed it strongly, as did Born, and to me it seemed convincing because of my sympathies with the ideas of Ernst Mach. I suppose that the discussion [about it] was more concentrated in conversations between the three of us, although it is well possible that it was also mentioned once in Born's Seminar. I don't believe that the idea was discussed in great detail by anyone who did not belong to our circle, while it appeared really very convincing to the three of us and often recurred in our discussions. (Jordan, AHQP Interview, Second Session, p. 31)

Though Jordan forgot to mention Pauli's involvement—probably because he did not know about the earlier discussions between Pauli, Born and Einstein—he was quite right in stressing the importance of the discussions in Göttingen on the observability postulate during 1923 and 1924. For it was in Göttingen, where one first started to use only observable quantities in atomic theory. Thus, Born, when in 1924 he discussed the terms governing the dispersion of light by atoms in classical theory and their replacement in quantum theory, remarked: 'In this connection, it does not appear to make sense to look for the corresponding quantities [in quantum theory], which describe the Fourier coefficients C_τ [representing the perturbation term in the classical Hamiltonian describing the atom plus incident radiation] . . . ; apparently only the quadratic combinations, $|C_\tau|^2 = C_\tau C_{-\tau}$, have quantum-theoretical significance' (Born, 1924b, p. 388). The quadratic expressions, $|C_\tau|^2$, were indeed proportional to the observed intensities of transitions or to the well-known absorption and emission coefficients of Einstein. Hence Born appealed directly to 'the Mach principle,' and this call was taken up immediately by Hendrik Kramers in Copenhagen, whose March letter to *Nature* on dispersion theory (Kramers, 1924a) had in turn stimulated Born's work on the quantum-theoretical reformulation of the classical perturbation scheme (Born, 1924b). Kramers, in a second letter to *Nature*, took the opportunity of elucidating his earlier note in reply to a subsequent letter of Gregory Breit on the same topic (Breit, 1924b). In his new letter, Kramers used Born's formalism to derive his dispersion formula, Eq. (50), for the coherent scattering of light by atoms, and then concluded:

> Apart from the problem of the validity of the underlying theoretical assumptions and of any eventual restriction in the physical application of formula [(50)], the dispersion formula thus obtained possesses the advantage over a formula such as proposed by Mr. Breit in that *it contains only such quantities as allow of a direct physical interpretation* [our italics] on the basis of the fundamental postulates of the quantum theory of spectra and atomic constitution, and exhibits no further reminiscence of the mathematical theory of multiple periodic systems. (Kramers, 1924b, p. 311)

The novelty of this statement was contained in two points. First, Kramers addressed himself to the direct physical interpretation of all quantities appearing in the quantum theory of dispersion. Second, he declared openly the abandonment of the mathematical theory of multiply periodic systems. The latter point played a central role in the discussions in Copenhagen at that time. For example, Niels Bohr recalled about the conversations he had had with Heisenberg during the latter's visit in spring 1924: 'The discussions on problems of atomic physics were devoted, above all, to the strangeness of the quantum of action for the formulation of concepts that were applied to describe all experimental results. In this context, we also talked about the possibility that, as in relativity theory, also here [in quantum theory] mathematical abstractions might perhaps turn out to be useful' (Bohr, 1961b, p. ix). Yet the tendency to use only observable quantities in atomic theory did not have deep roots in Copenhagen. The direct conceptual link between Kramers' new derivation of the dispersion formula (Kramers, 1924b) and Born's work on the quantum-theoretical perturbation scheme (Born, 1924b) had been forged at a meeting of the German Physical Society on 21 June 1924 in Hamburg, where Kramers presented his work on dispersion theory and Born gave a talk on his reformulation formalism. But, what was even more important: in Hamburg, Pauli, Ernst Mach's godson, was present and he might have emphasized the necessity of using only observable quantities in atomic theory. Max Born, in his *Vorlesungen über Atommechanik*, finally raised the idea to a principle. After discussing Bohr's theory of atomic structure, in which unobservable properties such as mechanical frequencies and distances had been ascribed to atomic systems, he stated: 'Thus we arrive at the conclusion that our procedure [i.e., Bohr's theory of atomic systems] *is at present only a computational scheme*, which allows one in certain cases to replace the still unknown, real quantum laws by calculations on the basis of classical theory. Of these real quantum laws, we must require that they involve only relations between observable quantities such as energies, light-frequencies, intensities and phases' (Born, 1925, p. 114).

So much for the preoccupation with observable quantities in Göttingen until fall 1924. At other places, too, it had become clear by then how questionable it was to use mechanics in the sense of the Bohr–Sommerfeld theory of atomic systems. Even Sommerfeld began his short note on the theory of periodic systems, submitted in December 1924, by saying:

> Without doubt the fundamental assumptions of Bohr's theory of the periodic system of elements are sound: the step-wise building-up of shells, the increase of the principal quantum numbers for penetrating [diving] orbits, the earlier occurrence of higher types of orbits and the later completion of the gaps in the previous shells. On the other hand, it is evident that the specialized model-conceptions [of Bohr's theory] concerning orbits can be formed only in relation to the spectroscopic experiments. (Sommerfeld, 1925a, p. 70)

Hence it occurred to Sommerfeld that the concepts of atomic theory were only as sound as could be verified experimentally. Pauli fully agreed with his former teacher's point of view; but he emphasized more drastically than Sommerfeld the insufficiency of the earlier atomic models. Thus, in his *Handbuch* article on quantum theory, he summarized the conclusions from spectroscopy in the following words:

> It is not possible to interpret, in a physically meaningful way, all the properties of an atom in a given stationary state by associating with each of its electrons just one and the same orbit. Rather we are forced to use, for the interpretation of different groups of atomic properties, different orbits of a given electron. Evidently, the latter procedure can be viewed only as a provisional device in the light of the failure of classical kinematics for the description of the stationary states of the atoms, so long as we do not know the concepts, which in the consistent quantum theory, would replace those of classical kinematics. (Pauli, 1926b, p. 209)

After this historical digression let us return to the situation of summer 1925 when Heisenberg looked for a guiding philosophical principle to assist him in the discovery of a consistent quantum-theoretical scheme. As we have mentioned earlier, Heisenberg and Pauli were agreed since December 1923 in the opinion that one could not properly describe the radiation from atoms by assuming definite electron orbits. A year later, Pauli confirmed this opinion by his investigation of doublet spectra (Pauli, 1925a) and of electron levels in complex atoms (Pauli, 1925b). At that time Heisenberg still talked about 'virtual electron orbits' (Heisenberg to Pauli, 8 October 1924), but nine months later he wanted 'to kill the concept of electron orbits completely.' When Pauli mentioned again the argument concerning the (electron) 'orbits dipping into the nucleus'—which Bohr had earlier used, for example, to eliminate Heisenberg's half quantum numbers from the description of the anomalous Zeeman effects—he replied:

> But I do not know what you mean by orbits that fall into the nucleus. We certainly agree that already the kinematics of quantum theory is totally different from that of classical theory ($h\nu$-relation), hence I do not see any geometrically-controllable sense in the statement "falling into the nucleus." It is really my conviction that an interpretation of the Rydberg formula [e.g., for hydrogen] in terms of circular and elliptical orbits (according to *classical* geometry) does not have the slightest physical significance. And all my wretched efforts are devoted to killing totally the concept of an orbit—which one cannot observe anyway—and replace it by a more suitable one. (Heisenberg to Pauli, 9 July 1925)

In the light of Heisenberg's previous work on atomic theory the last statements sound very revolutionary indeed. It was not so much the renunciation of the mechanical electron orbits in atoms; Heisenberg had arrived at this conclusion gradually during the course of several years. More serious was the radical denial of mechanical models, for they had provided the backbone of his previous calculations. And finally there was the reference to the philosophical guiding principle about using only observable quantities in physical theory. Had Heisenberg, the Saul who used to profess mechanical '*Ersatz*' models, become the Paul of Mach's philosophy? It was not quite so.

In June 1925, in the process of sharpening the correspondence principle Heisenberg had finally replaced the mechanical orbits by a set or pattern of transition amplitudes. The latter were none other than what he called the 'quantum-theoretical re-interpretation' of the classical Fourier coefficients of the orbits. Thus, the classical mechanical orbits, though abolished as a valid concept in the description of atomic systems, had been transformed into a quantum-theoretical form. This quantum-theoretical form represented the essential outcome of the previous attempts, by using mechanical '*Ersatz*' models, to bring the classical results on dispersion phenomena into harmony with the behaviour of atomic systems. In the process of re-interpretation and reformulation, the correspondence principle had played a much more crucial role than the principle of using only observable quantities in the theoretical description. Indeed, the latter principle made its way into Heisenberg's formalism *a posteriori*. That is, Heisenberg did not establish his new quantum-theoretical scheme on Ernst Mach's positivistic views; he had heard about them before from his friends and collaborators, but they had not guided his work directly and openly.

Altogether, it would be misleading to assume that the young Heisenberg concerned himself much with philosophy. True, as a schoolboy he had tried to read Immanuel Kant's *Critique of Pure Reason*, but he had not gone very far; again, in high school, he had succeeded in reading Plato's *Timaeus* for his course in Greek, and Plato's description of the physical world had left some impression on him. Yet, no impact of philosophical views, whatever their origin, could be found in Heisenberg's scientific work until summer 1925. Pauli had even complained to Bohr about the absence of any philosophical thinking in the conversations and publications of Heisenberg. Of course, Pauli did not have in mind that Heisenberg should apply the ideas of any particular philosopher; only that he should organize his presentations more systematically, logically and consistent with the fundamental assumptions. Following his visit to Copenhagen in spring 1924, Heisenberg had assumed a more philosophical outlook, at least in Pauli's opinion, but he had not really become philosophical. He felt the lack of philosophical thinking in himself, and wrote to Pauli as late as November 1925: 'To begin with, many thanks for your philosophical letter. It helped me much to hear your clear opinion about all these difficult questions. Unfortunately, my own private philosophy is far from being so clear; rather it is a mixture of all possible moral and aesthetic calculational rules, through which I do not often find my way anymore' (Heisenberg to Pauli, 24 November 1925). In spite of all the instruction which he had received from Niels Bohr, Heisenberg was aware that his work was not guided by a clear philosophy.[295]

Wherever else the discussions about philosophical influences may lead, it is evident that philosophical guidelines in physics are important only to the extent that they help in discovering new relationships about nature. For instance, prior to Heisenberg's quantum-theoretical reformulation of classical mechanical vari-

[295] In later years Heisenberg maintained that the philosopher who influenced him most was Plato. He had especially read Plato's *Timaeus*, which discussed the organization of the physical world. In his *Gifford Lectures*, delivered in winter 1955–1956 at St. Andrews University, Scotland, Heisenberg summarized Plato's theory of matter in the following words:

> Plato . . . constructs the regular solids from two basic triangles, which are put together to form the surface of solids . . . But the fundamental triangles cannot be considered as matter, since they have no extension in space. It is only when the triangles are put together to form a regular solid that a unit of matter is created. The smallest parts of matter are not the fundamental Beings, as in the philosophy of Democritus, but are mathematical forms. Here it is quite evident that the form is more important than the substance of which it is the form. (Heisenberg, 1958a, pp. 65–66)

It might seem that the process of abandoning the 'substantial' Bohr orbits and replacing them by 'formal' Fourier series could be interpreted as an application of Plato's philosophy. However, at the time that Heisenberg made this replacement, he thought of Bohr's or Born's procedure and not Plato's. In the 1950s, the older Heisenberg was more directly influenced by Plato's theory of matter, especially when he worked on his unified theory of matter in which the spinor field represented not the substance, like protons and electrons, but what the Greek philosopher called the 'form.'

ables, the principle of observability of physical quantities had been put to just such a practical use by Born in his paper '*Über Quantenmechanik*' ('On Quantum Mechanics') where he derived, among other things, Kramers' dispersion formula (Born, 1924b). While the classical formula (Born, 1924b, p. 385, Eq. (24)) contained the action variables and other reminders of mechanical orbits, Born showed that in its 'quantum mechanical' translation (Born, 1924b, p. 390, Eq. (34)), in which differential quotients had been replaced by the corresponding difference expressions, only observable quantities—namely, the frequencies and intensities of the emission and absorption lines—entered. Following this work Born looked for further applications of his replacement scheme. Finally, in early June 1925, at about the time when Heisenberg was ready to leave Göttingen for Helgoland to seek a cure for his hay fever, Born and Jordan completed a paper in which they sought to describe the influence of aperiodic fields on atoms (Born and Jordan, 1925a). In this paper they explicitly stated the principle of observability enunciated by Mach and Einstein, for they said: 'A fundamental axiom of large range and fruitfulness states that the true laws of nature involve only such quantities as can be observed and determined in principle' (Born and Jordan, 1925a, p. 493). At the end of this sentence they added the footnote: 'Thus relativity came about by the fact that Einstein recognized the impossibility in principle, to determine the absolute velocities of two events occurring at two different positions' (Born and Jordan, 1925a, p. 493, footnote 1). Born and Jordan then gave certain applications of the observability principle. They considered, in particular, the absorption and dispersion of radiation by atoms and found that the classical formulae for the absorbed and emitted energies did not contain the phases of the electronic motions. Since the latter were unobservable, in principle, it made sense to translate these classical formulae into quantum-theoretical ones. The authors thus decided 'to regard only those formulae of classical optics as final, i.e., as being valid in quantum theory also in the limit of large quantum numbers, that are independent of the motions of the electrons; these being the ones which are left after averaging over the phases [of electrons in different atoms]' (Born and Jordan, 1925a, p. 494).[296] By following this rule they were able to verify Einstein's derivation of Planck's radiation law (Einstein, 1916d), without invoking the concept of light-quanta.

All this had taken place before Heisenberg himself ever invoked the principle of the observability of physical quantities in the actual construction of a theory. One cannot find any reference to it in any of his earlier papers or scientific correspondence. Naturally, one should not exclude the possibility that he had discussed this principle in private conversations, both in Göttingen and Copenhagen. However, it is remarkable that he first mentioned it in a letter to Pauli as late as 24 June 1925, several days after his return from Helgoland to

[296]The fact that in the dispersion theory of Kramers and Heisenberg the phases had played a role, does not contradict the above statement of Born and Jordan, because in the treatment of incoherent scattering one had to deal with the amplitudes of spontaneous emission and absorption of the *same* electron.

Göttingen. Then he wrote: 'I have almost no desire to write about my own work, because to me everything is still unclear and I just vaguely anticipate how things will turn out, but perhaps the basic ideas are still correct. The fundamental axiom is: In calculating any quantities, like energy, frequency, etc., only relations between those quantities should occur, which can be controlled [i.e., determined experimentally] in principle' (Heisenberg to Pauli, 24 June 1925).

This statement by itself could not be considered as revolutionary at all, for it had already been expounded earlier several times. However, the date of the letter containing the statement and the person to whom he addressed it were of interest. That Heisenberg mentioned the '*Grundsatz*' so explicitly to Pauli, who had stated it in 1919 and recognized its importance, obviously meant that he had not discussed the relevance of the principle of observability of physical quantities as the philosophical basis of the new atomic mechanics when he met Pauli on his return from Helgoland to Göttingen. One may even conclude that he had not yet applied the principle consciously in formulating the quantum-theoretical scheme for the anharmonic oscillator. However, after returning to the Göttingen circle, he participated again in discussions with Born and Jordan, who spoke with enthusiasm about the philosophical principle which they had just employed in their treatment of dispersion phenomena. It was only then, i.e., only after 21 June 1925, when he wrote his first letter to Pauli after the trip to Helgoland, that Heisenberg became fully aware of the usefulness of the principle for his current work. He was not interested in philosophical questions *per se*, but he was very interested in the practical question of justifying the steps, which he had introduced earlier in order to formulate a more consistent mechanics of atomic systems. Indeed, he saw now that the principle concerning the use of observable quantities alone in the theoretical description agreed perfectly with what he had done. When he had represented, for example, the position coordinates of electrons in atomic systems by the Fourier series, and reformulated the latter as patterns of transition amplitudes, each associated with a periodic function containing the correct transition frequency, he had done just the right thing: he had removed a quantity representing the nonobservable motion of the electron, introducing instead a description in terms of observable quantities, i.e., the transition amplitudes (whose squares were the measurable transition probabilities) and the atomic radiation frequencies. Moreover, in deriving the multiplication law for the patterns of quantum-theoretical amplitudes, Heisenberg had made use of an observed fact, namely, of Walther Ritz' combination principle of spectroscopic terms. Hence he could claim rightly that his entire procedure of the quantum-theoretical reformulation scheme was more or less dictated by applying the '*Grundsatz*' which he had mentioned in the letter to Pauli. And he could argue that if quantum mechanics was to be built on this principle at all, then equations of motion such as the ones he had considered for the anharmonic oscillator should make sense, for they involved only quantities that were directly related to experimental data.

However, the philosophical principle proved to be even more valuable, as it justified the method which Heisenberg had used to solve the equation of motion of the anharmonic oscillator. As we have described in Section IV.3, the equation of motion could determine the transition amplitudes only up to a constant; this constant then became fixed with the help of a sum-rule, which Heisenberg obtained in Helgoland by an ingenious reasoning (see Section IV.4). He first differentiated the Bohr–Sommerfeld quantum condition for the phase-space integral with respect to the action variable, and then rewrote this classical expression according to Born's discretization prescription (51). This then led to the quantum-theoretical equation (90) which, together with the assumption of a lowest state of the oscillator system, enabled Heisenberg to calculate the desired constant. Now he motivated this procedure *a posteriori* from the point of view of the principle that all quantities in equations of atomic theory should be observable. The phase-space integral itself, he argued, did not represent an observable quantity, hence one was not permitted to use the old quantum condition anymore. However, differences of the phase-space integral taken for neighbouring states still made sense, for they were related to empirical data. The reason was that the energy terms of the oscillator depended uniquely on the action variable (whose values were given by those of the phase-space integral), and differences of terms could be observed in spectroscopy. In classical theory, the differences of the phase-space integrals associated with states denoted by the quantum numbers n and $n-1$ were given by the equation (see Eq. (88))

$$h = \frac{\pi m}{2} \sum_{\tau=-\infty}^{+\infty} \lambda^{2|\tau|-2}\left[a_\tau^2(n)\tau^2\omega_0 - a_\tau^2(n-1)\tau^2\omega_0\right], \tag{114}$$

where the $a_\tau(n)$ and $a_\tau(n-1)$ denote, respectively, the Fourier amplitudes describing the quantum states n and $n-1$ of the anharmonic oscillator. Heisenberg realized that the observability principle demanded that one had to reformulate Eq. (114) in quantum theory rather than the equation representing the quantum condition, i.e.,

$$nh = \frac{\pi m}{2} \sum_{\tau=-\infty}^{+\infty} \lambda^{2|\tau|-2}\left[a_\tau^2(n)\tau^2\omega_0\right]. \tag{115}$$

However, for convenience, he did not reformulate Eq. (114) itself, but rewrote the term in the square bracket on the right-hand side of it as $\partial[a_\tau^2(n)\tau^2\omega_0]/\partial n$—which was permitted in the limit of high quantum numbers n—and then used Born's discretization procedure to obtain Eq. (90).

With these applications of the old observability principle stemming from Ernst Mach and Albert Einstein, Heisenberg reached a higher level in his endeavours to formulate quantum mechanics. Thus, he was finally able to understand the various methods that had been used earlier to obtain quantum-theoretical equations from classical ones. During the period of guesswork one had applied such methods as Born's discretization prescription or the introduction of emission and absorption probabilities (instead of squares of Fourier amplitudes) without seeking a logical connection between them. The philosophical principle of admitting only observable quantities in atomic theory pointed to the one single deep reason behind all methods of reformulation; indeed, it represented the *unifying principle*, on which all quantum-theoretical reformulation was based. Thus in June 1925, Heisenberg really terminated the earlier period of clever guesswork by providing a clear prescription of how to proceed in order to arrive at a consistent quantum theory of atomic systems. This prescription required simply to drop the most questionable step in the dispersion-theoretic approach, i.e., the setting up of classical equations containing quantities like the frequency of revolution in electron orbits. Heisenberg proposed to skip that step completely and to concern oneself only with equations between observable quantities. It may be argued that his procedure did not practically change the description of atomic systems because one still had to start somehow from the classical description; and that Heisenberg had simply postulated the existence of an *a priori* description of atoms in terms of observable quantities. However, this argument does not do justice to the situation. The important point was that the postulate always to seek relations between observable quantities in atomic theory meant a *conceptual break with the past*. If one took the postulate seriously, one did not lose time anymore in investigating nonsensical equations; one could concentrate right away on finding the true equations governing atomic mechanics.

The conceptual breakthrough leading to quantum mechanics also explains Heisenberg's reaction, mentioned earlier, about Pauli's argument concerning 'electrons falling into the nucleus of an atom.' This argument had been raised first by Niels Bohr in connection with half-integral quantum numbers. Bohr had claimed that for nuclei of atomic number $Z > 67$ an electron with angular momentum $k = \frac{1}{2}h/2\pi$ would fall into the nucleus (Bohr, 1923c, pp. 265–266, footnote 1). Since such orbits were forbidden in the Bohr–Sommerfeld theory of atomic structure, Bohr denied the use of half-integral quantum numbers. But Bohr's verdict had not prevented the appearance of forbidden orbits in atomic systems. For example, Oskar Klein and Wilhelm Lenz had shown that an allowed electron orbit (which did not pass through the nucleus) in a hydrogen atom could assume a 'dangerous' pendulum-type motion, hitting the nucleus, when crossed electric and magnetic fields were switched on adiabatically (Klein, 1924a, b; Lenz, 1924). Pauli, who was very familiar with this kind of difficulty in the old atomic theory, thought that the new quantum mechanics should remove it and brought up the question again in summer 1925.[297] Heisenberg, instead of giving a detailed answer, just pointed out that the kinematics of quantum theory was so different from the one of classical theory that it made no sense to talk about 'the falling of an electron into the nucleus.' He may have been surprised that Pauli, who had been so keen to abolish electron orbits a few months earlier, now raised the question at all. For Heisenberg, the reformulation of atomic

[297] During winter 1922–1923, Pauli had helped Bohr to prepare his memoir on atomic structure containing the argument against half-integral quantum numbers (Bohr, 1923c); he had also been in Hamburg when Lenz treated the crossed-fields problem.

theory in terms of observable quantities removed the difficulty; it allowed him to consign to oblivion all the known troubles about electron orbits.[298] Pauli, on the other hand, still had to become fully acquainted with the consequences of the new situation in atomic mechanics.

In spite of the conceptual progress achieved by using the observability principle, the practical task of calculating, say, the energy states of a given atom remained extremely difficult. The only quantum-mechanical case, which Heisenberg had solved in an approximation, was that of the anharmonic oscillator in one space dimension. From it he could conclude how to proceed in more realistic cases; namely, by starting from the classical equation of motion in which the kinematical variables were reformulated as patterns of transition amplitudes. But even for the simplest case, that of the hydrogen atom, the formulation of the quantum-theoretical equation of motion and its integration appeared to be extremely complicated. Therefore, Heisenberg did not immediately attack the hydrogen problem, though his fundamental investigations had initially begun by considering it. He confined himself to writing up the general results on the structure of quantum mechanics, together with the specific results on the anharmonic oscillator and a few preliminary considerations on the rotating electron; he wanted to submit them for publication before the end of the summer semester.

V.3 Quantum-Theoretical Kinematics and Mechanics

On 9 July 1925 Heisenberg sent to Pauli the manuscript of a paper containing the results which he had obtained in the previous two months on the reformulation of atomic theory. In the covering letter he stated his complete rejection of the Bohr–Sommerfeld theory and continued:

> Therefore I venture to send you just the manuscript of my paper [without much additional commentary], for I believe that it contains—at least in the critical, i.e., negative, part,—[some] real physics. However, I have a very bad conscience, for I

[298] In summer 1925 Heisenberg was not alone in his insistence on getting rid of the electron orbits. He could count on some support both from Copenhagen and Göttingen. In Copenhagen, for example, Niels Bohr had already declared in his talk 'On the Law of Conservation of Energy' before the Royal Danish Academy on 20 February 1925 that 'the attempts to develop an atomistic interpretation of directly observable phenomena have led us to recognize the necessity of revising the ideas hitherto underlying the description of natural phenomena' (Report in *Nature* **116** (1925), p. 262). In Göttingen Max Born had stated in his *Vorlesungen über Atommechanik*, as quoted above, that electron orbits themselves could not be observed; and, in a footnote, he had added: 'Measurements of atomic radii and similar things do not yield a higher approximation to reality than, say, the [rough] agreement between orbital frequencies and [emitted] light frequencies' (Born, 1925, p. 114, footnote).

It was another question how far other people wanted to go in abolishing orbits. Born was probably prepared to follow Heisenberg in this, as his belief in the observability principle shows. Pauli, however, still found arguments from the old theory still quite useful, especially insofar as they highlighted peculiar important difficulties.

> must ask you to return the paper to me within two or three days, the reason being that I do wish either to complete it [for publication] in the last days of my presence here [in Göttingen] or to burn it. My own opinion about what I have written ["*das Geschreibsel*," playing down what he had done], and about which I am not happy at all, is this: that I am firmly convinced about the negative critical part, but that the positive part is fairly formal and meagre; however, perhaps people who know more can turn it into something reasonable. Therefore, please, read especially the introduction. [And he added in closing:] Now I request you again for a sharp criticism and to return the paper soon. (Heisenberg to Pauli, 9 July 1925)

Pauli read the manuscript carefully and critically; he was satisfied and returned it promptly to Heisenberg, who was thus enabled to complete his paper in the middle of July 1925 and hand it over to his professor. Max Born went over it, approved it, and submitted it to *Zeitschrift für Physik* by the end of the same month.

Heisenberg described the content of the paper, entitled '*Über die quantentheoretische Umdeutung kinematischer und mechanischer Beziehungen*' ('On the Quantum-Theoretical Re-interpretation of Kinematical and Mechanical Relations'), briefly as follows: 'In the present paper [we] seek to obtain the foundations of a quantum-theoretical mechanics based exclusively upon relationships between quantities which are observable in principle' (Heisenberg, 1925c, p. 879). As he confessed to Pauli, Heisenberg was not happy with his write-up, for he considered the 'positive' part of the paper as being 'poor.' What he meant with this qualification becomes clear if one compares the results of the new paper with those contained in earlier publications which had given him great satisfaction. In these 'successful' papers, say his very first publication on the doublet and triplet spectra and their associated Zeeman effects (Heisenberg, 1922a) or his doctoral thesis on the origin of turbulence in two-dimensional flows (Heisenberg, 1924f), he had been able to formulate and completely solve a given problem—or so he believed at the time he submitted the papers. Now, instead of a complete formulation of a consistent atomic theory, the future quantum mechanics, Heisenberg felt that he was able to present only a step in the direction of establishing its foundations, but by no means the final solution. He found that wherever he had tried to go beyond the 'negative' critical argument, i.e., the prescription to avoid unobservable quantities in the theory, he had not proceeded very far. He had just been able to write down a few formal equations, which were difficult to handle mathematically and even more difficult to interpret physically. He feared that the 'positive' or constructive part of his work might make as little sense as earlier formal schemes.

Besides these deep doubts about the validity of the reformulation scheme, Heisenberg had another reason to be dissatisfied with the product of his recent endeavours: the material assembled in the paper appeared to be very inhomogeneous. It contained a variety of topics: general considerations about the description of emission of radiation from atoms; the quantum-theoretical reformulation of classical Fourier series for the position coordinates and other variables describing atomic systems; the integration of the equation of motion of the anharmonic oscillator system in one space dimension; and a preliminary discussion of the rotating electron and certain consequences from it for the intensities of multiplet spectra. Heisenberg was perfectly aware that, in spite of the fact that he had enunciated a guiding principle which allowed one to look at these diverse topics from a unified point of view, he had succeeded only superficially in giving a logical presentation of his ideas. The diverse nature of the various topics may already be seen from a certain lack of unity in the notation he employed. For example, in Section 1 of the paper, Heisenberg used the Gothic script symbol $\mathfrak{A}$ to denote the complex vector describing the absorption and emission amplitudes of radiation, the same symbol as Kramers and Heisenberg had employed earlier in dispersion theory. Heisenberg retained the Gothic script letters for the transition amplitudes throughout Section 1, and denoted the quantum-theoretical amplitudes and their products by $\mathfrak{A}(n, n-\alpha)$, $\mathfrak{B}(n, n-\beta)$ and $\mathfrak{C}(n, n-\gamma)$. However, in Section 2, dealing with the equation of motion, he changed the notation. Now, for the classical Fourier coefficients of a system having one degree of freedom and the analogous quantum-theoretical amplitudes, he used the notation, $a_\alpha(n)$ and $a(n, n-\alpha)$, respectively. He had employed this notation previously when dealing with the anharmonic oscillator, as can be seen from his letter to Kronig, dated 5 June 1925. Heisenberg stuck to this notation in the rest of the paper; for example, in Section 2, he wrote the amplitudes of Kramers' dispersion formula consistently as $a(n, n \pm \alpha)$ (see Heisenberg, 1925c, p. 887). This lack of unity in the symbols denoting the same physical quantities in one paper seems bewildering on first inspection. However, if one keeps in mind the different origins of the various topics presented and carefully follows the physical content, the reading of Heisenberg's paper does not meet with great difficulty.

Heisenberg organized his paper into four parts, an introduction and three sections. In the introduction he summarized his critical analysis of the failure of the Bohr–Sommerfeld atomic theory and stated the principle of using only observable quantities as the philosophy guiding the formulation of quantum mechanics. He then developed his main theoretical step: the re-interpretation of the Fourier series representing the classical kinematical coordinates of atomic systems as quantum-theoretical quantities (Section 1). Then, in Section 2, followed the central core of the new mechanics: the use of the classical equation of motion for a given atomic system as a formal equation, involving the reformulated kinematic quantities; and the derivation of the quantum-theoretical relations which determined the arbitrary constants entering the solution of the equation of motion. The final section, 3, dealt with the application of the scheme to a few examples: the one-dimensional anharmonic oscillator and two cases of the rotating electron. These examples served to support Heisenberg's formal scheme; for instance, the preliminary results concerning the rotator system agreed with the generally accepted intensity rules for the multiplet lines of many-electron atoms and their Zeeman components. We shall first discuss the contents of the introduction and Sections 1 and 2, by which the general structure of quantum mechanics was established. The applications of the scheme and its confirmation will be reviewed in the following section.

In his letter of 9 July 1925 Heisenberg directed Pauli's attention in particular to the introduction of his paper. Compared to the introductory remarks in his earlier papers, this introducton was long and was intended to play a more important role. In Heisenberg's opinion, it constituted the most valuable part of the entire work; he thought that if anything of his attempts might remain correct in the future, it was the introduction. We shall therefore analyze it in detail.

In the first part Heisenberg dealt with the reasons for the failure of the previous atomic theory. He said:

> It is well known that the formal rules which are used in quantum theory for calculating observable quantities (such as the energy of the hydrogen atom) may be seriously criticized on the grounds that they contain, as an essential element, relationships between quantities that are apparently unobservable in principle (such as position, period of revolution of the electron, etc.); that these rules lack an evident physical foundation, unless one still retains the hope that the hitherto unobservable quantities may perhaps later become accessible to experimental determination. (Heisenberg, 1925c, p. 879)

Heisenberg thus addressed himself to what he regarded was the weakest point of the Bohr–Sommerfeld theory of atomic structure, i.e., the fact that its formulation involved apparently unobservable quantities. However, he did not forget to mention the possibility that quantities that were hitherto unobservable might become observable in future with improved techniques.[299] Still, Heisenberg gave this possibility a low probability, for he remarked:

> This hope might be regarded as justified if the above-mentioned rules were internally consistent and applicable to a clearly defined range of quantum-theoretical problems. However, experience shows that only the hydrogen atom and its Stark effect satisfy those formal rules of quantum theory, and that fundamental difficulties arise already in the problem of "crossed fields" (hydrogen atom in electric and magnetic fields of differing directions); that the reaction of atoms to periodically varying fields certainly cannot be described by the rules in question; and finally, an extension of the quantum rules to the treatment of atoms having several electrons has proved to be impossible. (Heisenberg, 1925c, p. 879)

This analysis of the failure of the Bohr–Sommerfeld theory of atomic structure had clearly emerged from the discussions of Heisenberg and Pauli. Heisenberg especially took into account Pauli's emphasis on the difficulty concerning the crossed-fields. At the same time he publicly admitted the inconsistencies existing

[299] The difficulty of deciding once and for all about which quantities are observable and which are not observable had been pointed out before by Hermann Weyl to Wolfgang Pauli in connection with the question whether electromagnetic fields within electrons made sense or not (see Weyl to Pauli, 9 December 1919).

in the description of many-electron atoms, such as the core model, which he had cherished over many years.

In the second part of the introduction, Heisenberg turned to deriving consequences from the above analysis.[300] 'It has become the practice,' he said, 'to denote this failure of the quantum-theoretical rules, which are essentially characterized by the application of classical mechanics, as deviations from classical mechanics' (Heisenberg, 1925c, p. 879). The question concerning the validity of classical mechanics in atomic physics had been a major problem since the early days of quantum theory, to which people had given different answers. For a long time, especially after Niels Bohr's successful treatment of the hydrogen atom, it was believed that classical mechanics would also provide the necessary foundation for atomic mechanics; all one had to do was to add to it the quantum conditions for multiply periodic systems. Only when the failure of the Bohr–Sommerfeld theory to account for the helium atom and similar several-body atomic systems had become definite in 1923, did people start to look for the deviations from classical mechanics that might remove the failures. Since at the same time other difficulties had also arisen in atomic and radiation theory—e.g., in connection with the observed Compton and Stern–Gerlach effects—it had not been possible to decide how much the classical mechanics and electrodynamics had to be changed in order to arrive at a more satisfactory description of the quantum phenomena.[301] The renewed critical analysis carried out immediately afterwards had revealed that both classical theories contributed their share to the difficulties of atomic theory. In an attempt to rescue the wave description of radiation (for the purpose of extending the application of the correspondence principle) Bohr, Kramers and Slater had been forced to adopt rather revolutionary changes in the mechanical description of atomic processes (Bohr, Kramers and Slater, 1924). However, the outcome of the Bothe–Geiger experiment concerning the nature of Compton scattering in April 1925 had restored the validity of energy and momentum conservation in individual processes involving atoms and radiation, which Bohr *et al.* had sought to deny in their radiation theory. Hence, in summer 1925 it was certain that the old atomic theory could not be saved by introducing new, unconventional coupling between electrons, or electrons and radiation. Heisenberg therefore remarked:

> This connotation [i.e., a "deviation from classical mechanics" to which the failure of atomic theory was formerly attributed] can, however, hardly be admitted as a reasonable one, if one considers that already the (generally valid) Einstein–Bohr frequency condition represents such a complete denial of classical mechanics or—more accurately from the point of view of the wave theory—of the kinematics underlying this mechanics, that even in the simplest quantum-theoretical problems it is impossible to think of the validity of classical mechanics. (Heisenberg, 1925c, pp. 879–880)

[300] In Heisenberg's paper the introduction formed one paragraph. B. L. van der Waerden split it in his translation, beginning the second paragraph at this point (van der Waerden, 1967, p. 261).

[301] See, e.g., Niels Bohr's statements in his review article on atomic structure (Bohr, 1923c, p. 228).

Thus, he interpreted the failure of the Bohr–Kramers–Slater theory in a very specific way: not to be blamed primarily was the electromagnetic theory of radiation, but rather the description of electrons in atoms responsible for the emission, absorption and scattering of radiation. By retaining the classical electromagnetic theory of radiation, Heisenberg secured the opportunity of still being able to apply correspondence arguments, for he regarded this possibility as being helpful in constructing the new atomic theory. On the other hand, he joined Pauli in the opinion, expressed by the latter several months previously, of attributing the failures of the earlier atomic theory to the fact that it involved wrong kinematic concepts. Again, as he had done frequently before, Heisenberg made a compromise between apparently irreconcilable points of view: between Bohr's correspondence arguments and Pauli's demand for the radical revision of classical mechanics.[302]

After pointing out the absoute necessity of developing new kinematic concepts in atomic theory, Heisenberg concluded his introduction by expounding the guiding principle of his reformulation scheme. In particular, he said: 'In this situation it seems more appropriate to discard all hope of observing the erstwhile unobservable quantities (like position, period of revolution of the electron) and simultaneously to concede that the partial agreement of the quantum rules in question is more or less fortuitous; one should try to establish a quantum-theoretical mechanics, analogous to classical mechanics, in which only the relations between observable quantities occur' (Heisenberg, 1925c, p. 880). This statement constituted Heisenberg's *credo* of atomic theory, and it was this belief in the unobservability of the electron's motion in atoms, which he thought would provide the foundation of the new theory. Still, the time would come, within about a year, when Heisenberg would have to relax this statement and admit that somehow the electron's motion in an atom must make sense. Then, however, the quantum-theoretical scheme would be firmly established in such a way that one could also relax the above claim that the position and periods of electrons in atomic systems were unobservable in principle. In July 1925 Heisenberg had just started to formulate quantum mechanics, and for this formulation the criterion of observability proved to be indispensable. Indeed, it allowed him to build his approach on the previously successful dispersion-theoretic treatment of atomic phenomena by Kramers (1924a), Born, (1924b), Kramers and Heisenberg (1925), and Born and Jordan (1925a). He mentioned these works, besides the frequency condition, as 'the most important first steps toward such a quantum-theoretical mechanics' (Heisenberg, 1925c, p. 880). Then he proceeded to develop the new quantum-mechanical relations.

Since he had declared the revision of classical kinematical relations to be the crucial step on the way towards a consistent quantum mechanics, Heisenberg

[302]Concerning Pauli's other demand, namely, the revision of the classical concept of force, Heisenberg did not yet state his opinion. It is very likely that he wanted to see first how far the change of kinematic concepts would lead before introducing still further changes in the mechanical description.

devoted the entire first section of his paper to that question.[303] He began by examining the classical expression for the radiation emitted by an accelerated electron, in which the electric and magnetic field vectors, **E** and **H**, at large distances were given by

$$\mathbf{E} = \frac{e}{r^3c^2}\left[\mathbf{r} \times (\mathbf{r} \times \dot{\mathbf{v}})\right] \tag{116a}$$

and

$$\mathbf{H} = -\frac{e}{r^2c^2}(\dot{\mathbf{v}} \times \mathbf{r}). \tag{116b}$$

Here $-e$ and r denote the charge of the electron and its distance from the point of observation; the vectors $\mathbf{r}$, $\mathbf{v}$ and $\dot{\mathbf{v}}$ represent the radius vector, velocity and acceleration, respectively; c denotes the velocity of light *in vacuo*, and the cross between the vectors indicates that the vector product has to be taken. To the right-hand sides of Eqs. (116a) and (116b) corrections have to be added, which are proportional to higher powers of the inverse of distance r and may be interpreted as the 'quadrupole,' 'octopole,' and other higher-order radiations from the electron. In classical theory all electromagnetic radiation from an accelerated electron (or another charged particle) can be expressed in terms of the electron's position (vector $\mathbf{r}$) and the change of its position in time (velocity $\mathbf{v}$ and acceleration $\dot{\mathbf{v}}$) by forming their appropriate combinations, as in Eqs. (116a) and (116b). Heisenberg assumed that the classical expressions for dipole and higher-order radiations remain valid in quantum theory. In particular, he argued: 'Since in the classical theory the higher approximations can be simply calculated if the motion of the electron or its Fourier representation, respectively, is given, one would expect a similar result in the quantum theory' (Heisenberg, 1925c, pp. 880–881). However, this postulated analogy of classical and quantum theory led to the conclusion, which Heisenberg had stated without proof in the introduction, that the problem of calculating the properties of atomic radiation did not require any change of the dynamical relations themselves. 'This question has nothing to do with electrodynamics, but is rather—and this seems to be particularly important—of a purely *kinematic* nature,' he said, 'for we may pose it in the simplest way as follows: If instead of a classical quantity $x(t)$ we have a quantum-theoretical quantity, then what quantum-theoretical quantity replaces $\{x(t)\}^2$?' (Heisenberg, 1925c, p. 881).

The observation that the radiation from electrons could also be determined in quantum theory alone from the specific kinematics of the systems under investigation struck Heisenberg's attention particularly. And yet, such an observation had already been made a few decades earlier as the main point in classical electron theory, where Hendrik Lorentz had connected all electromagnetic radia-

[303] This part of the paper grew out of Heisenberg's unsuccessful attempts to guess the intensities of the hydrogen lines, which we have discussed in Section IV.2.

tion phenomena with the motion of charged particles, especially the electrons. In atomic theory, Bohr, in 1913, had first discarded this intimate connection between kinematics and electromagnetic radiation by introducing the non-mechanical frequency condition besides the usual dynamical description of the electron motion in atoms, thereby treating radiation and electron motion in atoms as completely unrelated properties. The difficulties that then arose both with the mechanical and the electrodynamical parts of atomic theory in the early 1920s reminded the physicists that the separation had not solved the problem of atomic radiation and that one had to solve both difficulties at the same time. Heisenberg now proposed to seek the solution in re-establishing the classical connection between kinematics and electrodynamic radiation, but only after reformulating the kinematical relations in a proper quantum theory. To that end he considered the classical equation giving the energy loss of an atom in the frequency mode $(\alpha\nu)$ as a function of $\mathbf{A}_\alpha$, the vector Fourier amplitude associated with the three-dimensional electron motion (where the position vector is represented by a three-dimensional Fourier series with frequencies $(\alpha\nu)$, $\alpha = 0, 1, 2, \ldots$), the equation being

$$\left(\frac{dE}{dt}\right)_{\text{cl}} = \frac{2}{3}\frac{e^2}{c^3}\left[2\pi(\nu\alpha)\right]^4 |\mathbf{A}_\alpha|^2, \tag{117}$$

where $|\mathbf{A}_\alpha|^2$ denotes the scalar product of the vector $\mathbf{A}_\alpha$ with its conjugate complex, $\bar{\mathbf{A}}_\alpha$.[304] In quantum-theoretical reformulation this equation assumed the form

$$\left(\frac{dE}{dt}\right)_{\text{qu}} = \frac{2e^2}{3c^3}\left[2\pi\nu(n, n-\alpha)\right]^4 |\mathbf{A}(n, n-\alpha)|^2, \tag{117a}$$

with the quantum-theoretical (vector) transition amplitude $\mathbf{A}(n, n-\alpha)$ and the corresponding frequency $\nu(n, n-\alpha)$. Hence the quantum-theoretical amplitudes occurring in the reformulated Fourier series (which represents the position coordinate of the electron) turned out to be identical (but for a factor $-e$, the charge of the electron) with the transition amplitude, whose square gives the intensity of a spectral line. Although Heisenberg did not mention the above consideration in his paper, he knew the result, Eq. (117a), from his previous work with Kramers on dispersion theory.[305]

After indicating how to treat the problem of atomic radiation intensities as a problem of quantum-theoretical kinematics, Heisenberg discussed his reformula-

[304] According to the classical theory, the total energy loss by radiation from an electron (with acceleration vector $\dot{\mathbf{v}}$ is given by $\frac{2}{3}\, e^2/c^3\, |\dot{\mathbf{v}}|^2$, including a factor 2 from the two modes of polarization associated with each frequency of radiation. The substitution of the Fourier series for the position vector and subsequent second time derivative then yields Eq. (117) for the energy loss in the particular frequency $(\alpha\nu)$.

[305] See Kramers and Heisenberg, 1925, p. 683, Eq. (5). The factor 2 stemming from the two modes of polarization was not included in the Kramers–Heisenberg formula.

tion prescription for the Fourier series representing electron orbits, as well as the product of such series. He first reiterated the impossibility of observing the position of an electron in an atom. The only observable quantities, which one could associate with the electrons, were the frequencies $\nu(n, n \pm \alpha)$ and the amplitudes $A(n, n \pm \alpha)$ of the radiation absorbed or emitted by the atom. In classical theory the frequencies were determined by the equation

$$\nu(\alpha) = \alpha\nu(n) = \alpha \frac{1}{h} \frac{\partial W(n)}{\partial n}, \tag{118}$$

where $W(n)$ denotes the energy of the state n and h Planck's constant, and the frequencies are integral multiples ($\alpha = 1, 2, \ldots$) of a fundamental frequency ν. This was different from quantum theory, where each frequency was given by the difference of two energy terms, i.e.,

$$\nu(n, n-\alpha) = \frac{1}{h}\left[W(n) - W(n-\alpha)\right], \tag{118a}$$

and this difference had nothing to do with the motion of the electron in space and time. Thus, while in classical theory the Fourier series, i.e., the sum of all terms $A_\alpha(n)\exp[2\pi i(\alpha\nu)t]$, represented $x(n,t)$, a function in space and time describing the electron orbit, the analogous sum in quantum theory did not admit the same interpretation. Heisenberg noticed that: 'Such a combination of the corresponding quantum-theoretical quantities does not seem to be possible without arbitrariness—because of the equal status of the quantities $n, n-\alpha$—and is therefore not meaningful.' However, he immediately proposed another interpretation by saying: 'But one may readily regard the ensemble, $A(n, n-\alpha) \times \exp[i\omega(n, n-\alpha)t]$ as a representation of the quantity $x(t)$' (Heisenberg, 1925c, p. 882). That is, he considered the two-dimensional pattern of the quantities $A(n, n-\alpha)\exp[i\omega(n, n-\alpha)t]$ for all values of n and α (with $\omega(n, n-\alpha) = 2\pi\nu(n, n-\alpha)$) as the quantum-theoretical substitute for the position coordinate of an electron bound in an atom. In a similar way he reformulated the classical expression for $\{x(t)\}^2$, the square of a coordinate. If $x(t)$ were represented by the total set of quantities $A(n, n-\alpha)\exp[2\pi\nu(n, n-\alpha)t]$, then $\{x(t)\}^2$ had to be represented by the total set of quantities $B(n, n-\beta)\exp[2\pi i\nu(n, n-\beta)t]$, where

$$\begin{aligned} &B(n, n-\beta)\exp\left[2\pi i\nu(n, n-\beta)t\right] \\ &\quad = \sum_{\alpha=-\infty}^{+\infty} A(n, n-\alpha)A(n-\alpha, n-\beta)\exp\left[\pi i\nu(n, n-\beta)t\right], \end{aligned} \tag{119}$$

because in quantum theory the frequency $\nu(n, n-\beta)$ is the sum of the combining frequencies, $\nu(n, n-\alpha)$ and $\nu(n-\alpha, n-\beta)$. After stating the product rule, Eq. (119), Heisenberg drew attention to the fact that 'the phases of the quantum-theoretical A have as great a physical significance as in classical theory' (Heisenberg, 1925c, p. 883). This conclusion did not contradict the principle of having only observable quantities, for he noted: 'Only the origin of time and therefore a phase constant, which is common to all B, is arbitrary and possesses no physical significance; however, the phases of the individual A enter into the quantity B in an essential manner' (Heisenberg, 1925c, p. 883). At this point he referred to the importance of the relative phases in the problem of incoherent scattering of light by an atom, noticed earlier by Kramers and himself (Kramers and Heisenberg, 1925, p. 706). Again, there existed a difference between classical and quantum theory. While in classical theory phase relations could be visualized as interference effects in space and time, Heisenberg noted that: 'A geometrical interpretation of such quantum-theoretical phase relations in analogy with those of classical theory scarcely seems to be possible at the moment' (Heisenberg, 1925c, p. 883).

Besides this problem of interpreting the products of quantum-theoretical patterns, Heisenberg also mentioned the essential mathematical difficulty which occurred in his reformulation scheme, namely, that the factors in quantum-theoretical products would not commute in general. Hence the reformulation of a classical product of two variables, $x(t)y(t)$, did not seem to be uniquely determined in quantum theory, except when one took the products of quantity with itself, say $\{x(t)\}^2$; in all other cases one needed further information to go over from classical to quantum expressions. For example, in order to find the quantum-theoretical product corresponding to the classical product of velocity v and acceleration $\dot{v}$—occurring in the expressions for higher multipole radiation—one might use the form $\frac{1}{2}(v\dot{v} + \dot{v}v)$ because it represented the time derivative of the expression $\frac{1}{2}v^2$ with the noncommuting v and $\dot{v}$. 'Similarly, it would always seem to be possible to find natural expressions for the quantum-theoretical average values,' Heisenberg suggested, though he added that such expressions rested on less safe grounds than the product rule, Eq. (119) (Heisenberg, 1925c, p. 884). For the moment, however, he did not wish to concern himself further with the difficulty of noncommutativity of quantum-theoretical products.

The prescription for the reformulation of the classical Fourier series representing classical variables, such as the positions of electrons in atoms and their time derivatives, exhausted what Heisenberg called the kinematical part of quantum mechanics. In Section 2 of his paper he turned to the general problem of how to calculate the kinematical quantities, i.e., the quantum-theoretical amplitudes and associated transition frequencies. He first referred to the classical solution of the problem. For a given mechanical system, this solution consists of two steps: the integration of its equation of motion and the determination of the constants in the solution for the position variables. Heisenberg proposed to treat the simple and (in quantum theory) unique example of a system described by the equation of motion

$$\ddot{x} + f(x) = 0, \tag{120}$$

where $\ddot{x}$ denotes the second time derivative of x, the position variable, and $f(x)$ is a polynomial in x. Then he remarked:

> If one seeks to construct a quantum-theoretical mechanics, which is as closely analogous to classical mechanics as possible, then it is quite natural to take over the equation of motion [(120)] into quantum theory as directly as possible. For this purpose it is only necessary—so as not to depart from the firm foundation provided by quantities that are observable in principle—to replace the quantities $\ddot{x}$ and $f(x)$ by their quantum-theoretical representations, as given in section 1. (Heisenberg, 1925c, pp. 884–885)

In quantum theory, this procedure led to a set of infinitely many equations for the (in general) infinitely many quantum-theoretical amplitudes and transition frequencies. While one could sometimes solve the classical problem without inserting the Fourier series, Heisenberg observed that: 'In quantum-theory, however, we are so far dependent on this manner of obtaining a solution of Eq. [(120)], since . . . a quantum-theoretical function cannot be defined, which is directly analogous to the function $x(n, t)$' (Heisenberg, 1925, p. 885). He realized that until a proper method for simplifying the treatment of infinitely many equations were found, only a few selected problems could be solved in quantum mechanics.

The second step in the integration procedure was the determination of the constants of motion for the system under investigation. In a problem of classical mechanics, say the astronomical two-body system like the earth and moon, these constants were fixed by the initial conditions. In the old quantum theory, the initial conditions took the special form of quantum conditions stating that J, i.e., the f phase-space integrals of a periodic system having f degrees of freedom taken over full periods of the motion, assumed values given by integral multiples of Planck's constant. Heisenberg argued that: 'Not only does such a condition fit into the mechanical calculation by really being forced, it also appears from the erstwhile point of view—i.e., in the sense of the correspondence principle—quite arbitrary. The reason is that according to the correspondence principle the J are determined as integral multiples of h just up to an additive constant' (Heisenberg, 1925c, p. 885). He therefore suggested that the old quantum conditions should be replaced by properly modified ones. For example, for a periodic system with one degree of freedom the original quantum condition was

$$J = \oint m\dot{x}^2 dt = nh, \qquad n = 1, 2, \ldots, \tag{121}$$

where m denotes the mass and $\dot{x}$ the velocity of the moving object. Heisenberg obtained an equation, modified in accordance with the correspondence principle, by differentiating Eq. (121) with respect to the quantum number n (or with respect to J) as

$$\frac{d}{dn}(nh) = \frac{d}{dn} \oint m\dot{x}^2 \, dt. \tag{122}$$

But by substituting for $x(n,t)$, the position coordinate of the system in the state n, the Fourier series with Fourier coefficients $A_\alpha(n)$ and harmonic frequencies $\alpha\omega(n)/2\pi$, the modified quantum condition in classical theory assumed the form

$$h = 2\pi m \sum_{\alpha=-\infty}^{+\infty} \alpha \frac{d}{dn}\left[\alpha\omega(n)|A_\alpha(n)|^2\right]. \tag{122a}$$

This equation could be reformulated because it involved only the quantities $\alpha\omega(n)$ and $A_\alpha(n)$ and a differentiation with respect to the phase integral or action variable. The quantum-theoretical equation

$$h = 4\pi m \sum_{\alpha=1}^{\infty}\left[|A(n+\alpha,n)|^2\omega(n+\alpha,n) - |A(n,n-\alpha)|^2\omega(n,n-\alpha)\right], \tag{123}$$

derived by applying the usual prescriptions, then contained only the observable transition amplitudes, $A(n,n+\alpha)$, and the observable frequencies $\nu(n+\alpha,n) = (1/2\pi)\omega(n+\alpha,n)$. Equation (123) would pass over into Eq. (90) for the anharmonic oscillator if one put $A(n+\alpha,n) = \frac{1}{2}\lambda a(n+\alpha,n)$ and $A(n,n-\alpha) = \frac{1}{2}\lambda a(n,n-\alpha)$.[306] In classical theory, Eq. (122a) would only suffice to determine the amplitudes, $A_\alpha(n)$, up to an arbitrary additive constant. However, this arbitrariness was removed in quantum theory, because Heisenberg realized that: 'This relation [i.e., Eq. (123)] suffices here to determine A in a unique manner, because the constant, which is at first undetermined in the quantities A, becomes fixed automatically by the condition that a ground state should exist from which no radiation is emitted' (Heisenberg, 1925c, p. 886). As in the example of the anharmonic oscillator, that condition, i.e., the equations

$$A(n_0, n_0 - \alpha) = 0 \qquad \text{for } \alpha = 1, 2, \ldots, \tag{124}$$

where n_0 denotes the ground or the lowest state of the system, allowed one to fix the constant occurring in the integrated solution of the equation of motion for quantum systems having one degree of freedom.[307] The procedure involving Eqs. (123) and (124) thus removed the previous ambiguity in atomic theory, which had bothered the theoreticans for many years, for as Heisenberg concluded: 'The question whether to use half-integral or integral quantization does not therefore occur in a quantum-theoretical mechanics, which uses only the relations between observable quantities' (Heisenberg, 1925c, p. 886). Evidently, he was very satis-

[306] Note that Heisenberg's Eq. (16) (1925c, p. 886) contains an error, for he wrote the first terms of the sum as $|A(n,n+\alpha)|^2\omega(n,n+\alpha)$ rather than as $|A(n+\alpha,n)|^2\omega(n+\alpha,n)$. Since $|A(n,n+\alpha)|^2$ is identical with $|A(n+\alpha,n)|^2$ and $\omega(n,n+\alpha) = -\omega(n+\alpha,n)$, Heisenberg's terms differed from ours by a minus sign. However, this error was corrected in the applications. (See, e.g., Heisenberg, 1925c, the equation on p. 888 for the anharmonic oscillator or the equation on p. 891 for the rigid rotator.)

[307] We have changed Heisenberg's notation of the Fourier coefficients and the corresponding quantum-theoretical amplitudes from small to capital letters, i.e., we have used $A(n, n \pm \alpha)$ instead of Heisenberg's $a(n, n \pm \alpha)$. We have thus retained consistently the notation of capital letters for the transition amplitudes. In Section 2 of Heisenberg's paper the variables and their Fourier representations are scalars because he is dealing with the oscillator having one degree of freedom only.

fied with this result because it enabled him to reinvestigate the cases of multiply periodic systems, where half-integral quantum numbers had occurred previously. Though he restricted himself in the paper to systems having only one degree of freedom, he expected no fundamental difficulties in generalizing the procedure to systems having several degrees of freedom. Obviously one would then have to deal with several quantum conditions of the form of Eq. (123), and, in Eqs. (124), n_0 would denote the set of values assumed by the quantum numbers of the system in the lowest state, and n the set of values in any higher state.

Though Heisenberg did not concern himself with systems of several degrees of freedom, he did mention another generalization of his quantum-mechanical scheme from the very beginning. In classical theory one could treat not only periodic but also nonperiodic motions with the help of Fourier methods. The difference in the mathematical formulation consisted in the fact that $x(n,t)$, the position coordinate of an aperiodic system, had to be expressed by a Fourier integral rather than a Fourier series; say, in the case of a one-dimensional system (which, of course, has one degree of freedom) by

$$x(n,t) = \int_{-\infty}^{+\infty} d\alpha\, A_\alpha(n) \exp\left[i\alpha\omega(n)t \right]. \tag{125}$$

Some theoreticians had argued for many years that one should also be able to apply quantum theory to aperiodic motions. Just recently Max Born and Pascual Jordan had extended the correspondence approach to situations that occurred when free electrons interacted with atoms (Born and Jordan, 1925a). All one had to do in those cases, they had claimed, was again to reformulate the classical amplitudes, $A_\alpha(n)$, and frequencies, $(1/2\pi)\alpha\omega(n)$ (with $\alpha = \pm 1, \pm 2$, etc.), as quantum-theoretical transition amplitudes, $A(n, n-\alpha)$, and transition frequencies, $\nu(n, n-\alpha)$. Heisenberg was familiar with the results of his Göttingen colleagues, and he accepted them as being correct in quantum theory. He even incorporated them into his new scheme and claimed that it (the scheme) remained valid also for systems whose motion was aperiodic.

The inclusion of aperiodic motions in quantum mechanics presented no formal difficulties. As for the kinematic description, one had just to assume that the position coordinates were given by the set of amplitudes $A(n, n-\alpha)$, with α taking continuous values in the interval from $-\infty$ to $+\infty$. Of course, one had to rewrite Eq. (119), expressing the multiplication rule, and Eq. (123), giving the quantum condition; that is, one had to replace the α-sums on the right-hand sides of these equations by α-integrals. Heisenberg noticed a further complication in the description of quantum systems, such as the hydrogen atom, which exhibited both discrete stationary states as well as continuous states. In such cases four types of quantum-theoretical amplitudes would occur: transition amplitudes from a discrete stationary state to another discrete stationary state; transition amplitudes from a discrete to a continuous state; transition amplitudes from a continuous state to a discrete state; and transition amplitudes between two continuous states. In the Bohr–Sommerfeld theory an atom in a discrete stationary state represented a periodic system. In quantum mechanics, the kinematic description contained both periodic and aperiodic parts, the former characterized by amplitudes $A(n, n-\alpha)$, with α assuming discrete values, ± 1, $\pm 2, \ldots,$ and the latter denoted by amplitudes $A(n, n-\alpha')$, with α' assuming continuous values. Moreover, in the expressions for the products of physical quantities, the summation and integration terms would show up simultaneously; for instance,

$$\begin{aligned} B(n, n-\beta)\exp[i\omega(n, n-\beta)t] \\ = \sum_{\alpha=-\infty}^{+\infty} A(n, n-\alpha)A(n-\alpha, n-\beta)\exp[i\omega(n, n-\beta)t] \qquad (126) \\ + \int_{-\infty}^{+\infty} d\alpha' A(n, n-\alpha')A(n-\alpha', n-\beta)\exp[i\omega(n, n-\beta)t]. \end{aligned}$$

Heisenberg emphasized that for quantum systems having discrete and continuous states, the right-hand side of Eq. (123), the quantum condition, would also have to be replaced by a sum plus an integral. Without going into details, he remarked: 'In quantum mechanics one cannot separate in general "periodic" from "aperiodic" motions' (Heisenberg, 1925c, p. 886). This conclusion implied, of course, a drastic change in atomic theory. In the Bohr–Sommerfeld theory of atomic structure, only periodic systems could be treated, otherwise the quantum condition, Eq. (121), did not make sense. Heisenberg's new quantum-mechanical scheme made it possible to treat periodic and aperiodic systems on the same level. Hence only in quantum mechanics could the hydrogen atom be described completely. As a result of the general and unified treatment of systems in quantum mechanics, aperiodic systems had also to satisfy a quantum condition. Indeed, there seemed to exist no evident reason why quantum mechanics should not possess as wide a range of applicability as classical mechanics.

Like Einstein's work 'On the Electrodynamics of Moving Bodies' exactly twenty years earlier (Einstein, 1905d), Heisenberg's reformulation of kinematical and mechanical relations marked the birth of a new mechanical theory. If one compares the two fundamental papers, Einstein's and Heisenberg's, one discovers certain aspects which they had in common. The most important aspect was that both Einstein and Heisenberg had built their schemes on a few basic principles. In the case of Heisenberg this was not surprising at all, for he—together with Bohr, Born and Jordan, with whom he had close scientific relations—had for quite some time been deeply conscious of Einstein's successful approach to the mechanics and electrodynamics of fast-moving bodies. What Einstein had done was to have resolved the clash between mechanics and electrodynamics by redefining the kinematic concepts of mechanics and abandoning 'superfluous' concepts such as the 'luminiferous aether.' The latter had been employed in seeking to avoid the difficulties arising in the kinematical aspects of Newtonian mechanics due to the occurrence of a limiting speed, the velocity of light *in vacuo*, in Maxwell's electrodynamics. Einstein, in Part I of his memoir, entitled the 'Kinematical Part,' simply redefined the concepts of simultaneity (having to do with the measurement of time differences), of length and time intervals, as well as of space and time coordinates in general, and deduced the theorem of the addition of velocities for systems which move with great uniform velocities with respect to one another. Only after deriving 'the requisite laws of the theory of kinematics' (Einstein, 1905d, p. 907; English translation, p. 51), had Einstein proceeded to apply them to electrodynamics. The electrodynamics of moving bodies, which had been the central problem in the earlier treatments of Lorentz and Poincaré, had emerged in Einstein's special relativity theory as a natural consequence of the changed kinematic concepts.

Heisenberg, consciously or unconsciously, followed Einstein's footsteps in summer 1925. He declared that it was the kinematics in atomic theory, which had to be redefined. He also gave up 'superfluous' concepts like electron orbits and other unobservable quantities, which had been employed before in order to allow a simultaneous application of classical mechanics and the quantum postulates. Again, as in Einstein's case, he resolved the real difficulties by introducing new kinematic concepts, by redefining, for example, the position coordinate of an electron in an atom. With these redefined concepts Heisenberg arrived at a conclusion, which was at least as puzzling as the relativistic theorem of the addition of velocities: the noncommutativity of the products of quantum-theoretical variables. The new kinematics, then, had to be introduced in the usual mechanical laws and a consistent quantum-mechanical scheme was obtained.

The close parallel between Einstein's relativity kinematics and Heisenberg's quantum kinematics becomes more evident if one recalls that many physicists of the older generation believed that the specific interactions existing within atoms and molecules, i.e., the dynamics of the system, were somehow responsible for the difficulties in quantum theory.[308] Einstein, in particular, tried hard during the 1920s, and even afterwards, to derive the specific quantum-theoretical behaviour (i.e., the deviations from the classical mechanical behaviour) of atomic particles from a new dynamics of nonlinear field equations (see, for instance, Einstein, 1923d). Even Pauli was influenced by this opinion, which demanded a new dynamics, completely different from the classical one. Thus, he emphasized the need for 'a physical analysis of the concepts of motion *and of force* according to quantum theory' (Pauli to Bohr, 31 December 1924, our italics). At the same time Pauli had expressed his doubts that a rapid progress in solving the quantum problem could be made. Even after Heisenberg's discovery he remained uncer-

[308]The opinion that the specific quantum effects were caused by the particular dynamics of the system was expressed already in the early development of quantum theory (see, e.g., the mechanism which J. H. Jeans devised at the first Solvay Conference in order to explain the law of blackbody radiation, in Mehra, 1975c, Chapter Two, §4.) A similar difference of opinion existed in the case of electrodynamics of moving bodies. Einstein's predecessors, Hendrik Lorentz and Henri Poincaré, had tried to develop a consistent electrodynamics of moving electrons, i.e., *the dynamics* of the electron. (This goal even shines through the title of Poincaré's memoir '*Sur la Dynamique de l'Electron*' (Poincaré, 1906).) Einstein, on the other hand, put the emphasis on the *kinematic* aspects of the problem and thus created relativity theory.

tain whether the kinematic changes would suffice, or whether some new forces had also to be introduced. As late as November 1925, writing to Bohr, he spoke about the possibility of changing the dynamical laws:

> Now it is evidently the simplest *Ansatz*, as well as natural—in view of the asymptotic agreement with the classical theory in the limiting case of large quantum numbers—first to assume the classical form for the energy function $H(p,q)$. However, [later on], in view of the stress (*Zwang*) and the anomalous Zeeman effect, . . . one may think of introducing other functions $H(p,q)$ than the classical one. The problem would then arise, of course, to do this in a natural way, free from arbitrariness. (Pauli to Bohr, 17 November 1925)

It may be considered as an advantage of Heisenberg's procedure that, at the moment of realizing the new quantum-mechanical scheme, he left aside the complications brought in by the many-electron systems and restricted himself to the most elementary case, in which only the kinematics had to be changed. Eventually it would turn out to be Heisenberg's main triumph that merely the change of kinematics solved the quantum problem, just as twenty years earlier relativistic kinematics had sufficed to solve the problem of the electrodynamics of moving bodies.

Finally, we should point to the similarity of structure in the presentation of special relativity theory and quantum-theoretical kinematics. Both Einstein and Heisenberg started from a few fundamental principles: in one case the principles of relativity and the constancy of the velocity of light; in the other, the principle of using only observable quantities and the correspondence principle. And in both theories a fundamental constant was involved: in Einstein's theory, c, the velocity of light in vacuum, and in quantum mechanics, h, the quantum of action. These constants could not be derived from first principles, nor could they be calculated from the equations occurring in the schemes. But the occurrence of h determined the laws of quantum mechanics, just as c determined the laws of relativistic mechanics. And if proper limits were considered—e.g., when in quantum mechanics the quantum number n was taken to be very large and Planck's constant h as very small, or in relativistic mechanics the velocity constant c was taken to be infinite, then the new mechanical laws passed into the well-known laws of classical, Newtonian mechanics. Thus, by formulating new kinematic concepts, Heisenberg followed in the footsteps of Einstein to discover a new mechanical scheme.

V.4 Preliminary Tests and Applications of Quantum Mechanics

In embarking upon the development of a general quantum-mechanical formalism, which in many respects differed so radically from the semiclassical theory that had been applied to atomic problems since 1913, Heisenberg urgently needed some proof that he was on the right track. Such a proof could be provided if agreement were found between certain specific results obtained by the application of the new scheme and the well-established consequences of the old atomic theory, or if experiments confirmed the results directly. For this purpose, it seemed necessary to derive as many results as possible from the quantum-mechanical treatment of atomic systems. Heisenberg was aware of this necessity, but he immediately realized that calculations were rather difficult to perform with the new scheme. The impossibility of proceeding quickly with the computation of the properties of realistic atoms during the decisive weeks of June and July 1925 made Heisenberg unhappy; this was why he considered the positive aspects of his work on the reformulation of kinematics to be inferior to its negative aspect: the criticism of the Bohr–Sommerfeld atomic theory. Still he tried his best to assemble such evidence as he could in support of the quantum-mechanical scheme.

The tests of quantum mechanics, which Heisenberg presented in Sections 2 and 3 of his paper (1925c, pp. 886–893), did not cover the field systematically. They were quite diverse in nature; some dealt with the general, qualitative consequences that followed from the structure of the theory, while others dealt with specific properties of atomic systems. The most important problem facing Heisenberg was that the technical and mathematical apparatus at his disposal was too limited to carry out actual calculations in quantum mechanics. Because of the serious problem arising from the noncommutativity of the products of kinematical quantities, he was not able to deal immediately with systems other than the one-dimensional anharmonic oscillator. As he remarked to Pauli, perhaps more powerful mathematical methods were needed (Heisenberg to Pauli, 9 July 1925). These methods had to be selected in such a way that they could deal with the full complexity of realistic atomic systems, including their property of exhibiting periodic as well as aperiodic motions. For the moment, Heisenberg was not aware of any mathematical methods that would do the job; hence he restricted himself to reporting certain preliminary results demonstrating the correctness of his physical ideas. In particular, he expected that the solution of his quantum-mechanical equations, such as the reformulated Eq. (120) and Eq. (123), would indeed describe the behaviour of atomic systems, 'if one could demonstrate that this solution agrees with, or does not contradict, the presently known quantum mechanical relations' (Heisenberg, 1925c, p. 886).

Heisenberg did not hesitate in selecting those 'quantum mechanical relations' which he believed to be 'established.' He noted that there were three such relations that had to be satisfied under all circumstances within the quantum-mechanical scheme. The first was the requirement that 'a small perturbation of a mechanical problem gives rise to additional terms in the energy or in the frequency, respectively, which correspond exactly to the expressions found by Kramers and Born but are inconsistent with the ones provided by the classical theory' (Heisenberg, 1925c, pp. 886–887). Second, energy conservation had to hold in all processes involving atoms and radiation; this condition included the existence of stationary states, as we have discussed in Section IV.5. Third, the mechanical frequencies had to satisfy the Einstein–Bohr frequency condition, Eq. (118a). Besides these theoretical requirements, Heisenberg also demanded that the intensity ratios, observed for the components of multiplet spectra and their Zeeman effects, should follow from the application of his scheme to rotating electrons.

However, because of the difficulties mentioned above, Heisenberg did not carry out any of the tests completely. For example, he was not able to furnish the proofs of energy conservation and of the frequency condition for general atomic systems. He had to restrict himself to indicating a few steps of these proofs in the case of the one-dimensional anharmonic oscillator, the only system whose mathematics he could handle completely. Still, he succeeded in demonstrating that one general consequence of his quantum-mechanical scheme was correct, for he noted that 'from Eq. [(120)] (i.e., from its quantum-theoretical analogue) it follows, just as it does in classical theory, that the oscillating electron behaves in the presence of light—which possesses much shorter wavelengths than all eigen-oscillations of the system—like a free electron' (Heisenberg, 1925c, p. 887).[309] To prove this assertion Heisenberg took Kramers' dispersion formula for the induced electric moment in the case of coherent scattering of light by a one-electron atom, Eq. (55), in the limit that the frequency of incident radiation is much larger than every emission or absorption frequency ($\nu \gg \nu(n+\alpha, n)$, $\nu(n, n-\alpha)$), i.e.,

$$P_{\text{qu}}(\nu, t) = -\frac{2E_0 e^2 \cos(2\pi\nu t)}{\nu^2 h} \cdot \sum_{\alpha=1}^{\infty} \left[|A(n+\alpha, n)|^2 \nu(n+\alpha, n) - |A(n, n-\alpha)|^2 \nu(n, n-\alpha) \right]. \tag{127}$$

In Eq. (127), E_0 denotes the absolute value of the amplitude of the electric field and the products $-e \cdot A(n+\alpha, n)$ and $-e \cdot A(n, n-\alpha)$—with $-e$, the charge of the electron—stand for Kramers' absorption and emission amplitudes, i.e., A^a and A^e in Eq. (55). For the sake of completeness, it should be mentioned that on the right-hand side of Eq. (127) an integral has to be added that contains the absorption and emission terms of the electron into the continuum. However, the value of the α-sum—more accurately, of the sum and the integral—was given by the sum-rule of Willy Thomas and Werner Kuhn as being identical with $h/(4\pi^2 m_e)$, where m_e is the mass of the electron (Thomas, 1925a; Kuhn, 1925a). The equation for the quantum-theoretical moment thus reduced to

$$P_{\text{qu}}(\nu, t) = -\frac{e^2 E_0 \cos(2\pi\nu t)}{4\pi^2 \nu^2 m_e}. \tag{128}$$

[309] Evidently, in that case the interaction term, $f(x)$ in Eq. (120), which gives rise to the frequencies of the system, might be neglected in quantum as in classical theory.

This formula agreed completely with Joseph John Thomson's formula for the scattering of X-rays by free electrons (Thomson, 1903, p. 263). And, as Heisenberg quickly realized, exactly the same result followed from his quantum-theoretical reformulation of Eq. (120) and the quantization condition, Eq. (123). Thus, the scheme seemed to give a correct answer, at least in the problem of calculating the coherent scattering of high-frequency light by atoms.

The solution obtained by integrating the equation of motion for the one-dimensional anharmonic oscillator, Eq. (101), allowed Heisenberg to check all three consistency requirements mentioned above. As we have discussed in detail earlier, Heisenberg had arrived in Helgoland at the result that the formal energy expression, given by Eq. (107), did not exhibit any time dependence; hence energy conservation seemed to hold for the anharmonic oscillator, at least in the approximation up to which he had calculated the energy. But in his paper, Heisenberg also claimed that the explicit result for $W(n)$, the energy of the oscillator in the stationary state n—including terms of order λ^2, λ being the anharmonicity parameter—that is,

$$W(n) = \frac{h\omega_0}{2\pi}\left(n + \frac{1}{2}\right) + \lambda\frac{3h^2}{32\pi^2\omega_0^2 m}\left(n^2 + n + \frac{1}{2}\right) - \lambda^2\frac{h^3}{512\pi^3\omega_0^5 m^2}\left(17n^3 + \frac{51}{2}n^2 + \frac{59}{2}n + \frac{21}{2}\right), \tag{129}$$

did satisfy the other requirements as well. In the case of the frequency condition this claim was easy to prove by direct calculation. For this, one had just to calculate the difference of, say, $W(n)$ and $W(n-1)$; the result,

$$W(n) - W(n-1) = \frac{h\omega_0}{2\pi} + \lambda\frac{3nh^2}{16\pi^2\omega_0^2 m} - \lambda^2\frac{3h^2}{512\pi^3\omega_0^5 m^2}(17n^2 + 7), \tag{130}$$

indeed agreed with the expression for $(h/2\pi)\omega(n, n-1)$, as obtained from Eq. (106a).

It probably took Heisenberg slightly longer to prove his statement that: 'This energy [i.e., the right-hand side of Eq. (129)] can also be computed, according to the method of Kramers and Born, by treating the term $(m\lambda/4)x^4$ as a perturbation term of the harmonic oscillator. Then one really arrives again at exactly the result [(129)]; this fact seems to me to constitute a remarkable confirmation of the quantum-mechanical equations, on which the result is based' (Heisenberg, 1925c, p. 890). We shall outline here only the proof in the lowest approximation, i.e., keeping only the terms of the order λ. In classical perturbation theory the first-order correction to the energy of the anharmonic oscillator is given by averaging the perturbation term, $(\lambda/4)mx^4$, over a complete period of motion. This procedure (with $x = a_1\cos(\omega_0 t) +$ terms of order λ, λ^2, etc.) yields the result

$$W_1 = \frac{\lambda}{4}m\frac{3}{8}a_1^4. \tag{131}$$

According to the quantum-theoretical perturbation scheme of 1924 (see especially Born, 1924b) one must replace the product of Fourier amplitudes, a_1^4, by the quantum-theoretical product of transition amplitudes. Hence the perturbation energy becomes

$$\begin{aligned} W_1 = \frac{\lambda}{4} m \frac{1}{16} \big[& a(n,n-1)a(n-1,n-2)a(n-2,n-1)a(n-1,n) \\ & + a(n,n-1)a(n-1,n)a(n,n-1)a(n-1,n) \\ & + a(n,n-1)a(n-1,n)a(n,n+1)a(n+1,n) \\ & + a(n,n+1)a(n+1,n)a(n,n-1)a(n-1,n) \\ & + a(n,n+1)a(n+1,n)a(n,n+1)a(n+1,n) \\ & + a(n,n+1)a(n,n+2)a(n+2,n+1)a(n+1,n) \big]. \end{aligned} \tag{131a}$$

After substituting for $a(n,n-1)$ the quantum-theoretical expression given by Eq. (106b)—and for the other amplitudes the corresponding ones—and neglecting terms of order λ^2, λ^3, etc., one obtains indeed the λ-term in Eq. (129), in contrast to the classical result.[310]

Though he had calculated the anharmonic oscillator in some detail, Heisenberg did not expect experimental confirmation of his specific results in this case.[311] However, he carefully investigated whether the mechanical treatment of another class of systems, which included rotating electrons, would yield results in agreement with observation. At a very early stage in his work he had noted that the intensity formulae of multiplets and Zeeman lines seemed to follow from the quantum-mechanical scheme (see Heisenberg to Kronig, 5 June 1925). He had devoted the following several weeks entirely to the solution of the problem of the anharmonic oscillator, but then he turned again to the rotator system. As he reported to Pauli: 'I have calculated the rotator more exactly; there one can obtain the formulae of Kronig and Kemble' (Heisenberg to Pauli, 24 June 1925).

[310]The classical result is obtained by substituting for a_1 the expression $(hn/\pi\omega_0 m)^{1/2}$; then W_1 becomes equal to $3n^2\lambda(32\pi^2\omega_0^2 m)^{-1}$.

[311]The only consequence of the calculation, which one could hope to check, was the harmonic part of the energy expression, Eq. (129). Remarkably enough, neither Heisenberg nor Pauli quoted R. S. Mulliken's analysis of the boron oxide bands (Mulliken, 1924b), which became available in fall 1924, as a proof for this term. There occurred no reference to Mulliken also in the later writings of the Göttingen theoreticians (e.g., in Born and Jordan, 1925b; Born, Heisenberg and Jordan, 1926) nor in Pauli's *Handbuch* article (Pauli, 1926b). Probably Heisenberg and Pauli took Otto Stern's calculation of the chemical constant (Stern, 1913, 1919), which included the zero-point energy $\frac{1}{2}h\nu$, as an indication of the correctness of the energy term for the harmonic oscillator.

He reported the results of these calculations in Section 3 of his paper (Heisenberg, 1925c, pp. 891–893).

The explicit reference to the work of Kronig and Kemble provides the background of Heisenberg's quantum-mechanical treatment of the rotator. Edwin C. Kemble had, in a paper published in the January issue of *The Physical Review*, applied the correspondence principle to obtain the relative intensities of the spectral lines of diatomic molecules (Kemble, 1925). These molecules had been considered previously as degenerate systems because the rotator, by which the molecules were described, possessed two degrees of freedom, but only one quantum condition had been available. When Kemble evaluated Kramers' expression for a_{ik}, the Einstein emission coefficient of the line with frequency ν_{ik}, i.e.,

$$a_{ik} = \frac{16\pi^4 \nu_{ik}^3}{3hc^3} \overline{(X^2 + Y^2 + Z^2)} \tag{132}$$

—where X, Y and Z denote the corresponding Fourier coefficients of the electric moment of the system, and the bar over the term in parentheses indicates an averaging procedure between the quantities associated with the initial and final orbits of the system (see Kramers, 1919, p. 330)—he had found that the results did not agree with the experimental data on the intensities of band spectra. Kemble had therefore suggested an alternative mechanical treatment of the molecule; he had removed the degeneracy of the rotator system by assuming the presence of a weak magnetic field, which splits each rotational line into a triplet, and had then calculated the sum of the intensities of the magnetic multiplet arising from each band line. The results had now fitted the data satisfactorily. Evidently, Kemble's treatment of the rotator had attracted Heisenberg earlier in 1925, for it was very much along his own line of thinking at that time. First, he had independently used a similar argument to obtain the intensity of radiation emitted by a degenerate atomic system like Kemble's when, in fall 1924, he had discussed resonance fluorescence (Heisenberg, 1925a). And second, Kemble had advocated the use of half-integral quantum numbers to describe band spectra (Kemble, 1925, pp. 9–10, especially footnote 15). Kemble's work now encouraged him in formulating a quantum-mechanical theory of the rotating electron.

Ralph Kronig, on the other hand, had published, together with Samuel Goudsmit, a short note on the intensity of Zeeman components in January 1925 (Goudsmit and Kronig, 1925). A few weeks earlier, Helmut Hönl, a student of Sommerfeld's in Munich, had arrived at identical results (Hönl, 1925). Kronig had continued to pursue this problem in spring and summer 1925, maintaining a close connection with Heisenberg's attempts at sharpening the correspondence principle by its application to the intensity rules in line multiplets and their Zeeman components. In two papers, which he submitted in February and June 1925, Kronig had succeeded in working out subtle details of the theory of the multiplet and Zeeman effect intensities (Kronig, 1925a, b). All these results, including those published independently by Henry Norris Russell (1925), agreed quite well with the experimental data and their analysis, which had been obtained by Leonard Salomon Ornstein and his students.[312] It was to these intensity rules of Goudsmit and Kronig, Hönl and Russell that Heisenberg addressed himself in the last part of his paper (Heisenberg, 1925c, pp. 892–893).

Before treating the rotating electron in an external field, Heisenberg quickly sketched the theory of the one-dimensional rotator in quantum mechanics. In this system, a mass m is considered to move with uniform velocity in a circular orbit with radius a around a fixed point; hence its moment of inertia is given by ma^2. As noted above, this rotator had been used for the description of the rotational states of diatomic molecules in atomic theory, and results had been obtained that did not fit the data satisfactorily. Heisenberg now hoped to overcome the earlier difficulties by applying his quantum-mechanical scheme. He did not bother to write down the equation of motion, but right away gave its solution. The kinematic variables of the quantum-theoretical rotator, x, y, z—the analogues of the classical position variables—obviously had to satisfy the relation

$$x^2 + y^2 + z^2 = a^2, \tag{133}$$

where a is the radius of the orbiting mass m. There remained the task of computing the quantities $\omega(n, n-1)$ and $\omega(n+1, n)$, the analogues of the classical angular velocities. These quantities assumed, of course, the meaning of transition frequencies, apart from a factor of 2π, i.e., $\nu(n, n-1) = (1/2\pi) \times \omega(n, n-1)$, etc. Heisenberg determined them with the help of the quantum condition (123), which assumed the form[313]

$$h = 2\pi m\left[a^2\omega(n+1, n) - a^2\omega(n, n-1)\right]. \tag{134}$$

If, as in the case of the anharmonic oscillator, one assumed the existence of a lowest state with quantum number $n_0 = 0$ in the rotator system, Eq. (134) yielded an equation defining $\omega(n, n-1)$, i.e.,

$$\omega(n, n-1) = \frac{hn}{2\pi ma^2}. \tag{135}$$

Finally, Heisenberg calculated W_{qu}, the energy of the quantum-theoretical one-dimensional rotator in the stationary state n, by reformulating the classical

[312]The intensity rules for multiplets had been first given in a paper of Ornstein's students H. C. Burger and H. B. Dorgelo (1924) and in two papers by Ornstein and Burger (1924d, e) on the basis of experimental observations. Intensity rules for the Zeeman components were first predicted on the basis of theoretical conclusions by Ornstein and Burger (1924f, g), as well as by Hönl, and Goudsmit and Kronig, in the above-mentioned papers. The experimental verification of the rules was obtained only afterwards (Ornstein, Burger and van Geel, 1925).

[313]The phase integral of the classical motion is given by the product $2\pi ma^2\omega$ ($= \int p_\phi d\phi$ with $p_\phi = ma^2\omega$). The quantum-theoretical re-interpretation obviously yields Eq. (134).

equation, $W_{\text{cl}} = (m/2)a^2\omega^2$. Thus, he obtained the result

$$W_{\text{qu}} = \frac{m}{4}a^2\left[\omega^2(n+1,n) - \omega^2(n,n-1)\right]$$
$$= \frac{h^2}{8\pi^2 ma^2}\left[n^2+n+1\right], \qquad n = 0,1,2,\ldots . \tag{136}$$

He noted from the energy formula that the frequency condition followed at once.[314]

The new results on the rotator agreed completely with the ones found earlier by Kemble on the basis of detailed correspondence arguments. Kemble's main conclusion had been that the intensity of the band lines depended symmetrically on the rotational quantum numbers of the molecule involved in the transition. The quantum-mechanical expressions for the intensities, although they were not given explicitly by Heisenberg, satisfied the symmetry requirement automatically. The reason was obvious: in quantum mechanics the intensities were proportional to the absolute squares of the transition amplitudes, and this quantity remained unaltered when the initial and final rotational quantum numbers were exchanged. Hence Heisenberg's new scheme led to intensities that fitted the data as well as Kemble's. Moreover, Heisenberg was able to justify the assumption of half-integral quantum numbers, which had been introduced by Kratzer three years earlier and had also been taken over by Kemble. Kratzer had described the empirical terms of diatomic molecules by the formula

$$W_{\text{rot}} = B(m - \tfrac{1}{2}) + \ldots, \tag{137}$$

for the rotational energy W_{rot} of the system, with m denoting positive integers (Kratzer, 1922).[315] If Kratzer's rotational quantum numbers m were replaced by $n = m - 1$, then the right-hand side of Eq. (137) could be expressed by $B(n^2 + n + \frac{1}{4})$. Since the factor B may be identified with $h^2/8\pi^2 ma^2$, this expression deviates from Heisenberg's result, Eq. (136), only by the small quantity $\frac{1}{4}B$. Of course, one could decide whether Kratzer's or Heisenberg's energy expression was correct by measuring accurately the lowest rotational term of diatomic molecules. However, by mid-1925 it was still extremely difficult to identify this term; hence Heisenberg had good reason to believe that his formula accounted as perfectly for the observed band spectra as Kratzer's. Even a year later, when Schrödinger applied his wave mechanical methods to the rotator, the experimental situation had not become clearer. Schrödinger, who found the factor $(n^2 + n)$

[314]It should be noted, however, that the frequency condition may also be satisfied by assuming expressions different from the one of Eq. (136) for the quantum-theoretical energy $W(n)$.

[315]The terms of band spectra could be written in general as, $W = W_{\text{el}} + W_{\text{osc}}(n) + B_1(n) \cdot (m - \frac{1}{2})^2 - \beta(m - \frac{1}{2})^4$ + smaller terms, where the first term represents the electron energies in the single atoms constituting the molecule, the second term the (quantized) energy of the oscillation of the molecule, i.e., the energy associated with the combined oscillations of atoms in the molecule (quantum number n), and the following terms the purely rotational energy (quantum number m). (See Kratzer, 1925, p. 842.)

instead of Heisenberg's $(n^2 + n + \frac{1}{2})$ and Kratzer's $(n^2 + n + \frac{1}{4})$, remarked:

> This definition [i.e., $W_{rot} = B(n^2 + n)$] is different from all previous statements (except perhaps that of Heisenberg?). Yet from various arguments on the basis of experiment we are led to put "half-integral" values for n in formula (31) [i.e., $W_{rot} = Bn^2$]. It is easily seen that formula (34′) [i.e., $W_{rot} = B(n^2 + n)$] gives practically the same result as (31) with half-integral values of n. For we have, $n(n + 1) = (n + \frac{1}{2})^2 - \frac{1}{4}$. The discrepancy consists only of a small additive constant; the level *differences* in (34′) are the same as are obtained from "half-integral quantization". (Schrödinger, 1926d, p. 521)

In deriving the formula for the rotational energy terms, Heisenberg just followed his treatment of the one-dimensional anharmonic oscillator. The occurrence of half-integral quantization had exactly the same reason as before; namely, it was caused by taking into account (in the energy expression for the stationary state n) not only the transition from n to $n - 1$, but also from n to $n + 1$. Heisenberg could honestly treat only the rotator having one degree of freedom because he worked with a single quantum condition. In order to treat the system considered, say, by Kemble, he had to extend the quantum-mechanical formalism to describe also systems with several degrees of freedom. Since he did not perform this task in his paper, one should not really compare the result, Eq. (136), with the expression obtained later on by evaluating systematically the rigid rotator in space. Still Heisenberg's preliminary formula came very close to the final result.[316]

In spite of the initial limitation to systems having one degree of freedom, Heisenberg wanted to show how one could arrive at the intensity formulae for the Zeeman components of line multiplets, which Goudsmit and Kronig, or Hönl, had found earlier on the basis of correspondence arguments. The classical mechanical system describing the atomic situation possesses two degrees of freedom: i.e., an electron of mass m_e moves on a circular orbit of radius a with angular velocity ω_n, when no magnetic field is applied, and the orbit performs a slow precession with the frequency Ω around the axis of a (not too strong) magnetic field. Heisenberg had treated this system in detail with Sommerfeld three years earlier (Sommerfeld and Heisenberg, 1922b); hence he could simply take over the result of the previous paper.[317] If the vector of the external magnetic field lies in the z-direction, then x, y, z, the position coordinates of the rotating electron, are given by the equations

$$\begin{aligned} z &= a_0 \cos(\omega_n t), \\ x + iy &= \exp(i\Omega t)\left[b_{+1} \exp(i\omega_n t) + b_{-1} \exp(-i\omega_n t) \right], \end{aligned} \tag{138}$$

[316] A little later Erwin Fues calculated the energy of the diatomic molecule in wave mechanics. For the purely rotational term of the energy he found the result $(h^2/8\pi^2 A)(n + \frac{1}{2})^2$, A being the moment of inertia identical with Heisenberg's ma^2. (See Fues, 1926, p. 379, Eq. (42).) His result therefore completely confirmed Kratzer's semiempirical formula, Eq. (137).

[317] E. C. Kemble and H. Hönl, in their respective treatments of the rotator in an external field (Kemble, 1925; Hönl, 1925), referred to the paper of Sommerfeld and Heisenberg (1922b).

where a_0, b_{+1} and b_{-1} are, respectively, the coefficients associated with the undisturbed frequency ω_n (of the free atom), $\omega_n + \Omega$ and $\omega_n - \Omega$, the latter being the frequencies shifted by the Larmor precession (see Sommerfeld and Heisenberg, 1922b, p. 136, Eq. (9)). Since the coordinates of the rotating electron satisfy Eq. (133), the following relations hold[318]:

$$\begin{aligned} \tfrac{1}{2}a_0^2 + b_{+1}^2 + b_{-1}^2 &= a^2, \\ \tfrac{1}{2}a_0^2 + 2b_{+1}b_{-1} &= 0. \end{aligned} \tag{139}$$

In the Bohr–Sommerfeld theory, these equations, together with the quantum condition,

$$2\pi m_e(b_{+1}^2 - b_{-1}^2)\omega = (m + \text{const.})h, \tag{140}$$

allowed one to calculate the intensities of the components of the radiation emitted by the rotating electron in the state with the magnetic quantum number m.

In quantum theory Heisenberg represented the position coordinates by the patterns of amplitudes and their associated frequencies. For z in the state n he wrote the quantity $a(n,n-1;\ m,m)\cos[\omega(n,n-1)t]$, for $x + iy$ the two quantities $b(n,n-1;\ m,m-1)\exp\{i[\omega(n,n-1)+\Omega]t\}$ and $b(n,n-1;\ m-1,m) \times \exp\{i[-\omega(n,n-1)+\Omega]t\}$. The squares of the transition amplitudes, $a^2(n,n-1;m,m)$, $b^2(n,n-1;m,m-1)$ and $b^2(n,n-1;m-1,m)$, gave (up to a constant factor) the intensities of the components with frequencies $\omega(n,n-1)$, $\omega(n,n-1)+\Omega$, and $\omega(n,n-1)-\Omega$, respectively. The quantum-theoretical amplitudes could be obtained from equations corresponding to the classical equations (139) and (140). With the help of his multiplication rule, Eq. (119), Heisenberg reformulated the 'equation of motion,' Eq. (139), as[319]

$$\begin{aligned} &\tfrac{1}{2}\{a^2(n,n-1;\ m,m) + b^2(n,n-1;\ m,m-1) + b^2(n,n-1;\ m,m+1) \\ &\quad + \tfrac{1}{2}a^2(n+1,n;\ m,m) + b^2(n+1,n;\ m,m-1) \\ &\quad + b^2(n+1,n;\ m,m+1)\} = a^2, \\ &\tfrac{1}{2}a(n,n-1;\ m,m)a(n-1,n-2;\ m,m) \\ &\quad = b(n,n-1;\ m,m-1)b(n-1,n-2;\ m-1,m) \\ &\quad + b(n,n-1;\ m,m+1)b(n-1,n-2;\ m+1,m), \end{aligned} \tag{139a}$$

[318] In case the magnetic field has zero field strength, the electron moves in a plane; then the motion may be described by the equation $x + iy = a\exp(i\omega_n t)$.

In his paper, instead of the second Eq. (139), Heisenberg wrote the equation: $\frac{1}{4}a_0^2 = b_{+1}b_{-1}$ (Heisenberg, 1925c, p. 892, second Eq. (33)). The difference with our equation consists in the negative sign of the term $b_{+1}b_{-1}$, which may be absorbed in the definition of the amplitude. Since Heisenberg was interested in the intensities rather than amplitudes, this difference is not important.

[319] Concerning the change of the sign of the term on the right-hand side of the second Eq. (139a) with respect to the classical Eq. (139), see footnote 318.

and the quantum condition, Eq. (140), as

$$2\pi m_e\left[b^2(n,n-1;m,m-1) - b^2(n,n-1;m-1,m)\right]\omega(n,n-1)$$
$$= (m + \text{const.})h. \tag{140a}$$

One might wonder why Heisenberg reformulated the classical equation (140), instead of first transforming it into a difference equation. The reason was that he did not bother to carry out the details of a consistent calculation of the problem, which he could not do anyway; he was just interested in sketching a method of deriving the already established intensity formula. A later systematic treatment of the rotating electron in an external magnetic field would, of course, have to confirm the above steps.

Heisenberg now claimed that Eqs. (139a) and (140a) yielded as their 'simplest' solutions the following amplitudes:

$$b(n,n-1;m,m-1) = a\sqrt{\frac{(n+m+1)(n+m)}{4(n+\frac{1}{2})n}},$$

$$b(n,n-1;m-1,m) = a\sqrt{\frac{(n-m)(n-m+1)}{4(n+\frac{1}{2})n}}, \tag{141}$$

$$a(n,n-1;m,m) = a\sqrt{\frac{(n+m+1)(n-m)}{(n+\frac{1}{2})n}}.$$

These solutions indeed satisfied Eqs. (139a) and the quantization condition, Eq. (140a), if one substituted for the frequency $\omega(n,n-1)$ the expression from Eq. (135)—which should provide at least a reasonable approximation—and assumed no additive constant to occur on the right-hand side of Eq. (140a). Heisenberg justified the latter steps by their success, for he noticed that 'these expressions [Eqs. (141)] agree with the formulae of Goudsmit, Kronig and Hönl' (Heisenberg, 1925c, p. 893). Hönl had found for the same intensities expressions which looked slightly different but could be easily rewritten in Heisenberg's form.[320] In spite of this obvious success of the quantum-mechanical treatment, Heisenberg was not completely satisfied, for he remarked: 'However, one cannot easily understand why these expressions [i.e., the expressions for the transition amplitudes given by Eqs. (141)] should represent the *only* solution of Eq. [(139a)]

[320]In his paper, H. Hönl had only given explicit expressions for the transitions $\Delta j = 1$ in the rotational quantum number j (see Hönl, 1925, p. 350, Eq. (11a); note that he forgot a factor $\frac{1}{2}$ in the amplitudes of the shifted frequencies). However, he had given a rule for obtaining, from the expressions for the amplitudes $j \to j+1$, the expressions for the amplitudes $j \to j-1$. In this way one arrives at intensities proportional to $\frac{1}{4}(j+m)(j+m-1)$, $\frac{1}{4}(j-m+1)(j-m)$ and (j^2-m^2) for the transitions $(j \to j-1;\ m \to m-1)$, $(j \to j-1;\ m-1 \to m)$ and $(j \to j-1;\ m \to m)$, respectively. By substituting for Hönl's quantum numbers, j and m, Heisenberg's quantum numbers, $j = n + \frac{1}{2}$ and $m_{\text{Hönl}} = m_{\text{Heisenberg}} + \frac{1}{2}$, one reproduces Eqs. (141).

and Eq. [(140a)]' (Heisenberg, 1925c, p. 893). Still he hoped that the proof of uniqueness might follow eventually by taking into account the 'boundary conditions.' After all, the amplitudes $b(n, n-1; m-1, m)$ and $a(n, n-1; m, m)$ had to vanish in case the magnetic quantum number were to become equal to the rotational quantum number n; and similarly the amplitude $b(n, n-1; m, m-1)$ had to vanish when m became equal to $-m$.[321]

The agreement with the analysis of the intensities of multiplets and Zeeman components constituted the only sensible test of Heisenberg's new scheme, if one disregards the confirmation provided by the analysis of the mixed (rotation–oscillation) band spectra in terms of the energy states of the harmonic oscillator.[322] However, this success did not provide much reason for jubilation or for believing that the problems of quantum theory had now been resolved because the behaviour of complex spectra was so specific that the same intensity formulae could be deduced in many different ways, one being Heisenberg's scheme. More calculations of realistic atomic systems were needed with the new methods before one could decide about their value. The more fundamental aspects of the theory had also yet to be analyzed. For example, in his paper Heisenberg did not answer the question which he had mentioned earlier to Pauli, namely, 'what the equations of motion really mean if interpreted as relations between the transition probabilities' (Heisenberg to Pauli, 24 June 1925). Nevertheless, the first step in the right direction of formulating a consistent quantum mechanics had been taken, although Heisenberg concluded his paper with the cautious words:

> Whether a method of determining quantum-theoretical data using relations between observable quantities, such as the one proposed here, can be regarded as satisfactory in principle, or whether this method after all represents a still too rough attack on the physical problem of a quantum-theoretical mechanics—an obviously very involved problem at the moment—will be decided by a more thoroughgoing mathematical investigation of the method which has been employed here very superficially. (Heisenberg, 1925c, p. 893)

V.5 A Farewell to 'Term Zoology and Zeeman Botany'

On completing his paper on the quantum-theoretical re-interpretation of kinematic and mechanical relations in the middle of July 1925, Heisenberg handed it

[321] For a similar consideration of the boundary conditions, Heisenberg referred to the papers of Kronig (1925a), Sommerfeld and Hönl (1925) and Russell (1925). These authors had tried to determine the intensities with the help of correspondence arguments and sum-rules and had found that the resulting equations did not lead uniquely to expressions for the intensities. Extra physical conditions were needed to remove the ambiguities. For example, Kronig had argued that his solutions possessed 'the particularly remarkable property that they yield the value zero for all those single transitions, which originate from the levels with the largest and the smallest inner quantum numbers, if the final levels under consideration are not omitted due to the structure rule for J [the action variable in question]' (Kronig, 1925a, p. 892).

[322] Without going into details of calculation Heisenberg mentioned in his paper the result that the intensity formula for multiplets, which could be obtained by a similar consideration as the intensity formulae (141) for the Zeeman components, also agreed with the available data. (See Heisenberg, 1925c, p. 893, second paragraph.)

over to Max Born. 'I told him that I felt something real was in the paper, but that I was rather uncertain about it,' he recalled later (Heisenberg, Conversations, p. 269). In spite of the warm encouragement, which he had received from Pauli, Heisenberg also wanted to hear his professor's opinion as to whether he should publish the new ideas and preliminary results on quantum mechanics. And he said to Born: 'Do with it what you think is right' (Heisenberg, Conversations, p. 269). Since Born was very occupied at that time, Heisenberg did not even wait for Born's judgment, but asked his permission to leave Göttingen before the end of the summer semester in order to visit Leyden and Cambridge, where he had been invited. Thus, he did not witness how Born approved of his paper and started to develop a consistent mathematical formulation of quantum mechanics —the matrix formulation—before submitting his paper to *Zeitschrift für Physik* by the end of July 1925.[323]

It was quite natural that Heisenberg should have been invited to give lectures at Leyden and Cambridge. At both places there existed an interest in him, for his work had received high praise from acknowledged masters like Arnold Sommerfeld, Max Born and, especially, Niels Bohr. Moreover, with his publications on the theory of multiplets and the anomalous Zeemann effects, beginning in 1922, he had become by mid-1925 an acknowledged expert on these questions, whose ideas were exercising an influence on such people as Ralph Kronig and Friedrich Hund, who were concerned with the detailed analysis of spectroscopic data. Both in Leyden and Cambridge there existed a marked interest in spectroscopy. This interest was probably stronger in Leyden, where Samuel Goudsmit, a student of Ehrenfest's, was working on the empirical data on multiplets in close collaboration with the experimentalists at Utrecht. But also in Cambridge, where Ernest Rutherford and his nuclear research overshadowed most of the other activities in physics, Ralph Fowler had turned his attention to the study of spectroscopic data as a result of his interest in ionization problems and their application to astrophysics.[324]

Heisenberg had met Fowler in Copenhagen, where the latter spent about two months from the end of January to the beginning of March 1925. Fowler had certainly been impressed by the abilities of Bohr's young collaborator, who not only had just completed with Kramers the important work on dispersion theory, but also seemed to be able to calculate successfully the frequencies and intensities of most complex spectra. Thus he had decided to invite Heisenberg to Cambridge, especially since he [Fowler] was going to lecture on the 'recent developments of quantum theory' during the spring term.[325] 'I received a nice letter from Fowler with an invitation to visit Cambridge in July,' Heisenberg had

[323] Heisenberg's paper (1925c) was received by *Zeitschrift für Physik* on 29 July 1925.

[324] For example, in 1925 Ralph H. Fowler published two papers on spectroscopic problems, one about the intensities of band spectra (Fowler, 1925a), the other on the intensities of spectral lines in general (Fowler, 1925d), leaning heavily on correspondence arguments in his derivations. He also lectured regularly at Cambridge, at least since 1923, on quantum theory of atomic systems.

[325] A manuscript of the notes of Fowler's lectures, taken by Paul Dirac, is contained in the Dirac Collection at Churchill College, Cambridge.

already reported to Kronig in May (Heisenberg to Kronig, 8 May 1925). About two months later, Fowler's invitation was repeated by Peter Kapitza, one of Rutherford's closest collaborators at the Cavendish Laboratory, who visited Leyden and Göttingen in early July.[326] Kapitza was himself moderately interested in spectroscopy and had made the first use of his newly produced intense magnetic fields (Kapitza, 1924) in studying the Zeeman effect of ionized calcium and beryllium (Kapitza and Skinner, 1924). Thus, when he met Heisenberg at Göttingen, he thought that it would be a good idea to have him address the Kapitza Club at the Cavendish.

In inviting Heisenberg to Cambridge, Kapitza was probably also stimulated by his friend Paul Ehrenfest, whom he had just seen again in Leyden.[327] Ehrenfest had been in Göttingen for a few days during the second half of June and had given there a brilliant exposition of Einstein's new quantum theory of a monatomic ideal gas (Einstein, 1924c; 1925a, b).[328] Upon his return from Helgoland, Heisenberg saw Ehrenfest. As he recalled: 'Perhaps I had met him already in Göttingen, but certainly in Copenhagen. So I knew him well, and he invited me to come to Holland and see his young people' (Heisenberg, Conversations, p. 270). Ehrenfest himself did not work on spectroscopic problems; he was mainly concerned with the fundamental questions of quantum theory. However, he had always thought it to be important to assist the experimentalists at Leyden and the neighbouring Utrecht to understand their results theoretically. Therefore, he encouraged his students, like Goudsmit and young George Uhlenbeck, to investigate the details of atomic theory in order to describe properly the data on complex multiplets. Since Heisenberg was considered both in Göttingen and in Copenhagen as an expert on the theory of multiplets, Ehrenfest seized the opportunity of inviting him for a few days to Leyden. Heisenberg was very pleased to be invited to Leyden, where the great Hendrik Lorentz was still around and Einstein visited regularly. In those days one did not often have the opportunity of travelling abroad, especially from Germany. He therefore ac-

[326] According to the signatures on the wall of the living room in Ehrenfest's house at Leyden, Kapitza stayed as a guest of the Ehrenfests from 2 to 5 July 1925. On the other hand, Born wrote to Einstein on 15 July 1925 that 'last week Kapitza from Cambridge was here'; hence Kapitza passed through Göttingen during the week from 5 to 12 July 1925.

[327] Kapitza was well acquainted with Ehrenfest and his Russian-born wife Tatyana. For instance, he had already stayed with them at their home in July 1923. Again, this may be seen from the signatures on the wall of the living room in Ehrenfest's house in Leyden; all scientific visitors and house-guests had to write their names on the wall. Similarly, the Wednesday evening colloquium was a fixture at the Institute of Theoretical Physics at Leyden during Ehrenfest's time, and the speaker signed his name and date on the wall. This tradition has been maintained in the Wednesday evening 'Ehrenfest Colloquia,' where the speaker may be invited to sign his name on the wall after the lecture, though he would do so only once (i.e., not if he addresses the Colloquium again). Thus, I had the privilege of signing my name on the wall, at the invitation of P. Mazur, Director, *Instituut Lorentz*, after my lecture on 'The Historical Origins of General Relativity Theory,' 24 January 1974, but not after my second lecture on 'The Birth of Quantum Mechanics' on 6 December 1978. The invited Lorentz Professors sign their names on another portion of the wall reserved for them. (J. Mehra)

[328] Ehrenfest's visit to Göttingen and his talk there was mentioned by Born in his letter to Einstein, dated 15 July 1925.

cepted Ehrenfest's invitation as well as Fowler's. As soon as he had finished his duties for the summer semester under Born at Göttingen, he took off for Leyden and Cambridge.[329]

In Leyden, Heisenberg was well received by Paul Ehrenfest. 'I stayed in his house and was with his family. We had many discussions in the Institute. That was a very nice time, but not very long, just a few days.' he recalled later (Heisenberg, Conversations, p. 270).[330] As Ehrenfest had planned, Heisenberg discussed spectroscopic problems with his young disciples. He had the most interaction with Goudsmit, who also acted as his guide for sightseeing in the Netherlands. 'I remember very well Heisenberg in Leyden at Ehrenfest's home,' recalled Goudsmit. 'I was the tourist guide; I had to take Landé around, I had to take Sommerfeld around, I had to take Heisenberg around. He may even have come to The Hague with me' (Goudsmit, AHQP Interview, First Session, p. 27).[331]

It was Heisenberg's first encounter with Ehrenfest at really close quarters; he found him to be a very good physicist, but of a character quite different from those whom he had learned to appreciate and admire before. He recalled: 'Ehrenfest played a very interesting role in physics at that time because he had such a critical mind. By his criticism he always could stir up people and could get interesting discussions going. He had a kind of catalytic effect on physics to a very high degree. He was a "*Zauderer*" [irresolute, hesitant] of a man, who always had his scruples and his doubts and difficulties, but he had a very interesting and certainly very clear mind' (Heisenberg, Conversations, p. 270). More than his previous meetings with Ehrenfest in Copenhagen and Göttingen, he now had a really good opportunity of appreciating his scientific qualities. He talked with him about all the fundamental difficulties in quantum theory. 'Ehrenfest would always say how dreadful it was that one could not understand anything,' Heisenberg recalled. 'He was interested when I told him what I was trying to do in quantum mechanics' (Heisenberg, Conversations, p. 271). Thus, apart from Pauli, Kronig, Kramers and a few people in Göttingen, Ehrenfest and his band of young theoretical physicists in Leyden were among the first to learn about Heisenberg's new ideas. 'The arguments [in Heisenberg's scheme] appeared to be a little strange,' Goudsmit recalled. 'It was the emphasis on observable [quantities]—and that did not appeal too much to some people I heard talk

[329] Heisenberg did not participate in the meeting of the Lower Saxony Section (*Gauverein Niedersachsen*) of the German Physical Society at Hanover on Sunday, 19 July 1925. It is very likely that he left Göttingen for Leyden on that weekend.

[330] Heisenberg's signature on the wall of Ehrenfest's home is dated 26 July 1925. He probably signed on the day he left Leyden, for he was scheduled to talk in Cambridge on 28 July.

[331] For example, Goudsmit recalled: 'He came to see me from time to time, probably was at my home, the home of my parents, [had] dinner there, and I remember taking him to the fireworks at the seaside. It may have been [during] that particular visit, but I am not sure, because he came to Holland during that period, off and on' (Goudsmit, AHQP Interview, First Session, p. 27).

After leaving Leyden, Heisenberg wrote a letter to Goudsmit, dated 28 July 1925, thanking him for his help during his stay in Holland.

about it . . . But the mechanics was right' (Goudsmit, AHQP Interview, First Session, p. 27). To the experienced and scrupulous Ehrenfest, Heisenberg's ideas on the quantum-theoretical reformulation of atomic mechanics must surely have appeared to be very strange, and he needed time to get accustomed to their implications. In any case, the younger people in Leyden, especially those who, like Goudsmit, worked on spectroscopy, were inclined to take a more pragmatic view. Since Heisenberg had been able to reproduce the intensity formulae for multiplets and the Zeeman components, they found his quantum-theoretical scheme useful but did not concern themselves with the philosophical import of his ideas.

After spending several days in Leyden, Heisenberg arrived in Cambridge on the evening of 26 July and stayed with Fowler. Many years later he still remembered vividly the circumstances of his arrival in England. He said:

> By that time I was completely exhausted. First I had worked hard in Helgoland and then written the paper in Göttingen. I had been kept very busy in Holland. Then there was also a change of climate: England was very warm. So I was in a state of complete exhaustion when I arrived in Cambridge. After my first night at Fowler's place, he had to go away for a whole day for meetings in London. Apparently his wife was also away and he had asked the maid to provide me with breakfast and all the other meals during the day. I got up in the morning to have my breakfast and while I was at the table I just fell asleep. The maid took the breakfast away and I slept. At noon she came in and said that lunch was ready in another room. I did not hear a thing, I just kept on sleeping. So she took away the lunch and was a bit frightened. In the afternoon she told me that tea was ready. I said, "Yes," but I went on sleeping, and the same thing happened at dinner. At about 9 o'clock in the evening Fowler came home. The maid was terribly upset. She told him that I must be very ill, probably half dead, as I had slept through the whole day. Fowler came into my room and I realized that he was there, and I said, "Oh, hello Fowler." Fowler said, "What happened to you? Are you ill?" I replied, "No, I am perfectly all right and in the best of health." That was a strange experience. (Heisenberg, Conversations, pp. 271–272; also AHQP Interview)

Rest and sleep helped Heisenberg to recover from his tiredness. Fowler played golf the next day and he took his guest along. Heisenberg had a good time walking and talking with Fowler before they went to the meeting of the Kapitza Club in the evening. This meeting provided Heisenberg with a new experience, which he described as follows: 'The Kapitza Club was a peculiar thing. Kapitza had rooms at Trinity College. Some younger people used to come there periodically in the evening and sit by the fireplace. Somebody would report either on his own scientific work, or some other work of interest, and others would discuss it, but in an absolutely informal way. One sat on the floor, there were no chairs' (Heisenberg, Conversations, p. 272).[332] There, on 28 July 1925, at the 94th

[332] Although Kapitza was connected with Trinity College, it is not certain that the meeting of his Club took place in his rooms there. John Desmond Bernal, who joined the Kapitza Club a little later, recalled that the meetings were held in John Cockcroft's room in the Cavendish Laboratory (see Larsen, 1962, p. 62). Possibly Heisenberg, who was brought to the meeting by Fowler in the evening, did not carefully notice the location.

Meeting of the Kapitza Club, Heisenberg gave a talk, whose title he entered in the *Minute Book* as '*Termzoologie und Zeemanbotanik*.' Later on Heisenberg did not remember any details about his talk or the response of the young participants. He was not certain whether they were enthusiastic about the problems of atomic spectroscopy which he discussed, nor did he have closer interaction with any of them. This is the more surprising as, at that meeting, Patrick Maynard Stuart Blackett, Paul Adrien Maurice Dirac, and Llewellyn Hilleth Thomas should have been present.[333] Heisenberg knew Blackett from the time when the latter stayed at Franck's Institute in Göttingen. About Dirac he vaguely recalled that his host Fowler had told him that 'there was a young man who was an electrical engineer at that time and was an extremely good mathematician. But it seems that I did not meet Dirac this first time' (Heisenberg, Conversations, p. 273). Dirac, who was extremely interested in atomic theory and knew Heisenberg's work, did not remember much about the meeting of 28 July 1925 either. Heisenberg spoke about the specialized topic of the theory of multiplets and Zeeman effects, which did not particularly excite Dirac. In any case, he had heard nothing about Heisenberg's new quantum-mechanical ideas until he saw the proof-sheets of the paper several weeks later. It should be mentioned that at the 94th Meeting of the Kapitza Club, besides Heisenberg, another speaker gave a talk: the astrophysicist Harold Delos Babcock of Mt. Wilson Observatory discussed 'Symmetry Errors in the Measurements of Close Pairs of Spectral Lines.' Hence the two visitors had to share the evening, and there was not much time left for a longer discussion of Heisenberg's talk, independently of whether he mentioned his new quantum-mechanical ideas or not.[334]

In Cambridge, Heisenberg spoke for the last time on a subject that had been his favourite for several years: the explanation of multiplet spectra and anomalous Zeeman effects. The task had been to bring order into the puzzling variety of spectroscopic data; the physicists had tried to perform this task in a style similar to the one applied in biology by zoologists and botanists, when they organized the living species. Moreover, since no consistent atomic theory was available, one had been forced to use semiempirical rules as the theoretical basis for the organization of the spectroscopic data. Heisenberg himself had contributed importantly to providing some quantum-theoretical foundation for the semiempirical rules governing complex multiplets and anomalous Zeeman effects. This was the state of affairs, which he could discuss at the Kapitza Club, nothing more, because in July 1925 he possessed only a few preliminary ideas about how to connect the multiplet formulae to the quantum-mechanical scheme. Only several months later, after the discovery of electron spin and the

[333]Blackett, Dirac and Thomas, according to a list in the *Minute-Book*, were members of the Kapitza Club in the year 1924–1925. The attendance at Tuesday evening meetings of the Club was compulsory.

[334]Heisenberg himself claimed later that he had talked about the new quantum mechanics. He said: 'I was asked about this new paper of mine, and I explained all the details of the paper' (Heisenberg, Conversations, p. 273). However, from the title of the talk he inscribed in the *Minute-Book* of the Kapitza Club, it appears very unlikely that he made more than a few remarks about his latest work.

establishment of more powerful mathematical methods, would he and others be able to formulate a quantum-mechanical theory of complex atomic spectra.

While he had little opportunity at the meeting of the Kapitza Club, Heisenberg freely mentioned his recent attempts to formulate a consistent quantum-mechanical theory in private discussions with Ralph Fowler. Fowler, who had just been in Copenhagen and had worked on questions of the intensity of atomic and molecular spectra, was very excited by the report of his guest. As Heisenberg recalled: 'Fowler got very interested and told me, "As soon as you have something ready in print, could you send me the proofs of it?"' (Heisenberg, Conversations, p. 273). A few weeks later, when he received the proof-sheets of his paper, Heisenberg sent one copy to Cambridge, which Fowler passed on to his student Paul Dirac for closer examination. Dirac then began his own work on quantum mechanics, thus endowing Heisenberg's visit to Cambridge with an historic significance.

After his brief stay in Cambridge, Heisenberg went on a well-deserved vacation to Munich and the neighbouring mountains. This time he was not able to take off for the usual summer vacation of three months because he had to return to Copenhagen to complete the remaining tenure of his International Education Board Fellowship. He went to Copenhagen in the middle of September, from where he returned to Göttingen about a month later. However, in the meantime people in both places had become familiar with his new work. Kramers had brought the news from Göttingen to Copenhagen as early as the beginning of July, before Heisenberg had completed the paper on quantum-theoretical kinematics and mechanics. A few weeks later Pauli, who had, of course, seen the paper commented on Heisenberg's work to Kramers as follows:

> In particular I have greatly rejoiced in Heisenberg's bold attempts (of which you have probably heard in Göttingen). To be sure, one is still very far from saying something definite, and we stand at the very beginning of things. However, what has pleased me so very much in Heisenberg's considerations is the *method* of his procedure and the aspiration, with which he has embarked upon his considerations. Altogether I believe that, with respect to my scientific ideas, I have now come very close to Heisenberg, and that we now agree almost about everything, as much as this is possible at all for two independently thinking persons. I have also noticed it with pleasure that Heisenberg has learned some philosophical thinking from Bohr in Copenhagen, and largely turns his back now against the purely formal methods. Therefore, I wish him success in his endeavours with all my heart. (Pauli to Kramers, 27 July 1925)

Pauli knew that Heisenberg would be back in Copenhagen and wanted to make sure that people there would not interfere with the progress of his quantum-mechanical work. However, his warnings were quite unnecessary because Heisenberg's ideas received important support from the Göttingen theoreticians, who had studied his paper before it was submitted. Max Born and Pascual Jordan succeeded in establishing the matrix formulation of quantum mechanics. In August, Heisenberg learned about this progress, and from Copenhagen he began to collaborate with Born and Jordan on the theory of matrix mechanics.

References

Note: We have used the following abbreviations for journals and periodicals. Journals and periodicals not mentioned here are cited by their full titles. In the list of references we have given the dates, if available, of the particular issues in which the papers were published—except the *Sitzungsberichte* or *Comptes rendus* of the European academies. The references are arranged in alphabetical order according to the author, and in time-order according to the date of publication of the papers. Books or contributions to books (though not the yearbooks, which are treated as journals) have always been placed at the end of the annual list of a given author. The time-ordering letters (a, b, c, . . .) refer to all the papers of an author cited in Volumes 1 to 4.

American Journal of Physics:	*Amer. J. Phys.*
Annalen der Physik, fourth series:	*Ann. d. Phys. (4)*
Annales de chimie et physique:	*Ann. chim. & phys.*
Archives for the History of Quantum Physics (Interviews):	AHQP Interview
Encyklopädie der mathematischen Wissenschaften mit Einschluß ihrer Anwendungen:	*Encykl. d. math. Wiss.*
Jahrbuch der Bayerischen Akademie der Wissenschaften (München):	*Jahrb. d. Bayer. Akad. Wiss. (München)*
Journal of Mathematics and Physics of the M.I.T.:	*J. Math. & Phys. M.I.T.*
Journal of the Optical Society of America and Review of Scientific Instruments:	*J. Opt. Soc. America & Rev. Sci. Instr.*
Mathematisk-Fysiske Meddelelser, Det Kgl. Danske Videnskabernes Selskab:	*Kgl. Danske Vid. Selsk. Math.-fys. Medd.*
Nachrichten der Akademie der Wissenschaften in Göttingen:	*Nachr. Akad. Wiss. (Göttingen)*
Nachrichten von der Kgl. Gesellschaft der Wissenschaften zu Göttingen:	*Nachr. Ges. Wiss. (Göttingen)*
Naturwissenschaften:	*Naturwiss.*
Philosophical Magazine, fifth and sixth series:	*Phil. Mag. (5), (6)*
Philosophical Transactions of the Royal Society of London, series A	*Phil. Trans. Roy. Soc. (London)* A
The Physical Review, second series:	*Phys. Rev. (2)*

Physikalische Zeitschrift:	*Phys. Zs.*
Proceedings of the National Academy of Sciences (U.S.A.):	*Proc. Nat. Acad. Sci. (U.S.A.)*
Proceedings of the Royal Society of London, series A:	*Proc. Roy. Soc. (London)* A
Proceedings, Koninklijke Akademie van Wetenschappen te Amsterdam:	*Proc. Kon. Akad. Wetensch. (Amsterdam)*
Rendiconti del Circulo Matematico di Palermo:	*Rend. Circ. Mat. Palermo*
Sitzungsberichte der (Kgl.) Bayerischen Akademie der Wissenschaften (München), Mathematisch-physikalische Klasse:	*Sitz.ber. Bayer. Akad. Wiss. (München)*
Sitzungberichte der (Kgl.) Preussischen Akademie der Wissenschaften (Berlin), (from 1922) Physikalisch-mathematische Klasse:	*Sitz.ber. Preuss. Akad. Wiss. (Berlin)*
Skrifter, Det Kgl. Danske Videnskabernes Selskab:	*Kgl. Danske Videk. Selsk. Skrifter*
Verhandlungen der Deutschen Physikalischen Gesellschaft, series 2 (1899–1919) and 3 (since 1920):	*Verh. d. Deutsch. Phys. Ges. (2), (3)*
Verslag van de gewone Vergadering der wis- en natuurkundige Afdeeling, Koninklijke Akademie van Wetenschappen te Amsterdam:	*Versl. Kon. Akad. Wetensch. (Amsterdam)*
Zeitschrift für angewandte Mathematik und Mechanik:	*Z. angew. Math. & Mech.*
Zeitschrift für Elektrochemie und angewandte physikalische Chemie:	*Z. Elektrochem.*
Zeitschrift für Naturforschung:	*Z. Naturf.*
Zeitschrift für Physik:	*Z. Phys.*

Abraham, Max

1914 *Theorie der Elektrizität. Zweiter Band: Elektromagnetische Theorie der Strahlung*, third edition, Leipzig–Berlin: B. G. Teubner.

1918 *Theorie der Elektrizität. Erster Band: Einführung in die Maxwellsche Theorie der Elektrizität*, fourth edition, Leipzig: B. G. Teubner.

Ackermann, Walter

1915 Betrachtungen über Pyroelektrizität in ihrer Abhängigkeit von der Temperatur, *Ann. d. Phys. (4)* **46**, 197–220 (from the doctoral dissertation, University of Göttingen; received 27 October 1914, published in issue No. 2 of 28 January 1915).

Back, Ernst

1923b Der Zeemaneffekt des Bogen- und Funkenspektrums von Mangan, *Z. Phys.* **15**, 206–243 (received 5 March 1923, published in issue No. 4/5 of 26 May 1923).

1925a Über den Zeemaneffekt des Neons, *Ann. d. Phys. (4)* **76**, 317–332 (received 31 October 1924, published in issue 2/3 of January 1925, honouring F. Paschen's sixtieth birthday).

BLASIUS, H.

1913 *Das Ähnlichkeitsgesetz bei Reibungsvorgängen in Flüssigkeiten* (*Forschungsarbeiten aus dem Gebiete des Ingenieurwesens*, issue No. 131).

BOGUSLAWSKI, SERGEI

1914a Zur Theorie der Dielektrika. Temperaturabhängigkeit der Dielektrizitätskonstante. Pyroelektrizität, *Phys. Zs.* **15**, 283–288 (received 4 February 1914, published in issue No. 6 of 15 March 1914).

1914b Pyroelektrizität auf Grund der Quantentheorie, *Phys. Zs.* **15**, 569–572 (received 4 May 1914, published in issue No. 11 of 1 June 1914).

1914c Zu Herrn W. Ackermanns Messungen der Temperaturabhängigkeit der pyroelektrischen Erregung, *Phys. Zs.* **15**, 805–810 (received 29 July 1914, published in issue No. 17/18 of 15 September 1914).

BOHLIN, KARL

1888 Über eine neue Annäherungsmethode in der Störungstheorie, *Bihang till Kungl. Svenska Vetenskap Akademiens Handlinger (Stockholm)* **14**, *Afd. I*, No. 5 (communicated by H. Gyldén on 9 May 1888).

BOHR, NIELS

1906 *Physics Prize Essay* (submitted 30 October 1906 to the Royal Danish Academy of Sciences and Letters on the problem, posed in February 1905 to investigate the vibration of liquid jets), unpublished.

1909 Determination of the surface-tension of water by the method of jet-vibration, *Phil. Trans. Roy. Soc. (London)* **A209**, 281–317 (communicated by W. Ramsay; received 12 January 1909, read 21 January 1909); reprinted in *Collected Works 1*, pp. 29–80.

1910 On the determination of the tension of a recently formed water surface, *Proc. Roy. Soc. (London)* **A84**, 395–403 (communicated by Lord Rayleigh; received 22 August 1910, read 10 November and published in issue No. A572 of December 1910); reprinted in *Collected Works 1*, pp. 81–89.

1913b On the constitution of atoms and molecules. (Introduction and Part I), *Phil. Mag. (6)* **26**, 1–25 (dated 5 April 1913, published in issue No. 151 of July 1913); reprinted in *Constitution of Atoms and Molecules* (Bohr, 1963), pp. 1–25.

1918a On the quantum theory of line-spectra. Part I. On the general theory, *Kgl. Danske Vid. Selsk. Skrifter, 8. Raekke*, **IV.1**, 1–36 (dated November 1917, published in April 1918); reprinted in *Collected Works 3*, pp. 67–102.

1918b On the quantum theory of line-spectra. Part II. On the hydrogen spectrum, *Kgl. Danske Vid. Selsk. Skrifter, 8. Raekke*, **IV.1**, 37–100 (published in December 1918); reprinted in *Collected Works 3*, pp. 103–166.

1921a Atomic structure, *Nature* **107**, 104–107 (letter dated 14 February 1921, published in the issue of 24 March 1921); reprinted in *Collected Works 4*, pp. 72–82.

1921e Atomernes Bygning og Stoffernes fysiske og kemiske Egenskaber, *Fysisk Tidsskrift* **19**, 153–220 (presented at the meeting of 18 October 1921 of the Physical Society and the Chemical Society of Copenhagen, published in issue No. 5/6 of

30 December 1921); also published as a separate book with the same title by J. Gjellerups Forlag, Copenhagen 1922; reprinted in *Collected Works 4*, pp. 185–256. German translation: Der Bau der Atome und die physikalischen und chemischen Eigenschaften der Elemente, *Z. Phys.* **9**, 1–67 (received 3 January 1922, published in issue No. 112 of 15 March 1922); reprinted in *Drei Aufsätze über Spektren und Atombau* (Bohr, 1922e), pp. 70–146. English translation: The structure of the atom and the physical and chemical properties of the elements, in *The Theory of Spectra and Atomic Constitution* (second edition, 1924), pp. 61–126; reprinted in *Collected Works 4*, pp. 263–328.

1922d On the quantum theory of line-spectra. Part III. On the spectra of elements of higher atomic number, *Kgl. Danske Vid. Selsk. Skrifter, 8. Raekke*, **IV.1**, 101–118 (with an appendix dated October 1922, published 30 November 1922); reprinted in *Collected Works 3*, pp. 167–184. German translation: Über die Quantentheorie der Linienspektren. Teil III. Über die Spektren der Elemente mit höherer Atomzahl, in *Quantentheorie der Linienspektren* (Bohr, 1923f).

1922e *Drei Aufsätze über Spektren und Atombau* (No. 56 of the series *Tagesfragen aus den Gebieten der Naturwissenschaften und der Technik*), Braunschweig: Fr. Vieweg & Sohn. English publication (A. D. Udden, trans.): *The Theory of Spectra and Atomic Constitution*, Cambridge: Cambridge University Press, 1922 (second edition, 1924). French publication: *Les spectres et la structure de l'atome*, Paris: J. Hermann & Cie., 1922.

1923a Über die Anwendung der Quantentheorie auf den Atombau. I. Die Grundpostulate der Quantentheorie, *Z. Phys.* **13**, 117–165 (received 15 November 1922, published in issue No. 3 of 31 January 1923). English translation (by L. F. Curtiss): On the application of the quantum theory to atomic structure. Part I. The fundamental postulates, *Proc. Camb. Phil. Soc. Supplement* **22**, 1–44 (1924); reprinted in *Collected Works 3*, pp. 455–499.

1923c Linienspektren und Atombau, *Ann. d. Phys. (4)* **71**, 228–288 (received 15 March 1923, published in issue No. 9–12 of 23 May 1923, honouring H. Kayser's seventieth birthday); reprinted in *Collected Works 4*, pp. 549–610. English translation: Line-spectra and Atomic structure, in *Collected Works 4*, pp. 611–656.

1923f *Über die Quantentheorie der Linienspektren*, (translation of Bohr's papers 1918a, b and 1922d by P. Hertz), Braunschweig, Fr. Vieweg & Sohn.

1924 Zur Polarisation des Fluorescenzlichtes, *Naturwiss.* **12**, 1115–1117 (letter dated 1 November 1924, published in the issue of 5 December 1924).

1925a Über die Wirkung von Atomen bei Stößen, *Z. Phys.* **34**, 142–157 (received 30 March 1925, with a postscript of July 1925, published in issue No. 2/3 of 28 September 1925).

1925b Atomic theory and mechanics, *Nature* **116**, 845–852 (elaborated text of a lecture, presented on 30 August 1925 at the sixth Scandinavian Mathematical Congress, published in the supplement to the issue of 5 December 1925). German translation: Atomtheorie und Mechanik, *Naturwiss.* **14**, 1–10 (dated December 1925, published in the issue of 1 January 1926).

1961b Die Entstehung der Quantenmechanik, in *Werner Heisenberg und die Physik unserer Zeit* (F. Bopp, ed.), Braunschweig: Fr. Vieweg & Sohn, pp. IX–XII.

1972 *Collected Works. Volume 1: Early Work (1905–1911)* (J. Rud Nielsen, ed.), Amsterdam: North-Holland Publishing Company, and New York: American Elsevier Publishing Company, Inc.

1976 *Collected Works. Volume 3: The Correspondence Principle (1918–1923)* (J. Rud Nielsen, ed.), Amsterdam–New York–Oxford: North-Holland Publishing Company.

1977 *Collected Works. Volume 4: The Periodic System (1920–1923)* (J. Rud Nielsen, ed.) Amsterdam–New York–Oxford: North-Holland Publishing Company.

BOHR, NIELS, AND DIRK COSTER

1923 Röntgenspektrum und periodisches System der Elemente, *Z. Phys.* **12**, 342–374 (received 2 November 1922, published in issue No. 6 of 9 January 1923); reprinted in *Collected Works 4*, pp. 485–518. English translation: X-ray spectra and the periodic system of elements, in *Collected Works 4*, pp. 519–548.

BOHR, NIELS, HENDRIK ANTHONY KRAMERS, AND JOHN CLARKE SLATER

1924 The quantum theory of radiation, *Phil. Mag. (6)* **47**, 785–822 (dated January 1924, published in issue No. 281 of April 1924); reprinted in *Sources of Quantum Mechanics* (van der Waerden, 1967), pp. 159–176. German publication: Über die Quantentheorie der Strahlung, *Z. Phys.* **24**, 69–87 (received 22 February 1924, published in issue No. 2 of 22 May 1924); reprinted in *Collected Scientific Papers* (Kramers, 1956), pp. 271–289.

BORN, MAX

1909a Die träge Masse und das Relativitätsprinzip, *Ann. d. Phys. (4)* **28**, 571–584 (received 9 January 1909, published in issue No. 3 of 2 March 1909).

1916a Einsteins Theorie der Gravitation und der allgemeinen Relativität, *Phys. Zs.* **17**, 51–69 (received 13 February 1916, published in issue No. 4 of 15 February 1916).

1922 Über das Modell der Wasserstoffmolekel, *Naturwiss.* **10**, 677–678 (letter dated 27 June 1922, published in the issue of 4 August 1922).

1923a Quantentheorie und Störungsrechnung, *Naturwiss.* **11**, 537–542 (published in the issue of 6 July 1923).

1923b Atomtheorie des festen Zustandes (Dynamik der Kristallgitter), *Encykl. d. math. Wiss. V/3*, pp. 527–781 (dated 7 September 1923, published in issue No. 4 of 24 October 1923).

1924b Über Quantenmechanik, *Z. Phys.* **26**, 379–395 (received 13 June 1924, published in issue No. 6 of 20 August 1924); reprinted in *Ausgewählte Abhandlungen 2*, pp. 61–77, and in *Begründung der Matrizenmechanik* (Born, Heisenberg and Jordan, 1962), pp. 13–29. English translation: Quantum mechanics, in *Sources of Quantum Mechanics* (van der Waerden, 1967), pp. 181–198.

1925 *Vorlesungen über Atommechanik*, Berlin: J. Springer Verlag. English translation (by J. W. Fisher): *Mechanics of the Atom*, London: Bell, 1927.

1955b Die statistische Deutung der Quantenmechanik (Nobel lecture delivered on 11 December 1954 at Stockholm), in *Les Prix Nobel en 1954*, Stockholm: Nobel Foundation, pp. 79–90; reprinted in *Ausgewählte Abhandlungen 2*, pp. 430–441. English translation: The statistical interpretation of quantum mechanics, in *Nobel Lectures: Physics 1942–1962* (Nobel Foundation, ed.), Amsterdam–New York: Elsevier Publishing Company, 1964, pp. 256–267.

1962 *Ausgewählte Abhandlungen*, 2 volumes, Göttingen: Vandenhoeck & Ruprecht.

1968 *My Life and My Views*, New York: Charles Scribner's Sons.

1978 *My Life: Recollections of a Nobel Laureate*, London: Taylor & Francis Ltd., New York: Charles Scribner's Sons. German translation: *Mein Leben: Die Erinnerungen eines Nobelpreisträgers*, Munich: Nymphenburger Verlagshandlung GmbH, 1975.

BORN, MAX, AND E. BRODY

1921b Über die Schwingungen eines mechanischen Systems mit endlicher Amplitude und ihre Quantelung, *Z. Phys.* **6**, 140–152 (received 9 July 1921, published in issue No. 2 of 27 August 1921); reprinted in *Ausgewählte Abhandlungen 1* (Born, 1962), pp. 418–430.

BORN, MAX, AND WERNER HEISENBERG

1923a Über Phasenbeziehungen bei den Bohrschen Modellen von Atomen und Molekeln, *Z. Phys.* **14**, 44–55 (received 16 January 1923, published in issue No. 1 of 5 March 1923).

1923b Die Elektronenbahnen im angeregten Heliumatom, *Z. Phys.* **16**, 229–243 (received 11 May 1923, published in issue No. 4 of 9 July 1923); reprinted in *Ausgewählte Abhandlungen 2* (Born, 1962), pp. 23–37.

1924a Zur Quantentheorie der Molekeln, *Ann. d. Phys. (4)* **74**, 1–31 (received 31 December 1923, published in issue No. 9 of April 1924).

1924b Über den Einfluß der Deformierbarkeit der Ionen auf optische und chemische Konstanten. I., *Z. Phys.* **23**, 388–410 (received 22 March 1924, published in issue No. 6 of 10 May 1924); reprinted in *Ausgewählte Abhandlungen 2* (Born, 1962), pp. 38–60.

BORN, MAX, WERNER HEISENBERG AND PASCUAL JORDAN

1926 Zur Quantenmechanik II., *Z. Phys.* **35**, 557–615 (received 16 November 1925, published in issue No. 8/9 of 4 February 1926); reprinted in *Ausgewählte Abhandlungen 2* (Born, 1962), pp. 155–213, and in *Begründung der Matrizenmechanik* (Born, Heisenberg and Jordan, 1962), pp. 77–135. English translation: On quantum mechanics II, in *Sources of Quantum Mechanics* (van der Waerden, 1967), pp. 321–385.

1962 *Zur Begründung der Matrizenmechanik (Dokumente der Naturwissenschaften—Abteilung Physik*, Vol, 2; A. Hermann, ed.), Stuttgart: E. Battenberg Verlag.

BORN, MAX, AND ERICH HÜCKEL

1923 Zur Quantentheorie mehratomiger Molekeln, *Phys. Zs.* **24**, 1–12 (received 1 November 1922, published in issue No. 1 of 1 January 1923).

BORN, MAX, AND PASCUAL JORDAN

1925a Zur Quantentheorie aperiodischer Vorgänge, *Z. Phys.* **33**, 479–505 (received 11 June 1925, published in issue No. 7 of 15 August 1925); reprinted in *Ausgewählte Abhandlungen 2* (Born, 1962), pp. 97–123.

1925b Zur Quantenmechanik, *Z. Phys.* **34**, 858–888 (received 27 September 1925, published in issue No. 11/12 of 28 November 1925); reprinted in *Ausgewählte Abhandlungen 2* (Born, 1962), pp. 124–154, and in *Begründung der Matrizenmechanik* (Born, Heisenberg and Jordan, 1962), pp. 47–76. English translation (except Chapter 4): On quantum mechanics, in *Sources of Quantum Mechanics* (van der Waerden, 1967), pp. 277–306.

BORN, MAX, AND WOLFGANG PAULI

1922 Über die Quantelung gestörter mechanischer Systeme, *Z. Phys.* **10**, 137–158 (received 29 May 1922, published in No. 3 of 7 July 1922); reprinted in *Ausgewählte Abhandlungen 2* (Born, 1962), pp. 1–22, and in *Collected Scientific Papers 2* (Pauli, 1964), pp. 48–69.

BOTHE, WALTHER, AND HANS GEIGER

1925a Experimentelles zur Theorie von Bohr, Kramers und Slater, *Naturwiss.* **13**, 440–441 (letter dated 18 April 1925, published in the issue of 15 May 1925).

BREIT, GREGORY

1924a The polarization of resonance radiation, *Phil. Mag. (6)* **47**, 832–842 (published in issue No. 281 of May 1924).

1924b The quantum theory of dispersion, *Nature* **114**, 310 (letter published in the issue of 30 August 1924).

BURGER, HERMAN CAREL, AND HENDRIK BEREND DORGELO

1924 Beziehungen zwischen inneren Quantenzahlen und Intensitäten von Mehrfachlinien, *Z. Phys.* **23**, 258–266 (received 8 March 1924, published in issue No. 3/4 of 19 April 1924).

CATALÁN, MIGUEL A.

1922 Series and other regularities in the spectrum of manganese, *Phil. Trans. Roy. Soc. (London)* **A223**, 127–173 (communicated by A. Fowler; received 22 February 1922, read 23 March and published 20 July 1922).

CHARLIER, CARL LUDWIG

1902, 1907 *Die Mechanik des Himmels. Vorlesungen*, 2 volumes, Leipzig: Veit & Companie.

COUETTE, M.

1890 Études sur le frottement des liquides, *Ann. chim. & phys. (6)* **21**, 433–510 (published December 1890).

COURANT, RICHARD, AND DAVID HILBERT

1924 *Methoden der mathematischen Physik. (Erster Band)*, Berlin: J. Springer Verlag.

DEBYE, PETER

1914 Zustandsgleichung und Quantenhypothese mit einem Anhang über Wärmeleitung (Wolfskehl lecture, April 1913), in *Vorträge über die kinetische Theorie, etc.* (Planck, Debye, Nernst, Smoluchowski, Sommerfeld, Lorentz, et al, 1914), pp. 17–60.

1916a Quantenhypothese und Zeeman-Effekt, *Nachr. Ges. Wiss. Göttingen*, pp. 142–153 (presented at the meeting of 3 June 1916).

1916b Quantenhypothese und Zeeman-Effekt, *Phys. Zs.* **17**, 507–512 (received 7 September 1916, published in issue No. 20 of 15 October 1916).

DEWEY, JANE M.

1926 Intensities in the Stark effect of helium, *Phys. Rev. (2)* **28**, 1108–1124 (dated 28 August 1926, published in issue No. 6 of December 1926).

1927 Intensities in the Stark effect of helium: II, *Phys. Rev. (2)* **30**, 770–780 (dated July 1927, published in issue No. 6 of December 1927).

DIRAC, PAUL ADRIEN MAURICE

1978 The development of quantum mechanics (lecture presented on 25 August 1975 at the University of New South Wales, Kensington, Sydney, Austrialia), in Dirac: *Directions in Physics* (H. Hora and J. R. Shepanski, eds.), New York–London–Syndey–Toronto: J. Wiley & Sons.

DÖLGER, FRANZ

1933 August Heisenberg. Geboren 13. November 1869, gestorben 22. November 1930, *Jahresberichte über die Fortschritte der klassischen Altertumswissenschaft* **59**, 25–55.

DORGELO, HENDRIK BEREND

1924 Die Intensität mehrfacher Spektrallinien, *Z. Phys.* **22**, 170–177 (received 23 January 1924, published in issue No. 3 of 10 March 1924).

EDDINGTON, ARTHUR STANLEY

1923b *The Mathematical Theory of Relativity*, Cambridge (England): Cambridge University Press.

1925 Hugo von Seeliger, *Monthly Notices of the Royal Astronomical Society (London)* **85**, 316–318 (published in issue No. 4 of 13 February 1925).

EINSTEIN, ALBERT

1905c Über die von der molekulartheoretischen Theorie der Wärme geforderte Bewegung von in ruhenden Flüssigkeiten suspendierten Teilchen, *Ann. d. Phys. (4)* **17**, 549–560 (received 11 May 1905, published in issue No. 8 of 15 July 1905); reprinted in *Untersuchungen über die Theorie der Brownschen Bewegungen (Ostwalds Klassiker der exakten Naturwissenschaften*, No. 199), Leipzig: Akademische Verlagsgesellschaft, 1922. English translation (by A. D. Cowper): On the movement of small particles suspended in a stationary liquid demanded by the molecular theory of heat, in *Investigations on the Theory of Brownian Movement*, London: Methuen, 1926.

1905d Zur Elektrodynamik bewegter Körper, *Ann. d. Phys. (4)* **17**, 891–921 (received 30 June 1905, published in issue No. 10 of 26 September 1905); reprinted in *Das Relativitätsprinzip* (Lorentz, Einstein and Minkowski, 1913), pp. 27–52. English translation: On the electrodynamics of moving bodies, in *The Principle of Relativity* (Lorentz, Einstein, Minkowski and Weyl, 1923), pp. 35–65.

1916a Ernst Mach, *Phys. Zs.* **17**, 101–104 (received 14 March 1916, published in issue No. 7 of 1 April 1916).

1916d Strahlungs-Emission und -Absorption nach der Quantentheorie, *Verh. d. Deutsch. Phys. Ges. (2)* **18**, 318–323 (received 17 July 1916, presented at the meeting of 21 July and published in issue No. 13/14 of 30 July 1916).

1919 Spielen Gravitationsfelder im Aufbau der materiellen Elementarteilchen eine wesentliche Rolle?, *Sitz.ber. Preuss. Akad. Wiss. (Berlin)*, pp. 349–356 (communicated to the meeting of 10 April 1919); reprinted in *Das Relativitätsprinzip* (third edition, 1920). English translation: Do gravitational fields play an essential part in the structure of elementary particles?, in *The Principle of Relativity* (Lorentz, Einstein, Minkowski and Weyl, 1923), pp. 189–198.

1922b Zur Theorie der Lichtfortpflanzung in dispergierenden Medien, *Sitz.ber. Preuss. Akad. Wiss. (Berlin)*, pp. 18–22 (presented at the meeting of 2 February 1922).

1923d Bietet die Feldtheorie Möglichkeiten zur Lösung des Quantenproblems?, *Sitz.ber. Preuss. Akad. Wiss. (Berlin)*, pp. 359–364 (presented at the meeting of 13 December 1923).

1924c Quantentheorie des einatomigen idealen Gases, *Sitz.ber. Preuss. Akad. Wiss. (Berlin)*, pp. 261–267 (presented at the meeting of 10 July 1924).

1925a Quantentheorie des einatomigen idealen Gases. 2. Abhandlung, *Sitz.ber. Preuss. Akad. Wiss. (Berlin)*, pp. 3–14 (presented at the meeting of 8 January 1925).

1925b Quantentheorie des idealen Gases, *Sitz.ber. Preuss. Akad. Wiss. (Berlin)*, pp. 18–25 (presented at the meeting of 29 January 1925).

1949 Autobiographisches—Autobiographical Notes, in *Albert Einstein: Philosopher Scientist*, (P. A. Schilpp, ed.), New York: Tudor Publishing Company, pp. 1–95.

EINSTEIN, ALBERT, HEDWIG BORN AND MAX BORN

1969 *Briefwechsel 1916–1955, kommentiert von Max Born*, München: Nymphenburger Verlagshandlung. English translation (I. Born): *The Born–Einstein Letters. Correspondence between Albert Einstein and Max and Hedwig Born from 1916 to 1955 with commentaries by Max Born*, London and Basingstoke: The Macmillan Press Ltd., 1971.

EINSTEIN, ALBERT, AND PAUL EHRENFEST

1922 Quantentheoretische Bemerkungen zum Experiment von Stern und Gerlach, *Z. Phys.* **11**, 31–34 (received 21 August 1922, published in issue No. 1 of 16 September 1922); reprinted in Ehrenfest: *Collected Scientific Papers* (M. J. Klein, ed.), Amsterdam: North-Holland Publishing Company, 1959, pp. 452–455.

EINSTEIN, ALBERT, AND WANDER JOHANNES DE HAAS

1915 Experimenteller Nachweis der Ampèreschen Molekularströme, *Verh. d. Deutsch. Phys. Ges. (2)* **17**, 152–170 (presented at the meeting of 19 February 1915, extended manuscript received 10 April 1915, published in issue No. 8 of 30 April 1915). Dutch translation: Profondervindelijk bewijs voor het bestaan der moleculaire stroomen van Ampère, *Versl. Kon. Akad. Wetensch. (Amsterdam)* **23**, 1449–1464.

EINSTEIN, ALBERT, AND ARNOLD SOMMERFELD

1968 *Briefwechsel* (A. Herman, ed.), Basle–Stuttgart: Schwabe & Co.

ELDRIDGE, JOHN A.

1924b Theoretical interpretation of the polarization experiment of Wood and Ellett, *Phys. Rev. (2)* **24**, 234–242 (dated 29 March 1924, published in issue No. 3 of September 1924).

EPSTEIN, PAUL SOPHUS

1916a Zur Theorie des Starkeffekts, *Phys. Zs.* **17**, 148–150 (received 29 March 1916, published in issue No. 8 of 15 April 1916).

1916c Zur Theorie des Starkeffektes, *Ann. d. Phys. (4)* **50**, 489–521 (received 9 May 1916, published in issue No. 13 of 25 July 1916).

1916e Zur Quantentheorie, *Ann. d. Phys. (4)* **51**, 168–188 (received 9 August 1916, published in issue No. 18 of 10 October 1916).

FERMI, LAURA

1954 *Atoms in the Family: My Life with Enrico Fermi*, Chicago: The University of Chicago Press.

FOSTER, JOHN STUART

1927a Stark patterns observed in helium, *Proc. Roy. Soc. (London)* **A114**, 47–66 (communicated by A. S. Eve; received 1 November 1926, published in issue No. A766 of 1 February 1927).

1927b Theory of the Stark effect in the arc spectra of helium, *Phys. Rev. (2)* **29**, 916 (abstract of a paper presented at the Washington meeting of the American Physical Society, 22–23 April 1927, published in issue No. 6 of June 1927).

1927c Application of quantum mechanics to the Stark effect in helium, *Proc. Roy. Soc. (London)* **A117**, 137–163 (communicated by N. Bohr; received 8 August 1927, published in issue No. A776 of 1 December 1927).

FOWLER, RALPH HOWARD

1925a Applications of the correspondence principle to the theory of line-intensities in band-spectra, *Phil. Mag. (6)* **49**, 1272–1288 (published in issue No. 294 of June 1925).

1925d A note on the summation rules for the intensities of spectral lines, *Phil. Mag. (6)* **50**, 1079–1083 (published in issue No. 299 of November 1925).

FUES, ERWIN

1926 Das Eigenschwingungsspektrum zweiatomiger Moleküle in der Undulationsmechanik, *Ann. d. Phys. (4)* **80**, 367–396 (received 27 April 1926, published in issue No. 12 of 22 June 1926).

GAVIOLA, E., AND PETER PRINGSHEIM

1924 Über die Polarisation der Natrium-Resonanzstrahlung in magnetischen Feldern, *Z. Phys.* **25**, 367–377 (received 24 May 1924, published in issue No. 4–6 of 4 August 1924).

GIESELER, HILDE

1922 Serienzusammenhänge im Bogenspektrum des Chroms, *Ann. d. Phys. (4)* **69**, 147–160 (received 19 June 1922, published in issue No. 18 of 2 November 1922).

GÖTZE, RAIMUND

1921 Liniengruppen und innere Quanten, *Ann. d. Phys. (4)* **66**, 285–292 (received 6 October 1921, published in issue No. 20 of 20 December 1921).

GOUDSMIT, SAMUEL, AND RALPH DE LAER KRONIG

1925 Die Intensität der Zeemankomponenten, *Naturwiss.* **12**, 90 (letter dated 17 December 1924, published in the issue of 30 January 1925).

HANLE, WILHELM

1923 Über den Zeemaneffekt bei der Resonanzfluoreszenz, *Naturwiss.* **11**, 690–691 (letter dated 28 June 1923, published in the issue of 10 August 1923).

1924 Über magnetische Beeinflussung der Polarisation der Resonanzfluoreszenz, *Z. Phys.* **30**, 93–105 (received 1 November 1924, published in issue No. 2 of 12 December 1924).

1926 Die elektrische Beeinflussung der Polarisation der Resonanzfluoreszenz von Quecksilber, *Z. Phys.* **35**, 346–364 (received 12 November 1925, published in issue No. 5 of 9 January 1926).

1927 Die Polarisation der Resonanzfluoreszenz von Natriumdampf bei Anregung mit zirkular polarisiertem Licht, *Z. Phys.* **41**, 164–183 (received 22 December 1926, published in issue No. 2/3 of 10 February 1927).

HANSEN, HANS MARIUS, TOSHIO TAKAMINE AND SVEN WERNER

1923 On the effect of magnetic and electric fields on the mercury spectrum, *Kgl. Danske Vid. Selsk. Math.-fys. Medd.* **5**, No. 3 (dated October 1922, ready for printing 3 August 1923).

HEISENBERG, WERNER

1922a Zur Quantentheorie der Linienstruktur und der anomalen Zeemaneffekte, *Z. Phys.* **8**, 273–297 (received 17 December 1921, published in No. 5 of 15 February 1922).

1922b Die absoluten Dimensionen der Kármánschen Wirbelbewegung, *Phys. Zs.* **23**, 363–366 (received 18 July 1922, published in issue No. 18 of 15 September 1922).

1924a Quantitatives über die Deformierbarkeit edelgasähnlicher Ionen, *Verh. d. Deutsch. Phys. Ges. (3)* **5**, 7–8 (abstract of a paper presented at the Baunschweig meeting of the *Gauverein Niedersachsen* of the German Physical Socitey, 9–10 February 1924, published in issue No. 1 of 31 March 1924).

1924b Bemerkungen zu einer Arbeit von F. v. Wiśniewski: "Zur Theorie des Heliums," *Z. Phys.* **25**, 175–176 (received 21 May 1924, published in No. 2 of 2 July 1924).

1924c Über die Stabilität und Turbulenz von Flüssigkeitsströmen, *Ann. d. Phys. (4)* **74**, 577–627 (doctoral dissertation; received 20 February 1924, published in issue No. 15 of July 1924).

1924d Über den Einfluß der Deformierbarkeit der Ionen auf optische Konstanten. II. Stabilität und Bildungswärme dreiatomiger Molekeln und Ionen, *Z. Phys.* **26**, 196–204 (received 4 June 1924, published in issue No. 3 of 11 August 1924).

1924e Über eine Abänderung der formalen Regeln der Quantentheorie beim Problem der anomalen Zeemaneffekte, *Z. Phys.* **26**, 291–307 (received 13 June 1924, published in issue No. 4/5 of 14 August 1924).

1924f Nichtlineare Lösungen der Differentialgleichungen für reibende Flüssigkeiten, in *Vorträge aus dem Gebiete der Hydro- und Aerodynamik (Innsbruck 1922)* (T. v. Kármán and T. Levi-Cività, eds.), Berlin: J. Springer Verlag, pp. 139–142 (presented in September 1922 at the Innsbruck meeting).

1925a Über eine Anwendung des Korrespondenzprinzips auf die Frage nach der Polarisation des Fluoreszenzlichtes, *Z. Phys.* **31**, 617–626 (received 30 November 1924, published in issue No. 7/8 of 17 March 1925).

1925b Zur Quantentheorie der Multiplettstruktur und der anomalen Zeemaneffekte, *Z. Phys*, **32**, 841–860 (received 10 April 1925, published in issue No. 11/12 of 30 June 1925).

1925c Über die quantentheoretische Umdeutung kinematischer und mechanischer Beziehungen, *Z. Phys.* **33**, 879–893 (received 29 July 1925, published in issue No. 12 of 18 September 1925); reprinted in *Begründung der Quantenmechanik* (Born, Heisenberg and Jordan, 1962), pp. 31–46. English translation: Quantum-theoretical re-interpretation of kinematic and mechanical relations, in *Sources of Quantum Mechanics* (van der Waerden, 1967), pp. 261–276.

1943a Die "beobachtbaren Größen" in der Theorie der Elementareilchen, *Z. Phys.* **120**, 513–538 (received 8 September 1942, published in issue No. 7–10 of 25 March 1943).

1943b Die beobachtbaren Größen in der Theorie der Elementarteilchen. II, *Z. Phys.* **120**, 673–702 (received 30 October 1942, published in issue No. 11–12 of 6 April 1943).

1944 Die beobachtbaren Größen in der Theorie der Elementarteilchen. III, *Z. Phys.* **123**, 93–112 (received 12 May 1944, published in issue No. 1/2 of 10 October 1944).

1948a Zur statistischen Theorie der Turbulenz, *Z. Phys.* **124**, 628–657 (received 16 December 1946, published in issue No. 7–12 of 8 September 1948).

1948b Bemerkungen zum Turbulenzproblem, *Z. Naturf.* **3a**, 434–437 (received 15 June 1948, published in issue No. 8–11 of August–October 1948).

1948c On the theory of statistical and isotropic turbulence, *Proc. Roy. Soc. (London)* **A195**, 402–406 (communicated by G. I. Taylor; received 15 June 1948, published in issue No. A1042 of 22 December 1948).

1949 Über die Enstehung der Mesonen in Vielfachprozessen, *Z. Phys.* **126**, 569–582 (received 28 May 1949, published in issue No. 6 of 15 July 1949).

1952 On the stability of laminar flow, in *Proceedings of the International Congress of Mathematicians, Volume II*, Providence (R.I.): American Mathematical Society, pp. 292–296 (presented in the Section 'Conference on Applied Mathematics,' 30 August–6 September 1950).

1953 Zur Quantisierung nichtlinearer Gleichungen, *Nachr. Akad. Wiss. Göttingen, Math.-phys.-chem. Abteilung*, IIa, pp. 111–127 (presented at the meeting of 6 November 1953).

1955 *Das Naturbild der heutigen Physik*, Hamburg: Rowohlt.

1958a *Physics and Philosophy. The Revolution in Modern Science* (Gifford Lectures, University of St. Andrews, winter 1955–1956), New York: Harper and Row (World Perspectives, No. 19). In German: *Physik und Philosophie*, Stuttgart: Hirzel Verlag, 1959.

1958b *Festrede*, (presented on the occasion of the 800th anniversary of Munich), Munich: Münchner Zeitungsverlag.

1958c *The Physicist's Conception of Nature*, New York: Harcourt, Brace & Company (translation by A. J. Pomerans of *Das Naturbild der heutigen Physik*, Heisenberg, 1955).

1969b *Der Teil und das Ganze: Gespräche im Umkreis der Atomphysik*, Munich: R. Piper & Co.

1971 *Physics and Beyond: Encounters and Conversations*, (English translation of Heisenberg, 1969b, by A. J. Pomerans), New York–Evanston–London: Harper & Row Publishers, Inc.; paperback edition in Harper Torchbooks, 1972.

1977 *Tradition in der Wissenschaft: Reden und Aufsätze*, Munich: R. Piper & Co.

HEISENBERG, WERNER, AND WOLFGANG PAULI

1929 Zur Quantentheorie der Wellenfelder, *Z. Phys.* **56**, 1–61 (received 19 March 1929, published in issue No. 1/2 of 8 July 1929).

1930 Zur Quantentheorie der Wellenfelder. II, *Z. Phys.* **59**, 168–190 (received 7 September 1929, published in issue No. 3/4 of 2 January 1930).

HELLER, K. D.

1964 *Ernst Mach: Wegbereiter der modernen Physik*, Vienna & New York: Springer-Verlag.

HERMANN, ARMIN

1976 *Werner Heisenberg in Selbstzeugnissen und Bilddokumenten* (*Rowohlts Monographien*, No. 240) Hamburg: Rowohlt Taschenbuch Verlag.

1977 *Die Jahrhundertwissenschaft: Werner Heisenberg und die Physik seiner Zeit*, Stuttgart: Deutsche Verlagsanstalt.

HÖNL, HELMUT

1925 Die Intensitäten der Zeemankomponenten, *Z. Phys.* **31**, 340–354 (received 26 November 1924, published in No. 1 – 4 of 11 February 1925).

HOPF, LUDWIG

1910 Turbulenz bei einem Flusse, *Ann. d. Phys. (4)* **32**, 777–808 (from the doctoral dissertation, University of Munich, 1909; received 18 February 1910, published in issue No. 9 of 21 June 1910).

1914 Der Verlauf kleiner Schwingungen auf einer Strömung reibender Flüssigkeit, *Ann. d. Phys. (4)* **44**, 1–60 (*Habilitation* thesis, Aachen; received 20 January 1914, published in issue No. 9 of 28 April 1914).

HOPF, LUDWIG, AND ARNOLD SOMMERFELD

1911 Über komplexe Integraldarstellungen der Zylinderfunktionen, *Archiv der Mathematik und Physik* **18**, 1–16, reprinted in *Gesammelte Schriften I* (Sommerfeld, 1968), pp. 256–271.

HUND, FRIEDRICH

1923 Theoretische Betrachtungen über die Ablenkung von freien langsamen Elektronen in Atomen, *Z. Phys.* **13**, 241–263 (received 21 December 1922, published in issue No. 4 of 10 February 1923).

1924 Rydbergkorrektionen und Radien der Atomrümpfe, *Z. Phys.* **22**, 405–415 (received 25 Febrary 1924, published in issue No. 6 of 4 April 1924).

1925a Die Gestalt mehratomiger polarer Molekeln. I., *Z. Phys.* **31**, 81–106 (received 4 December 1924, published in issue No. 1–4 of 11 February 1925).

1925b Die Gestalt mehratomiger polarer Molekeln. II. Molekeln, die aus einem negativen Ion und aus Wasserstoffkernen bestehen, *Z. Phys.* **32**, 1–18 (received 26 February 1925, published in issue No. 1 of 22 April 1925).

1925c Zur Deutung verwickelter Spektren, insbesondere der Elemente Scandium bis Nickel, *Z. Phys.* **33**, 345–371 (received 22 June 1925, published in issue No. 5/6 of 8 August 1925).

1925e Zur Deutung verwickelter Spektren. II., *Z. Phys.* **34**, 296–308 (received 20 August 1925, published in issue No. 4 of 5 October 1925).

1961 Göttingen, Kopenhagen, Leipzig im Rückblick, in *Werner Heisenberg und die Physik unserer Zeit* (F. Bopp, ed.), Braunschweig: Fr. Vieweg & Sohn, pp. 1–7.

JAFFÉ, GEORGE

1920 Bemerkung über die Entstehung von Wirbeln in Flüssgkeiten, *Phys. Zs.* **21**, 541–543 (received 29 July 1920, published in issue No. 20 of 15 October 1920).

JAHNKE E., AND FRITZ EMDE

1909 *Funktionentafeln mit Formeln und Kurven*, Leipzig & Berlin: B. G. Teubner.

JOOS, GEORG

1924 Der Einfluß eines Magnetfeldes auf die Polarisation des Resonanzlichts, *Phys. Zs.* **25**, 130–134 (received 11 March 1924, published in issue No. 6 of 15 March 1924).

JORDAN, PASCUAL

1925c Bemerkungen zur Theorie der Atomstruktur, *Z. Phys.* **33**, 563–570 (received 8 July 1925, published in issue No. 8 of 18 August 1925).

KAPITZA, PETER

1924 A method of producing strong magnetic fields, *Proc. Roy. Soc. (London)* **A105**, 691–710 (communicated by E. Rutherford; received 9 April 1924, published in issue No. A734 of 2 June 1924).

KAPITZA, PETER AND H. W. B. SKINNER

1924 The Zeeman effect in strong magnetic fields, *Nature* **114**, 273 (letter dated 20 July 1924, published in the issue of 23 August 1924).

KÁRMÁN, THEODOR VON

1911 Über den Mechanismus des Widerstandes, den ein bewegter Körper in einer Flüssigkeit erfährt, *Nachr. Ges. Wiss. Göttingen*, pp. 509–517 (presented by F. Klein at the meeting of 14 September 1911).

KÁRMÁN, THEODOR VON, AND H. RUBACH

1912 Über den Mechanismus des Flüssigkeits- und Luftwiderstandes, *Phys. Zs.* **13**, 49–59 (received 21 December 1911, published in issue No. 2 of 15 January 1912).

KELVIN, WILLIAM THOMSON, LORD

1887 Stability of fluid motion—rectilineal motion of viscous fluid between two parallel planes, *Phil. Mag. (5)* **24**, 188–196 (read before the Royal Society of Edinburgh on 18 July 1887, published in issue No. 147 of August 1887); reprinted in *Mathematical and Physical Papers 4*, pp. 321–330.

KEMBLE, EDWIN CRAWFORD

1925 The application of the correspondence principle to degenerate systems and the relative intensities of band lines, *Phys. Rev. (2)* **25**, 1–22 (dated 11 October 1924, published in issue No. 1 of January 1925).

KLEIN, OSKAR

1924a The simultaneous action on a hydrogen atom of crossed homogeneous electric and magnetic fields, *Phys. Rev. (2)* **23**, 308 (abstract of a paper presented at the Cincinnati meeting of the American Physical Society, 27–29 December 1923, published in issue No. 2 of February 1924).

1924b Über die gleichzeitige Wirkung von gekreuzten homogenen elektrischen und magnetischen Feldern auf das Wasserstoffatom. I, *Z. Phys.* **22**, 109–118 (dated 31 December 1923, received 23 January 1924 and published in issue No. 1/2 of 6 March 1924).

KRAMERS, HENDRIK ANTHONY

1919 Intensities of spectral lines: On the application of the quantum theory to the problem of the relative intensities of the components of the fine structure and of the Stark effect of these lines of the hydrogen spectrum, *Kgl. Danske Vid. Selsk. Skrifter, 8. Raekke, III.3*; reprinted in *Collected Scientific Papers*, pp. 3–108.

1920a On the application of Einstein's theory of gravitation to a stationary field of gravitation, *Proc. Kon. Akad. Wetensch. (Amsterdam)* **23**, 1052–1073 (communicated at the meeting of 25 September 1920); reprinted in *Collected Scientific Papers*, pp. 134–155.

1920b Über den Einfluß eines elektrischen Feldes auf die Feinstruktur der Wasserstofflinien, *Z. Phys.* **3**, 199–223 (received 1 October 1920, published in issue No. 4 of November 1920); reprinted in *Collected Scientific Papers*, pp. 109–133.

1923a Über das Modell des Heliumatoms, *Z. Phys.* **13**, 312–341 (received 31 December 1922, published in issue No. 5 of 19 February 1923), reprinted in *Collected Scientific Papers*, pp. 192–221.

1923b Über die Quantelung rotierender Moleküle, *Z. Phys.* **13**, 343–350 (received 3 January 1923, published in issue No. 6 of 26 February 1923); reprinted in *Collected Scientific Papers*, pp. 223–230.

1923d On the theory of X-ray absorption and of the continuous X-ray spectrum, *Phil. Mag. (6)* **46**, 836–871 (published in No. 275 of November 1923); reprinted in *Collected Scientific Papers*, pp. 156–191.

1924a The law of dispersion and Bohr's theory of spectra, *Nature* **113**, 673–674 (letter dated 25 March 1924, published in the issue of 10 May 1924); reprinted in *Collected Scientific Papers*, pp. 290–291, and in *Sources of Quantum Mechanics* (van der Waerden, 1967), pp. 177–180.

1924b The quantum theory of dispersion, *Nature* **114**, 310–311 (letter dated 22 July 1924, published in the issue of 30 August 1924); reprinted in *Collected Scientific Papers*, p. 292, and in *Sources of Quantum Mechanics* (van der Waerden, 1967), pp. 199–201.

1924c Die chemischen Eigenschaften der Atome nach der Bohrschen Theorie, *Naturwiss.* **12**, 1050–1054 (presented on 23 September 1924 at the 88th *Naturforscherversammlung*, Innsbruck, published in the issue of 21 November 1924).

1925a Om Vekselvirkningen mellem Lys og Stof, *Fysisk Tidsskrift* **23**, 26–40 (published in issue No. 1/2 of February 1925).

1925b On the behaviour of atoms in an electromagnetic wave field, *6e Skand. Mat. Kongress*, pp. 143–153; reprinted in *Collected Scientific Papers*, pp. 321–331.

1956 *Collected Scientific Papers*, Amsterdam: North-Holland Publishing Company.

KRAMERS, HENDRIK ANTHONY, AND WERNER HEISENBERG

1925 Über die Streuung von Strahlung durch Atome, *Z. Phys.* **31**, 681–708 (received 5 January 1925, published in issue No. 9 of 17 March 1925); reprinted in *Collected Scientific Papers* (Kramers, 1956), pp. 293–320. English translation: On the dispersion of radiation by atoms, in *Sources of Quantum Mechanics* (van der Waerden, 1967), pp. 223–257.

KRAMERS, HENDRIK ANTHONY, AND WOLFGANG PAULI

1923 Zur Theorie der Bandenspektren, *Z. Phys.* **13**, 351–367 (received 3 January 1923, published in issue No. 6 of 26 February 1923); reprinted in *Collected Scientific Papers* (Kramers, 1956), pp. 231–247, and in *Collected Scientific Papers* (Pauli, 1964), pp. 134–150.

KRATZER, ADOLF

1920a Eine ultraroten Rotationsspektren der Halogenwasserstoffe, *Z. Phys.* **3**, 289–307 (received 14 October 1920, published in issue No. 5 of December 1920).

1920b Die spektroskopische Bestätigung der Isotopen des Chlors, *Z. Phys.* **3**, 460–465 (received 28 November 1920, published in issue No. 6 of December 1920).

1922 Störungen und Kombinationsprinzip im System der violetten Cyanbanden, *Sitz. ber. Bayer. Akad. Wiss. (München)*, pp. 107–118 (presented by A. Sommerfeld at the meeting of 4 March 1922).

1925 Die Gesetzmäßigkeiten in den Bandenspektren, *Encykl. d. math. Wiss. V/3*, pp. 821–859 (dated May 1925, published in issue No. 5 of 19 December 1925).

KRATZER, ADOLF, AND ELISABETH SUDHOLT

1925 Die Gesetzmäßigkeiten im Resonanzspektrum des Joddampfes und die Bestimmung des Trägheitsmoments, *Z. Phys.* **33**, 144–152 (received 22 May 1925, published in issue No. 1/2 of 18 July 1925).

KRONIG, RALPH DE LAER

1925a Über die Intensität der Mehrfachlinien und ihrer Zeemankomponenten, *Z. Phys.* **31**, 885–897 (received 18 February 1925, published in issue No. 12 of 14 April 1925).

1925b Über die Intensität der Mehrfachlinien und ihrer Zeemankomponenten. II, *Z. Phys.* **33**, 261–272 (received 2 June 1925, published in issue No. 4 of 1 August 1925).

1960 The turning point, in *Theoretical Physics in the Twentieth Century* (M. Fierz and V. F. Weisskopf, eds.), New York: Interscience Publications, pp. 5–39.

KUHN, WERNER

1925a Über die Gesamtstärke der von einem Zustande ausgehenden Absorptionslinien, *Z. Phys.* **33**, 408–412 (received 14 May 1925, published in issue No. 5/6 of 8 August 1925). English translation: On the total intensity of absorption lines emanating from a given state, in *Sources of Quantum Mechanics* (van der Waerden, 1967), pp. 253–257.

1926b Die Stärke der anomalen Dispersion in nichtleuchtenden Dämpfen von Thallium und Cadmium, *Kgl. Danske Vid. Selsk. Math.-fys. Medd.* **7**, No. 12, 1–87 (dated 20 April 1926, ready for print 30 June 1926).

1927a Absorptionsvermögen von Atomkernen für γ-Strahlen, *Z. Phys.* **43**, 56–65 (received 1 April 1927, published in issue No. 1/2 of 23 May 1927).

1927b Polarisierbarkeit der Atomkerne und Ursprung der γ-Strahlen, *Z. Phys.* **44**, 32–35 (received 25 May 1927, published in issue No. 1/2 of 27 July 1927).

LADENBURG, RUDOLF

1921a Die quantentheoretische Deutung der Zahl der Dispersionselektronen, *Z. Phys.* **4**, 451–468 (received 8 February 1921, published in issue No. 4 of March 1921). English translation: The quantum-theoretical interpretation of the number of dispersion electrons, in *Sources of Quantum Mechanics* (van der Waerden, 1967), pp. 139–157.

LADENBURG, RUDOLF, AND FRITZ REICHE

1923 Absorption, Zerstreuung und Dispersion in der Bohrschen Atomtheorie, *Naturwiss.* **11**, 584–598 (published in the issue of 6 July 1923).

LANDÉ, ALFRED

1921c Über den anomalen Zeemaneffekt (Teil I), *Z. Phys.* **5**, 231–241 (received 16 April 1921, published in issue No. 4 of 23 June 1921).

1921d Anomaler Zeemaneffekt und Seriensysteme bei Ne und Hg, *Phys. Zs.* **22**, 417–422 (received 27 June 1921, published in issue No. 15 of 1 August 1921).

1921f Über den anomalen Zeemaneffekt (II. Teil), *Z. Phys.* **7**, 398–405 (received 5 October 1921, published in issue No. 6 of 30 November 1921).

1922a Zur Theorie der anomalen Zeeman- und magneto-optischen Effekte, *Z. Phys.* **11**, 353–363 (received 16 September 1922, published in issue No. 6 of 30 November 1922).

1923a Termstruktur und Zeemaneffekt der Multipletts, *Z. Phys.* **15**, 189–205 (received 5 March 1923, published in issue No. 4/5 of 26 May 1923).

1923b Zur Theorie der Röntgenspektren, *Z. Phys.* **16**, 391–396 (received 12 June 1923, published in issue No. 5/6 of 19 July 1923).

1923c Das Versagen der Mechanik in der Quantentheorie, *Naturwiss.* **11**, 725–726 (letter dated 15 July 1923, published in the issue of 24 August 1923).

1923d Zur Struktur des Neonspektrums, *Z. Phys.* **17**, 292–294 (received 5 July 1923, published in No. 4/5 of 8 September 1923).

1923e Schwierigkeiten in der Quantentheorie des Atombaues, besonders magnetischer Art, *Phys. Zs.* **24**, 441–444 (presented on 19 September 1923 at the Bonn meeting of the German Physical Society, published in issue No. 20 of 15 October 1923).

1923f Termstruktur und Zeemaneffekt der Multipletts. Zweite Mitteilung, *Z. Phys.* **19**, 112–123 (received 16 August 1923, published in issue No. 2 of 9 November 1923).

1924a Das Wesen der relativistischen Röntgendubletts, *Z. Phys.* **24**, 88–97 (received 15 March 1924, published in issue No. 2 of 22 May 1924).

1924d Über den quadratischen Zeemaneffekt, *Z. Phys.* **30**, 329–340 (received 4 November 1924, published in issue No. 4/5 of 29 December 1924).

1925a Zeemaneffekt bei Multipletts höherer Stufe, *Ann. d. Phys. (4)* **76**, 273–283 (received 19 October 1924, published in issue No. 2/3 of January 1925, dedicated to F. Paschen on his sixtieth Birthday).

1926c Neue Wege in der Quantentheorie, *Naturwiss.* **14**, 455–458 (published in the issue of 14 May 1926).

LANDÉ, ALFRED, AND WERNER HEISENBERG

1924 Termstruktur der Multipletts höherer Stufe, *Z. Phys.* **25**, 279–286 (received 18 May 1924, published in issue No. 4–6 of 4 August 1924).

LANDSBERG, GRIGORII, AND LEONID MANDELSTAM

1928 Eine neue Erscheinung bei der Lichtzerstreuung in Krystallen, *Naturwiss.* **16**, 557–558 (letter dated 6 May 1928, published in the issue of 13 July 1928).

LAPORTE, OTTO

1923a Multipletts im Spektrum des Vanadiums, *Naturwiss.* **11**, 779–780 (letter dated 20 July 1923, published in the issue of 14 September 1923).

LARSEN, EGON

1962 *The Cavendish Laboratory: Nursery of Genius*, London: Edmund Ward Ltd., New York: Franklin Watts, Inc.

LENZ, WILHELM

1924 Über den Bewegungsverlauf und die Quantenzustände der gestörten Keplerbewegung, *Z. Phys.* **24**, 197–207 (received 2 April 1924, published in issue No. 3/4 of 4 June 1924).

LIN, CHIA-CHIAO

1944 On the stability of two-dimensional parallel flows, *Proc. Nat. Acad. Sci. (U.S.A.)* **30**, 316–324 (communicated on 24 August 1944, published in issue No. 10 of 15 October 1924).

1945 On the stability of two-dimensional parallel flows. Parts I, II, III, *Quarterly of Applied Mathematics* **3**, 117–142, 218–234, 277–301 (received 3 March, 18 March and 18 July 1945, published in the issues No. 2 of July 1945, No. 3 of October 1945, and No. 4 of January 1946).

1955 *The Theory of Hydrodynamic Stability*, Cambridge: Cambridge University Press.

LOOMIS, FRANCIS WHEELER

1927 Correlation of the fluorescent and absorption spectra of iodine, *Phys. Rev. (2)* **29**, 112–134 (dated 18 September 1926, published in issue No. 1 of January 1927).

LORENTZ, HENDRIK ANTOON, ALBERT EINSTEIN AND HERMANN MINKOWSKI

1913 *Das Relativitätsprinzip* (O. Blumenthal, ed.), Leipzig: B. G. Teubner.

1923 *The Principle of Relativity: A Collection of Original Memoirs on the Special and General Theory of Relativity, with notes by A. Sommerfeld* (translated from the fourth German edition of *Das Relativitätsprinzip*, 1922, by W. Perret and G. B. Jeffery), London: Methuen and Company, Ltd.; paperback reprinted by Dover Publications, Inc., New York.

LYMAN, THEODORE

1922 The spectrum of helium in the extreme ultra-violet, *Nature* **110**, 278–279 (letter dated 3 August 1922, published in the issue of 26 August 1922).

MACH, ERNST

1883 *Die Mechanik in ihrer Entwicklung*, Leipzig: F. A. Brockhaus (7th edition 1912). English translation: *The Science of Mechanics*, Chicago, 1893.

MAUE, A. W.

1940 Zur Stabilität der Kármánschen Wirbelstraße, *Z. angew. Math. & Mech.* **20**, 129–137 (published in issue No. 3 of June 1940).

MEHRA, JAGDISH

1975c *The Solvay Conference on Physics: Aspects of the Development of Physics since 1911*, Dordrecht–Boston: D. Reidel Publishing Company.

MILLIKAN, ROBERT ANDREWS, AND IRA SPRAGUE BOWEN

1924a Extreme ultra-violet spectra, *Phys. Rev. (2)* **23**, 1–34 (dated 15 September 1923, published in issue No. 1 of January 1924).

1924b Some conspicuous successes of the Bohr atom and a serious difficulty, *Phys. Rev. (2)* **24**, 223–228 (dated 10 May 1924, published in issue No. 3 of September 1924).

MINKOWSKI, HERMANN

1909 Raum und Zeit, *Phys. Zs.* **10**, 104–111 (presented on 21 September 1908 at the 80th *Naturforscherversammlung*, Cologne; received 23 December 1908, published in issue No. 3 of 1 February 1909); reprinted with supplementary notes by A. Sommerfeld in *Das Relativitätsprinzip* (Lorentz, Einstein and Minkowski, 1913), pp. 56–73. English translation (by W. Perrett and G. B. Jeffery): Space and time, in *The Principle of Relativity* (Lorentz, Einstein, Minkowski and Weyl, 1923), Dover reprint, pp. 73–91.

MISES, RICHARD VON

1912 Beitrag zum Oszillationsproblem, in *Heinrich-Weber-Festschrift*, Leipzig and Berlin: B. G. Teubner.

MÜLLER, CONRAD HEINRICH, AND GEORG PRANGE

1923 *Allgemeine Mechanik*, Hanover: Helwigsche Verlagsbuchhandlung.

MULLIKEN, ROBERT SANDERSON

1924a Isotope effects in the band spectra of boron monoxide and silicon nitride, *Nature* **133**, 423–424 (letter published in the issue of 22 March 1924).

1924b The band spectrum of boron monoxide, *Nature* **114**, 349–350 (published in the issue of 6 September 1924).

1925a The isotope effect in band spectra, Part I, *Phys. Rev. (2)* **25**, 119–138 (dated 21 August 1924, published in issue No. 2 of February 1925).

1925b The isotope effect in band spectra. Part II: The spectrum of boron monoxide, *Phys. Rev. (2)* **25**, 259–294 (dated 12 September 1924, revised 11 December 1924, published in issue No. 3 of March 1925).

NOETHER, FRITZ

1921 Das Turbulenzproblem, *Z. angew. Math. & Mech.* **1**, 125–138 (published in issue No. 2 of April 1921).

1926 Zur asymptotischen Behandlung der stationären Lösungen im Turbulenzproblem, *Z. angew. Math. & Mech.* **6**, 232–243 (revised version of a paper presented at the Marburg meeting of the German Mathematical Association, 20–25 September 1923, published in issue No. 3 of June 1926).

ORNSTEIN, LEONARD SALOMON, AND HERMAN CAREL BURGER

1924d Strahlungsgesetz und Intensität von Mehrfachlinien, *Z. Phys.* **24**, 41–47 (received 25 March 1924, published in issue No. 1 of 15 May 1924).

1924e Die Feinstruktur der gelben Heliumlinie 5876 Å, *Z. Phys.* **26**, 57–58 (received 5 June 1924, published in issue No. 1 of 5 August 1924).

1924f Intensitäten der Komponenten im Zeemaneffekt, *Z. Phys.* **28**, 135–141 (received 20 August 1924, published in issue No. 3/4 of 10 October 1924).

1924g Nachschrift zu der Arbeit "Intensität der Komponenten im Zeemaneffekt," *Z. Phys.* **29**, 241–242 (received 20 September 1924, published in issue No. 3/4 of 28 October 1924).

ORNSTEIN, LEONARD SALOMON, HERMAN CAREL BURGER, AND WILLEM CHRISTIAAN VAN GEEL

1925 Intensitäten der Komponenten im Zeemaneffekt, *Z. Phys.* **32**, 681–683 (received 30 April 1925, published in No. 9 of 12 June 1925).

ORR, WILLIAM MCFADDEN

1907 The stability or instability of the steady motions of a perfect liquid and of a viscous liquid, *Proc. Roy. Irish Acad.* **A27**, 9–68, 69–138 (read 12 November 1906 and 24 June 1907, published in issues No. 2 of March 1907 and No. 3 of October 1907).

PASCHEN, FRIEDRICH, AND ERNST BACK

1921 Liniengruppen magnetisch vervollständigt, *Physica* **1**, 261–273 (published in issue No. 8–10 of 31 October 1921).

PAULI, WOLFGANG

1919a Über die Energiekomponenten des Gravitationsfeldes, *Phys. Zs.* **20**, 25–27 (received 22 September 1918, published in issue No. 2 of 15 January 1919); reprinted in *Collected Scientific Papers 2*, pp. 10–12.

1919c Merkurperihelbewegung und Strahlenablenkung in Weyls Gravitationtheorie, *Verh. d. Deutsch. Ges. (2)* **21**, 742–750 (received 3 November 1919, published in issue No. 21/22 of 5 December 1919); reprinted in *Collected Scientific Papers 2*, pp. 1–9.

1921b Relativitätstheorie, *Encykl. d. math. Wiss. V/2*, pp. 539–775 (dated December 1920, published in issue No. 4 of 15 September 1921); also published as a special monograph, *Relativitätstheorie*, with a preface of A. Sommerfeld, Leipzig: B. G. Teubner, 1921; reprinted in *Collected Scientific Papers 1*, pp. 1–237. English translation: *Theory of Relativity* (Pauli, 1958a).

1922 Über das Modell des Wasserstoffmolekülions, *Ann. d. Phys. (4)* **68**, 177–240 (improved and enlarged doctoral dissertation; received 4 March 1922, published in issue No. 11 of 3 August 1922); reprinted in *Collected Scientific Papers 2*, pp. 70–133.

1923a Über die Gesetzmäßigkeiten des anomalen Zeemaneffektes, *Z. Phys.* **16**, 155–164 (received 26 April 1923, published in issue No. 3 of 29 June 1923); reprinted in *Collected Scientific Papers 2*, pp. 151–160.

1924a Zur Frage der Zuordnung der Komplexstrukturterme in starken und schwachen äußeren Feldern, *Z. Phys.* **20**, 371–387 (received 20 October 1923, published in issue No. 6 of 11 January 1924); reprinted in *Collected Scientific Papers 2*, pp. 176–192.

1924c Zur Frage der theoretischen Deutung der Satelliten einiger Spektrallinien und ihrer Beeinflussung durch magnetische Felder, *Naturwiss.* **12**, 741–743 (letter dated 17 August 1924, published in the issue of 12 September 1924); reprinted in *Collected Scientific Papers 2*, pp. 198–200.

1925a Über den Einfluß der Geschwindigkeitsabhängigkeit der Elektronenmasse auf den Zeemaneffekt, *Z. Phys.* **31**, 373–385 (received 2 December 1924, published in issue No. 5/6 of 19 February 1925); reprinted in *Collected Scientific Papers 2*, pp. 201–213.

1925b Über den Zusammenhang des Abschlusses der Elektronengruppen im Atom mit der Komplexstruktur der Spektren, *Z. Phys.* **31**, 765–783 (received 16 January 1925, published in issue No. 10 of 21 March 1925); reprinted in *Collected Scientific Papers 2*, pp. 214–232. English translation: On the connexion between the completion of electron groups in an atom with the complex structure of spectra, in D. ter Haar: *The Old Quantum Theory*, Oxford–London–Edinburgh–New York–Toronto–Sydney–Paris–Braunschweig: Pergamon Press, 1967, pp. 184–203.

1925c Über die Absorption der Reststrahlen in Kristallen, *Verh. d. Deutsch. Phys. Ges. (3)* **6**, 10–11 (abstract of a paper presented on 8 February 1925 at the meeting of the *Gauverein Niedersachsen* of the German Physical Society at Göttingen, published in issue No. 1 of 31 March 1925); reprinted in *Collected Scientific Papers 2*, p. 251.

1925d Über die Intensitäten der im elektrischen Feld erscheinenden Kombinationslinien, *Kgl. Danske Vid. Selsk. Math. Mat.-fys. Medd.* **7**, No. 3 (ready for printing on 9 November 1925); reprinted in *Collected Scientific Papers 2*, pp. 233–250.

1926b Quantentheorie, in *Handbuch der Physik* (H. Geiger and K. Scheel, eds.) *23, Part 1*, Berlin: J. Springer Verlag, pp. 1–278; reprinted in *Collected Papers 1*, pp. 269–548.

1958a *Theory of Relativity*, translation of the encyclopedia article (Pauli, 1921b) with 'Supplementary Notes'; reprint of the 'Supplementary Notes' in *Collected Scientific Papers 1*, pp. 238–263.

1964 *Collected Scientific Papers* (R. Kronig and V. F. Weisskopf, eds.), 2 volumes, New York–London–Sydney: Interscience Publishers.

1979 *Wissenschaftlicher Briefwechsel mit Bohr, Einstein, Heisenberg, u. a., Volume I: 1919–1929* (A. Hermann, K. von Meyenn and V. F. Weisskopf, eds.), New York–Heidelberg–Berlin: Springer-Verlag.

PEKERIS, CHAIM LEIB

1948 Stability of the laminar parabolic flow of a viscous fluid between parallel fixed walls, *Phys. Rev. (2)* **74**, 191–199 (received 19 March 1948, published in issue No. 2 of 15 July 1948).

Planck, Max

1911a Eine neue Strahlungshypothese, *Verh. d. Deutsch. Phys. Ges. (2)*, **13**, 138–148 (presented at the meeting of 3 February 1911, published in issue No. 3 of 15 February 1911); reprinted in *Physikalische Abhandlungen und Vorträge*, Braunschweig: Fr. Vieweg & Sohn, 1958, Vol. II, pp. 249–259.

Planck, Max, Peter Debye, Walther Nernst, Marian von Smoluchowski, Arnold Sommerfeld, Hendrik Antoon Lorentz, et al.

1914 *Vorträge über die kinetische Theorie der Materie und der Elektrizität, gehalten in Göttingen auf Einladung der Wolfskehlstiftung* (with a preface of D. Hilbert), Leipzig: B. G. Teubner.

Poincaré, Henri

1893 *Les Méthodes Nouvelles de la Mécanique Céleste. Tome 2: Méthodes de MM. Newcomb, Lindstedt et Bohlin*, Paris: Gauthier-Villars & Cie.; paperback reprint, New York: Dover Publications, Inc., 1957.

1906 Sur la dynamique de l'électron, *Rend. Circ. Mat. Palermo* **21**, 129–175 (dated July 1905, published in issue No. 1 of January/February 1906).

Prandtl, Ludwig

1922a Bemerkungen über die Entstehung der Turbulenz, *Phys. Zs.* **23**, 19–25 (presented at the Jena meeting of the German Physical Society, 18–24 September 1921, published in issue No. 1 of 1 January 1922).

1922b Bemerkung zur vorstehenden Arbeit [of Heisenberg (1922b)], *Phys. Zs.* **23**, 366 (dated 29 July 1922, published in issue No. 18 of 15 September 1922).

1926a Zum Turbulenzproblem, *Z. angew. Math. & Mech.* **6**, 339–340 (letter dated 9 July 1926, published in issue No. 4 of August 1926).

1926b Zum Turbulenzproblem, *Z. angew. Math. & Mech.* **6**, 428 (letter dated 28 July 1926, published in issue No. 5 of October 1926).

Pringsheim, Peter

1924a Über polarisierte Resonanzfluoreszenz, *Naturwiss.* **12**, 247–248 (letter dated 21 February 1924, published in the issue of 28 March 1924).

1924b Über die Polarisation der Resonanzstrahlung von Dämpfen, *Z. Phys.* **23**, 324–332 (received 9 March 1924, published in issue No. 5 of 2 May 1924).

Raman, Chandrasekhara Venkata

1928 A new radiation, *Indian Journal of Physics* **2**, 387–398 (presented on 16 March 1928 at the Science Congress in Bangalore).

Rayleigh, John William Strutt, Lord

1880 On the resultant of a large number of vibrations of the same pitch and of arbitrary phase, *Phil Mag. (5)* **10**, 73–78 (dated June 1880, published in issue No. 60 of August 1880).

1892 On the question of the stability of the flow of fluids, *Phil. Mag. (5)* **34**, 59–70 (published in issue No. 206 of July 1892).

REHM, A.

1926 Dr. Nikolaus Wecklein (obituary), *Jahrbuch d. Bayer. Akad. Wiss. (München)*, pp. 21–24.

REICHE, FRITZ

1919b Bemerkungen zur Lebensdauer der Serienlinien, *Phys. Zs.* **20**, 296–298 (received 2 May 1919, published in issue No. 13 of 1 July 1919).

REYNOLDS, OSBORNE

1883 An experimental investigation of the circumstances which determine whether the motion of water shall be direct or sinuous, and the law of resistance in parallel channels, *Phil. Trans. Roy. Soc. (London)* **174**, 935–982 (received and read 15 March 1883, published in Part III).

1895 On the dynamical theory of incompressible viscous fluids and the determination of the criterion, *Phil. Trans. Roy. Soc. (London)* **A186**, 123–164 (received 25 April 1894, read 24 May 1894, published in Part I).

RITZ, WALTHER

1908b Über ein neues Gesetz der Serienspektren. (Vorläufige Mitteilung), *Phys. Zs.* **9**, 521–529 (received 6 June 1908, published in issue No. 16 of 15 August 1908).

ROSENTHAL, ARTUR

1913 Beweis der Unmöglichkeit ergodischer Gassysteme, *Ann. d. Phys. (4)* **42**, 796–806 (received 16 June 1913, published in issue No. 14 of 4 November 1913).

1914 Aufbau der Gastheorie mit Hilfe der Quasiergodenhypothese, *Ann. d. Phys. (4)* **43**, 894–904 (received 31 December 1913, published in issue No. 6 of 20 March 1914).

ROZENTAL, STEFAN

1967 *Niels Bohr: His Life and Work as Seen by his Friends and Colleagues*, Amsterdam: North-Holland Publishing Company.

RUBINOWICZ, ADALBERT

1918b Bohrsche Frequenzbedingung und Erhaltung des Impulsmomentes. II. Teil, *Phys. Zs.* **19**, 465–474 (received 22 May 1918, published in issue No. 21 of 1 November 1918).

RUNGE, CARL

1907 Über die Zerlegung von Spektrallinien im magnetischen Felde, *Phys. Zs.* **8**, 232–237 (received 3 April 1907, published in issue No. 8 of 15 April 1907).

RUSSELL, HENRY NORRIS

1925 The intensities of lines in multiplets, *Nature* **115**, 835–836 (letter dated 20 April 1925, published in the issue of 30 May 1925).

RUSSELL, HENRY NORRIS, AND FREDERICK ALBERT SAUNDERS

1925 New regularities in the spectra of the alkaline earths, *Astrophysical Journal* **61**, 38–69 (dated 3 October 1924, published in issue No. 1 of January 1925).

SCHRÖDINGER, ERWIN

1925b Die wasserstoffähnlichen Spektren vom Standpunkte der Polarisierbarkeit des Atomrumpfes, *Ann. d. Phys. (4)* **77**, 43–70 (received 7 April 1925, published in issue No. 9 of June 1925).

1926d Quantisierung als Eigenwertproblem. (Zweite Mitteilung), *Ann. d. Phys. (4)* **79**, 489–527 (received 23 February 1926, published in issue No. 6 of 6 April 1926). English translation: Quantization as a problem of proper values. Part II, in *Collected Papers on Wave Mechanics* (translated by J. F. Shearer and W. M. Deans), Glasgow: Blackie & Son, 1928, pp. 13–40.

SCHÜLER, HERMANN

1924 Das Spektrum des einfach ionisierten Lithiums, *Naturwiss.* **12**, 579 (letter dated 4 June 1924, published in the issue of 11 July 1924).

SEELIG, CARL

1954 *Albert Einstein. Eine dokumentarische Biographie*, Zurich: Europa Verlag. English translation (by M. Savill): *Albert Einstein. A Documentary Biography*, London: Staples. 1956.

SLATER, JOHN CLARKE

1924a Radiation and atoms, *Nature* **113**, 307–308 (letter dated 28 January 1924, published in the issue of 1 March 1924).

SMEKAL, ADOLF

1923c Zur Quantentheorie der Dispersion, *Naturwiss.* **11**, 873–875 (letter dated 15 September 1923, published in the issue of 26 October 1923).

1925c M. Born and W. Heisenberg: Die Elektronenbahnen des angeregten Heliumatoms, *Physikalische Berichte* **6**, 1258 (review published in issue No. 19 of 1 October 1925).

SOMMERFELD, ARNOLD

1909a Ein Beitrag zur hydrodynamischen Erklärung der turbulenten Flüssigkeitsbewegung, *Atti del 4. Congresso Internationale dei Matematici* (6–11 April 1908, Rome) *Vol. 3*, Rome: Accademia dei Lincei, pp. 116–124; reprinted in *Gesammelte Schriften I*, pp. 599–607.

1915b Zur Theorie der Balmerschen Serie, *Sitz.ber. Bayer. Akad. Wiss. (München)*, pp. 425–458 (presented at the meeting of 6 December 1915).

1915c Die Feinstruktur der Wasserstoff- und der Wasserstoff-ähnlichen Linien, *Sitz.-ber. Bayer. Akad. Wiss. (München)*, pp. 459–500 (presented at the meeting of 8 January 1916, included in the volume of 1915).

1915d Die allgemeine Dispersionsformel nach dem Bohrschen Modell, in *Arbeiten aus dem Gebiete der Physik, Mathematik und Chemie* (Elster-Geitel-Festschrift), Braunschweig: Fr. Vieweg & Sohn, pp. 549–584; reprinted in *Gesammelte Schriften III*, pp. 136–171.

1916b Zur Quantentheorie der Spektrallinien, *Ann. d. Phys. (4)* **51**, 1–94 (received 5 July 1916, published in issue No. 17 of 22 September 1916); reprinted in *Gesammelte Schriften III*, pp. 172–265.

1916c Zur Quantentheorie der Spektrallinien. (Fortsetzung), *Ann. d. Phys. (4)* **51**, 125–167 (received 5 July 1916, published in issue No. 18 of 10 October 1916); reprinted in *Gesammelte Schriften III*, pp. 266–308.

1916d Zur Theorie des Zeeman-Effekts der Wasserstofflinien mit einem Anhang über den Stark-Effekt, *Phys. Zs.* **17**, 491–507 (received 7 September, published in issue No. 20 of 15 October 1916); reprinted in *Gesammelte Schriften III*, pp. 309–325.

1918a Die Drudesche Dispersionstheorie vom Standpunkte des Bohrschen Modelles und die Konstitution von H_2, O_2 und N_2, *Ann. d. Phys. (4)* **53**, 497–550 (received 2 August 1917, published in issue No. 15 of 24 January 1918); reprinted in *Gesammelte Schriften III*, pp. 378–431.

1919 *Atombau und Spektrallinien*, Braunschweig: Fr. Vieweg & Sohn.

1920a Ein Zahlenmysterium in der Theorie des Zeemaneffekts, *Naturwiss.* **8**, 61–64 (published in the issue of 23 January 1920); reprinted in *Gesammelte Schriften III*, pp. 511–514.

1920b Bemerkungen zur Feinstruktur der Röntgenspektren, *Z. Phys.* **1**, 135–146 (received 12 January 1920, published in issue No. 1 of February 1920); reprinted in *Gesammelte Schriften III*, pp. 566–577.

1920f Allgemeine spektroskopische Gesetze, insbesondere ein magnetooptischer Zerlegungssatz, *Ann. d. Phys. (4)* **63**, 221–263 (received 23 March 1920, published in issue No. 19 of 7 October 1920): reprinted in *Gesammelte Schriften III*, pp. 523–565.

1921a Bemerkungen zur Feinstruktur der Röntgenspektren. II, *Z. Phys.* **5**, 1–16 (received 23 February 1921, published in issue No. 1 of 10 May 1921); reprinted in *Gesammelte Schriften III*, pp. 578–593.

1921b Über den Starkeffekt zweiter Ordnung, *Ann. d. Phys. (4)* **65**, 36–40 (received 27 January 1921, published in issue No. 9 of 23 May 1921); reprinted in *Gesammelte Schriften III*, pp. 601–605.

1921e *Atombau und Spektrallinien*, second edition, Braunschweig: Fr. Vieweg & Sohn.

1922a Quantentheoretische Umdeutung der Voigt'schen Theorie des anomalen Zeemaneffektes vom D-Linientypus, *Z. Phys.* **8**, 257–272 (received 12 December 1921, published in issue No. 5 of 15 February 1922); reprinted in *Gesammelte Schriften III*, pp. 609–624.

1922d *Atombau und Spektrallinien*, third edition, Fr. Vieweg & Sohn. English translation (by H. L. Brose): *Atomic Structure and Spectral Lines*, London and New York: Methuen and Dutton, 1923; French translation (by H. Bellenot): *La Constitution de l'Atome et les Raies Spectrales*, Paris: Albert Blanchard, 1923.

1923a Über die Deutung verwickelter Spektren (Mangan, Chrom usw.) nach der Methode der inneren Quantenzahlen, *Ann. d. Phys. (4)* **70**, 32–62 (received 20 August 1922, published in issue No. 1 of 18 January 1923); reprinted in *Gesammelte Schriften III*, pp. 675–705.

1923c The model of the neutral helium atom, *J. Opt. Soc. America & Rev. Sci. Instr.* **7**, 509–515 (published in issue No. 7 of July 1923); reprinted in *Gesammelte Schriften III*, pp. 706–712.

1924a Zur Theorie der Multipletts und ihrer Zeemaneffekte, *Ann. d. Phys. (4)* **73**, 209–227 (published in issue No. 3/4 of January 1924); reprinted in *Gesammelte Schriften III*, pp. 713–731.

1924c Grundlagen der Quantentheorie und des Bohrschen Atommodelles, *Naturwiss.* **12**, 1047–1049 (presented on 23 September 1924 at the 88th *Naturforscherversamm-*

lung, Innsbruck, published in the issue of 21 November 1924); reprinted in *Gesammelte Schriften IV*, pp. 535–543.

1924d *Atombau und Spektrallinien*, fourth edition, Braunschweig: Fr. Vieweg & Sohn.

1925a Zur Theorie des periodischen Systems, *Phys. Zs.* **26**, 70–74 (received 22 December 1924, published in issue No. 1 of 1 January 1925); reprinted in *Gesammelte Schriften III*, pp. 757–761.

1929 *Atombau und Spektrallinien: Wellenmechanischer Ergänzungsband*, Braunschweig: Fr. Vieweg & Sohn. English translation (by H. L. Brose): *Wave Mechanics*, London and New York: Methuen and Dutton, 1930.

1943 *Vorlesungen über theoretische Physik. Band I: Mechanik*, Leipzig: Akademische Verlagsgesellschaft. English translation (by M. O. Stern): *Mechanics*, New York: Academic Press Inc., 1952.

1945a *Vorlesungen über theoretische Physik. Band II: Mechanik der deformierbaren Medien*, Leipzig: Akademische Verlagsgesellschaft; second edition 1949. English translation of the second edition (by G. Kuerti): *Mechanics of Deformable Bodies*, New York: Academic Press Inc., 1950.

1945b *Vorlesungen über theoretische Physik. Band VI: Partielle Differentialgleichungen der Physik*, Leipzig: Akademische Verlagsgesellschaft. English translation (by G. Strauss): *Partial Differential Equations in Physics*, New York: Academic Press Inc., 1949.

1948 *Vorlesungen über theoretische Physik. Band III: Elektrodynamik*, Wiesbaden: Dieterich; also Leipzig: Akademische Verlagsgesellschaft, 1949. English translation (by E. G. Ramberg): *Electrodynamics*, New York: Academic Press Inc., 1952.

1949 Some reminiscences of my teaching career, *Amer. J. Phys.* **12**, 315–316 (published in issue No. 5 of May 1949).

1950 *Vorlesungen über theoretische Physik. Band IV: Optik*, Wiesbaden: Dieterich. English translation (by O. Laporte and P. A. Moldauer): *Optics*, Academic Press Inc., 1954.

1952 *Vorlesungen über theoretische Physik. Band V: Thermodynamik und Statistik* (F. Bopp & J. Meixner, eds.), Wiesbaden: Dieterich. English translation (by J. Kestin): *Thermodynamics and Statistical Mechanics*, New York: Academic Press Inc., 1956.

1968 *Gesammelte Schriften*, 4 volumes (*I–IV*), Braunschweig: Fr. Vieweg & Sohn.

SOMMERFELD, ARNOLD, AND ERNST BACK

1921 Fünfundzwanzig Jahre Zeemaneffekt, *Naturwiss.* **9**, 911–916 (published in the issue of 11 November 1921); reprinted in *Gesammelte Schriften IV* (Sommerfeld, 1968), pp. 531–534.

SOMMERFELD, ARNOLD, AND WERNER HEISENBERG

1922a Eine Bemerkung über relativistische Röntgendubletts und Linienschärfe, *Z. Phys.* **10**, 393–398 (received 3 August 1922, published in issue No. 6 of 13 September 1921); reprinted in *Gesammelte Schriften III* (Sommerfeld, 1968), pp. 649–654.

1922b Die Intensität der Mehrfachlinien und ihrer Zeemankomponenten, *Z. Phys.* **11**, 131–154 (received 26 August 1922, published in No. 3 of 24 October 1922); reprinted in *Gesammelte Schriften III* (Sommerfeld, 1968), pp. 625–648.

SOMMERFELD, ARNOLD, AND HELMUT HÖNL

1925 Über die Intensität von Multiplett-Linien, *Sitz.ber. Preuss. Akad. Wiss. (Berlin)*, pp. 141–161 (communicated by A. Sommerfeld to the meeting of 12 March 1925); reprinted in *Gesammelte Schriften III* (Sommerfeld, 1968), pp. 736–756.

STERN, OTTO

1913 Zur kinetischen Theorie des Dampfdrucks einatomiger fester Stoffe und über die Entropiekonstante einatomiger Gase, *Phys. Zs.* **14**, 629–632 (received 22 May 1913, published in issue No. 14 of 15 July 1913).

1919 Zusammenfassender Bericht über die Molekulartheorie des Dampfdruckes fester Stoffe und ihre Bedeutung für die Berechnung chemischer Konstanten, *Z. Elektrochem.* **25**, 66–80 (received 3 November 1918, published in issue No. 5/6 of March 1919).

STONER, EDMUND CLIFTON

1924 The distribution of electrons among atomic levels, *Phil. Mag. (6)* **48**, 719–736 (dated July 1924, published in issue No. 286 of October 1924).

THOMAS, LLEWELLYN HILLETH

1952 The stability of plane Poisseuille flow, *Phys. Rev. (2)* **86**, 812–813 (letter received 17 April 1952, published in issue No. 5 of 1 June 1952).

THOMAS, WILLY

1925a Über die Zahl der Dispersionselektronen, die einem stationären Zustande zugeordnet sind, *Naturwiss.* **13**, 627 (letter dated 28 June, published in the issue of 10 July 1925).

THOMSON, JOSEPH JOHN

1903 *Conduction of Electricity through Gases*, Cambridge: Cambridge University Press.

TOLLMIEN, WALTER

1929 Über die Entstehung der Turbulenz, 1. Mitteilung, *Nachr. Ges. Wiss. (Göttingen)*, pp. 21–44 (presented by L. Prandtl at the meeting of 22 March 1929).

VAN DER WAERDEN, BARTEL LEENDERT

1967 *Sources of Quantum Mechanics*, Amsterdam: North-Holland Publishing Company; paperback reprint in the *Classics of Science Series*, Vol. V, New York: Dover Publications, Inc., 1968.

VOIGT, WOLDEMAR

1913a Über die anormalen Zeemaneffekte der Wasserstofflinien, *Ann. d. Phys. (4)* **40**, 368–380 (dated January 1913, published in issue No. 2 of 4 February 1913).

1913c Weiteres zum Ausbau der Kopplungstheorie der Zeemaneffekte, *Ann. d. Phys. (4)* **41**, 403–440 (received 2 April 1913, published in issue No. 7 of 3 June 1913).

1913d Die anormalen Zeemaneffekte der Spektrallinien vom D-Typus, *Ann. d. Phys. (4)* **42**, 210–230 (dated 16 June 1913, published in issue No. 11 of 26 August 1913).

WALLER, IVAR

1926 Der Starkeffekt zweiter Ordnung beim Wasserstoff und die Rydbergkorrektion der Spektra von He und Li^+, *Z. Phys.* **38**, 635–646 (received 21 June 1926, published in issue No. 8 of 21 August 1926).

WEIGERT, FRITZ

1924 Über den Polarisationszustand der Resonanzstrahlung und über seine Beeinflussung durch schwache magnetische Felder, *Naturwiss.* **12**, 38–39 (letter dated 19 November 1923, published in the issue of 11 January 1924).

WEYL, HERMANN

1918a Gravitation und Elektrizität, *Sitz.ber. Preuss. Akad. Wiss. (Berlin)*, pp. 465–480 (communicated by A. Einstein on 2 May 1918, presented at the meeting of 30 May 1918); reprinted in *Das Relativitätsprinzip* (Lorentz, Einstein, Minkowski and Weyl, 1920). English translation (by W. Perret and G. B. Jeffery): Gravitation and electricity, in *The Principle of Relativity* (Lorentz, Einstein, Minkowski and Weyl, 1923), Dover reprint, pp. 191–198.

1918c *Raum-Zeit-Materie: Vorlesungen über allgemeine Relativitätstheorie*, Berlin: J. Springer Verlag.

1920 Elektrizität und Gravitation, *Phys. Zs.* **21**, 649–651 (presented at the 86th *Naturforscherversammlung*, Bad Nauheim, 19–25 September 1920, published in issue No. 23/24 of 1/15 December 1920).

1921 *Raum-Zeit-Materie*, fourth edition, Berlin: J. Springer Verlag. English translation (by H. L. Brose): *Space–Time–Matter*, London: Methuen & Company; paperback reprint New York: Dover Publications, Inc., 1952.

WIEN, WILLY

1919 Über Messungen der Leuchtdauer der Atome und der Dämpfung der Spektrallinien. I, *Ann. d. Phys. (4)* **60**, 597–637 (received 18 June 1919, published in issue No. 23 of 9 December 1919).

1921 Über Messungen der Leuchtdauer der Atome und der Dämpfung der Spektrallinien. II, *Ann. d. Phys. (4)* **66**, 229–236 (received 5 October 1921, published in issue No. 20 of 20 December 1921).

WIŚNIEWSKI, FELIX JOACHIM VON

1924 Zur Theorie des Heliums, *Phys. Zs.* **25**, 135–137 (received 23 February 1924, published in issue No. 6 of 15 March 1924).

WOOD, ROBERT WILLIAMS, AND ALEXANDER ELLET

1923 On the influence of magnetic fields on the polarization of resonance radiation, *Proc. Roy. Soc. (London)* A**103**, 396–403 (received 16 April 1923, published in issue No. A722 of 1 June 1923).

1924 Polarized resonance radiation in weak magnetic fields, *Phys. Rev. (2)* **24**, 243–254 (dated 13 June 1924, published in issue No. 3 of September 1924).

Author Index

Note: The principal character in this volume is Werner Heisenberg. His name occurs on most of the pages, hence we have omitted it from inclusion in the Author Index. Pages where biographical data are given are indicated in italics.

CPSIA information can be obtained at www.ICGtesting.com
Printed in the USA
LVOW12s1334240114

370854LV00002B/287/A